Optical Electromagnetic Field Analysis Using Python

In this book, Kajikawa and Okamoto explain how to use Python to calculate and visualize the optical response of microscopic structures and systems. Throughout, the book the authors provide varied examples to instruct readers in the application of theoretical knowledge to real-world scenarios.

Electromagnetic field analysis is often necessary to determine the optical response of materials with microscopic structures. Although the principles are widely described, the actual calculation and visualization of results are not and remain challenging. Python is the ideal language to use for this as it has a large functional library for visualizing analysis results and is suitable for programming beginners to use at low cost, so it has many advantages over languages like Fortran, BASIC, and C. Here, the authors introduce the application of Python to various electromagnetic field analysis scenarios in the field of nanophotonics. The first half of the book describes cases in which there is an analytical solution for the structure. It addresses scenarios such as scattering and absorption in spherical and cylindrical structures and complex structures such as rotating ellipsoids, sphere-aggregated structures, and hemispherical structures. The second half describes methods including rigorous coupling wave analysis, finite-difference time-domain method and discrete dipole approximation for numerically solving varied structures. This book enables readers to conduct their own electromagnetic field analysis quickly, cheaply, and accurately without in-depth study of other complicated and time-consuming approaches or programs.

This book is invaluable for researchers and postgraduate students working in the fields of optics and photonics. Additionally, the contents are useful not only for those conducting electromagnetic field analysis but also for those simulating physical, chemical, and biological phenomena.

Kotaro Kajikawa is Professor at Institute of Science Tokyo (formerly Tokyo Institute of Technology). He obtained his Bachelor's, Master's, and Doctorate degrees from Tokyo Institute of Technology in 1987, 1989, and 1992, respectively. Professor Kajikawa is a member and a fellow of The Japanese Society of Applied Physics. He is the author of around 140 journal articles and 10 books.

Takayuki Okamoto retired from RIKEN in 2022, where he worked as a research scientist since 1986. He obtained his Bachelor's, Master's and Doctorate degrees from Osaka University in 1981, 1983, and 1986, respectively. He is the author of over 80 refereed papers.

Optical Electromagnetic Field Analysis Using Python

Practical Application in Metallic and Dielectric Nanostructures

Kotaro Kajikawa and Takayuki Okamoto

CRC Press
Taylor & Francis Group
Boca Raton London New York

CRC Press is an imprint of the
Taylor & Francis Group, an **informa** business

First edition published 2025
by CRC Press
2385 NW Executive Center Drive, Suite 320, Boca Raton FL 33431

and by CRC Press
4 Park Square, Milton Park, Abingdon, Oxon, OX14 4RN

CRC Press is an imprint of Taylor & Francis Group, LLC

ISBN: 978-1-032-41351-8 (hbk)
ISBN: 978-1-032-41352-5 (pbk)
ISBN: 978-1-003-35767-4 (ebk)

DOI: 10.1201/9781003357674

Typeset in CMR10 font
by KnowledgeWorks Global Ltd.

Contents

Preface

Electromagnetic field analysis is often necessary when discussing the optical response of nanostructured materials. Although the principles are described in optics textbooks, computer-based calculations are usually required to understand the optical response better. Electromagnetic field analysis using commercially available software offers good programing prospects and provides excellent insight with little effort, but it is not easily accessible to everyone.

Python is a computer language with good program code descriptiveness; anyone can write prospective programs. In addition, it is easy to use due to its extensive libraries of various functions for scientific and technical calculations. As they are widely used in machine learning and statistics, a wealth of information is available on the Internet. In many cases, they provide sufficient computational speed. This program language is suitable for non-programing experts such as the authors who need access to computing facilities to interpret the experimental results or design optical structures. Best of all, it is open source and widely available on various platforms, such as Linux, Windows, and Mac OS, free of charge.

This book is a practical guide to calculating and visualizing the optical response of nanostructures and systems, with examples of Python programs. We aim to bridge the gap between understanding language and actual programing. While the programs may be not as optimal, we have strived to explain the background electromagnetic field analysis straightforwardly, translate it into programs, and present valuable content in real research. This practical approach will give you the confidence in programming.

Kajikawa, the author of chapters 1–4 and 7, delves into analytical calculations, while Okamoto, the author of chapters 5 and 6, focuses on numerical calculations. Both methods have merits and limitations, and using them appropriately is crucial, depending on the problems. We hope this book will inspire new findings and breakthroughs in optics and nanostructured materials. All programs in this book are based on Python version 3.

About the Authors

Kotaro Kajikawa is Professor at Institute of Science Tokyo (formerly Tokyo Institute of Technology). He obtained his Bachelor's, Master's, and Doctorate degrees from Tokyo Institute of Technology in 1987, 1989, and 1992, respectively. Professor Kajikawa is a member and a fellow of The Japanese Society of Applied Physics. He is the author of around 140 journal articles and 10 books.

Takayuki Okamoto retired from RIKEN in 2022, where he worked as a research scientist since 1986. He obtained his Bachelor's, Master's and Doctorate degrees from Osaka University in 1981, 1983, and 1986, respectively. He is the author of over 80 refereed papers.

1

Calculation of Reflectivity and Transmittance of Layered Structures

Calculating reflectance and transmittance when light is incident to layered structures is necessary in optics and various fields such as physics, chemistry, biology, materials science, and electronics. For example, it is used to measure the thickness of thin films. This chapter uses Python to describe the program for calculating light transmission and reflection at an interface and multilayers.

1.1 Introduction

A brief introduction to light as an electromagnetic wave is given as preparation.[1] Light has both particle and wave natures. Reflection, transmission, scattering, etc., are considered to have a wave nature, where light is an electromagnetic wave like radio waves. Wavelength λ_0 and the frequency ν in a vacuum have the following relationship:

$$\lambda_0 = \frac{c}{\nu} \tag{1.1}$$

where c is the velocity of light in a vacuum. The refractive index n, which determines the phase velocity of the electromagnetic wave in a medium, is the most fundamental optical constant. If light travels in a medium with a refractive index n, the wavelength λ changes from λ_0.

$$\lambda = \frac{\lambda_0}{n} = \frac{c}{n\nu}. \tag{1.2}$$

The refractive index n is related to the dielectric constant of the medium ϵ by $n = \sqrt{\epsilon}$.[2]

[1] Textbooks in this field are listed in References [1, 2, 3].

[2] In this book, when simply referring to "dielectric constant", we will refer to the relative permittivity. The same applies to magnetic permeability.

DOI: 10.1201/9781003357674-1

Instead of the frequency ν, the angular frequency ω, ν multiplied by 2π, is often used. The angular frequency ω corresponds to the wavenumber k, representing the number of waves per 2π. The wavenumber vector $\boldsymbol{k}$ is a vector whose length is the wavenumber k in the direction of the wavefront. It is a quantity related to the momentum vector $\boldsymbol{p}$ of light and has the relation $\boldsymbol{p} = \hbar\boldsymbol{k}$, using the parameter $\hbar$ that is Plank's constant h divided by 2π. On the other hand, the angular frequency ω is a quantity related to the energy of light U and has the relation $U = \hbar\omega$. The relation between ω and k is called the dispersion relation. In a medium with refractive index n without boundary (free space), the relation is $k = nk_0$. Here, k_0 is the wavenumber in a vacuum, and $k_0 = \omega/c$. However, the dispersion relation is sometimes complicated when light propagates through photonic crystals, waveguides, or other bounded structures.

Consider the case where light with a wavenumber vector $\boldsymbol{k}$ propagates in a uniform medium with refractive index n. The electric field $\boldsymbol{E}$ of the electromagnetic wave propagating in free space at position $\boldsymbol{r}$ is expressed as

$$\boldsymbol{E} = \boldsymbol{E}_0 \exp(i(\boldsymbol{k} \cdot \boldsymbol{r} - \omega t)). \tag{1.3}$$

Here, $\boldsymbol{E}_0$ is an amplitude vector of the light electric field. The direction of the electric field $\boldsymbol{E}$ is called polarization. A magnetic field $\boldsymbol{H}$ exists perpendicular to the electric field $\boldsymbol{E}$ in free space. The magnetic field is also a vector, and is expressed similarly to Eq. (1.3). It is related to the electric field $\boldsymbol{E}$ as

$$\boldsymbol{H} = \frac{\boldsymbol{k} \times \boldsymbol{E}}{\mu_0 \mu \omega} \tag{1.4}$$

$$\boldsymbol{E} = -\frac{\boldsymbol{k} \times \boldsymbol{H}}{\epsilon_0 \epsilon \omega}, \tag{1.5}$$

where $\times$ is the outer product, μ is the relative magnetic permeability, and ϵ_0 and μ_0 are the vacuum dielectric constant and magnetic permeability, respectively. We have the relationship between the electric and magnetic fields, $c = 1/\sqrt{\epsilon_0 \mu_0}$. Using the vacuum impedance Z_0,

$$H = \frac{nE}{Z_0}, \tag{1.6}$$

where

$$Z_0 = \sqrt{\frac{\mu_0}{\epsilon_0}}. \tag{1.7}$$

Light is generally observed as intensity by a photodetector or other means. The optical energy flow (Pointing vector), $\boldsymbol{S}$, observed in unit time is expressed as follows:

$$\boldsymbol{S} = \boldsymbol{E} \times \boldsymbol{H} \tag{1.8}$$

In an isotropic medium, the direction of the Pointing vector is the direction of energy flow, which is in the same direction as the wavenumber vector $\boldsymbol{k}$. The time average of the Pointing vector is defined as the intensity (irradiance), which is expressed as I as follows[3]:

$$I = \int_0^{2\pi} \boldsymbol{S}\, dt = \int_0^{2\pi} \boldsymbol{E} \times \boldsymbol{H}\, dt = \frac{1}{2}|\boldsymbol{E}_0||\boldsymbol{H}_0| = \frac{n}{2Z_0}|\boldsymbol{E}_0|^2 \tag{1.9}$$

We often ignore the proportionality constant. Then, $I = |\boldsymbol{E}|^2$. In this book, unless otherwise noted, we consider the square of the electric field to be the intensity. Since $\boldsymbol{E}$ is a complex vector, $I = \boldsymbol{E}\boldsymbol{E}^*$ in general, where $\boldsymbol{E}^*$ is the complex conjugate of $\boldsymbol{E}$.

1.2 Reflection and transmission at interface

As the most basic example, consider the case where light is incident on an interface between two media, as shown in Figure 1.1. Let the incident side be Medium 1, and the transmitted side be Medium 2. In this case, the light has two polarization directions: one is p-polarization, in which the light electric field oscillates in the plane of incidence,[4] and the other is s-polarization, in which the electric field oscillates in the direction perpendicular to the plane of incidence. The former is also called TM (transverse magnetic) polarization, and the latter is TE (transverse electric) polarization. There are two possible definitions of the direction of p-polarization, but in this book, it is defined as Figure 1.1(a). Reflection and refraction occur when light passes through an interface with different refractive indices. The incident angle θ_1 and the refracted angle θ_2 are related by

$$n_1 \sin\theta_1 = n_2 \sin\theta_2. \tag{1.10}$$

This is called Snell's law, which means that the tangential component of the wavenumber vector is conserved across the interface and is derived from the principle of least action. Thus, Snell's law can also be written as $k_{1x} = k_{2x}$, where $k_{ix} (i = 1 \text{ or } 2)$ is the tangential (x-direction) component of the wavenumber vector of light traveling through medium i.

The electric field is a vector, but when considering transmission and reflection problems, we usually distinguish p- and s-polarization and treat it as a scalar quantity E for each polarization.

[3]Light intensity I is proportional to the number of photons falling on a unit area in a unit time. On the other hand, the power of light is its total quantity, which is the light intensity I integrated over the area.

[4]The plane containing the wavenumber vector of the incident light and the normal vector of the surface is called the plane of incidence.

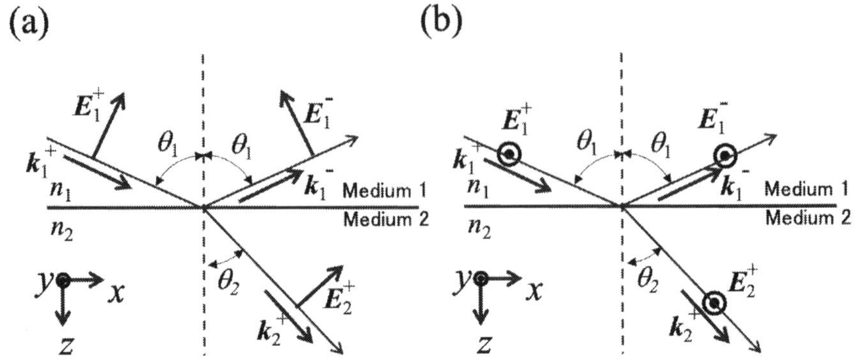

FIGURE 1.1
Optical geometry: (a) p-polarization and (b) s-polarization.

Let the magnitude of the electric field of incident light be E_1^+, the magnitude of the electric field of reflected light is E_1^-, and the magnitude of the electric field of transmitted light is E_2^+. The subscript is the number of the medium, and the superscripts + and − represent downward and upward propagating light, respectively. The ratio of the magnitude of the electric field of reflected light to that of incident light is called the reflection coefficient, and the reflection coefficient for incident light from Medium 1 into Medium 2 is written r_{12} with subscripts. Similarly, the ratio of the electric field of the transmitted light to incident light is called the transmission coefficient, and the transmission coefficient from Medium 1 to Medium 2 is written as t_{12}. The reflection and transmission coefficients are complex.

The reflection coefficient r_{12} and transmission coefficient t_{12} depend on polarization. For s-polarized light, they are denoted as r_{12}^{s} and t_{12}^{s}. For p-polarized light, they are denoted as r_{12}^{p}, t_{12}^{p}. With the angle of incidence θ_1, the refraction angle θ_2, refractive index of the Medium 1, n_1, and that of Medium 2, n_2, they are expressed as

$$r_{12}^{\mathrm{s}} = \frac{n_1 \cos\theta_1 - n_2 \cos\theta_2}{n_1 \cos\theta_1 + n_2 \cos\theta_2} = \frac{k_{1z} - k_{2z}}{k_{1z} + k_{2z}} \tag{1.11}$$

$$t_{12}^{\mathrm{s}} = \frac{2n_1 \cos\theta_1}{n_1 \cos\theta_1 + n_2 \cos\theta_2} = \frac{2k_{1z}}{k_{1z} + k_{2z}} \tag{1.12}$$

$$r_{12}^{\mathrm{p}} = \frac{n_2 \cos\theta_1 - n_1 \cos\theta_2}{n_2 \cos\theta_1 + n_1 \cos\theta_2} = \frac{n_2^2 k_{1z} - n_1^2 k_{2z}}{n_2^2 k_{1z} + n_1^2 k_{2z}} \tag{1.13}$$

$$t_{12}^{\mathrm{p}} = \frac{2n_1 \cos\theta_1}{n_2 \cos\theta_1 + n_1 \cos\theta_2} = \frac{2n_1 n_2 k_{1z}}{n_2^2 k_{1z} + n_1^2 k_{2z}}, \tag{1.14}$$

where k_{iz} is the z-direction component of the wave vector in Medium i, and $k_{iz} = n_i k_0 \cos\theta_i$. The ratio of the reflected light intensity to the incident light intensity is the reflectance R, and the ratio of the transmitted light intensity is the transmittance T, where they are real numbers ranging from 0 to 1. The intensity of light is given by Eq. (1.9),

$$R = rr^* \tag{1.15}$$

$$T = \frac{n_2 \cos\theta_2}{n_1 \cos\theta_1} tt^*. \tag{1.16}$$

The ratio of $\cos\theta$ is taken in the transmittance calculation because the reflection angle differs from the incident angle. If there is no absorption in the media, the energy conservation law

$$R + T = 1 \tag{1.17}$$

is hold.

Based on the discussion above, we calculate the reflection coefficient r and transmission coefficient t for each polarization when light is incident from Medium 1 with a refractive index of 1.0 to Medium 2 with a refractive index of 1.5. An example is shown in Program 1.1, which loads the numerical library scipy in the first line. The matplotlib library is loaded in the second line to graph the calculation result. First, the refractive index of Medium 1 is assigned to variable n1 in Line 7, and the refractive index of Medium 2 is assigned to variable n2 in line 8. Next, the linspace command on Line 10 is used to specify the array of incident angles. The argument of the linspace command is (first value, last value, number of divisions). The array t1 is the radian of the first value, the last value, and the number of divisions. In a simple calculation using arrays, we do not need to write a code assigning the components individually, as in Fortran or C. In Line 12, we use Snell's law, as shown in Eq. (1.10), to create t2 as an array of refraction angles. The calculation is made for the array of transmission coefficients tp and reflection coefficients rp in p-polarization, transmission coefficients ts, and reflection coefficients rs in s-polarization, using Eqs. (1.11)–(1.14) in Lines 13–16. We calculate the array of transmission coefficients tp and reflection coefficients rp for p-polarization, as well as transmission coefficients ts and reflection coefficients rs for s-polarization. The next part is the plotting of the results. The arguments of the plot command of the matplotlib library are (an array of variables for the x-axis, an array of variables for the y-axis, and a definition of the graph name). After Line 18, the label names and font sizes for the x- and y-axes are specified. Superscripts, Greek letters, etc. can be written as the r "$...$", with TeX commands. After specifying the graph's title, font size, grid, plotting range, etc., with the title command, the graph is displayed with the show() command.

Program 1.1

```
import scipy as sp
import matplotlib as mpl
import matplotlib.pyplot as plt
from scipy import pi,sin,cos,tan,arcsin,linspace
from matplotlib.pyplot import plot,show,xlabel,ylabel,title,legend,grid,axis,tight_layoutht_layout

n1 = 1    # refractive index of medium 1
n2 = 1.5  # refractive index of medium 2

t1Deg = linspace(0, 90, 91)  # Generate array of incident angle t1
t1 = t1Deg /180*pi              # Convert angle of incidence into radians.
t2 = arcsin((n1/n2)*sin(t1))    # Find the refraction angle t2
tp = 2*n1*cos(t1)/(n2*cos(t1)+n1*cos(t2))                  # tp
rp = (n2*cos(t1)-n1*cos(t2))/(n2*cos(t1)+n1*cos(t2))    # rp
ts = 2*n1*cos(t1)/(n1*cos(t1)+n2*cos(t2))                  # ts
rs = (n1*cos(t1)-n2*cos(t2))/(n1*cos(t1)+n2*cos(t2))    # rs

plt.figure(figsize=(8,6))                          # Set figure size
plot(t1Deg,rp, label=r"$r_{12}^{\rm{p}}$",linewidth = 3.0, color='black', linestyle='dashed')          # Plot rp
plot(t1Deg,tp, label=r"$t_{12}^{\rm{p}}$",linewidth = 3.0, color='black')                                # Plot tp
plot(t1Deg,rs, label=r"$r_{12}^{\rm{s}}$",linewidth = 3.0, color='gray', linestyle='dashed')           # Plot rs
plot(t1Deg,ts, label=r"$t_{12}^{\rm{s}}$",linewidth = 3.0, color='gray')                                 # Plot ts
xlabel(r"$\theta_1$ (deg.)",fontsize=20) # Label x-axis
ylabel(r"$r, t$",fontsize=20)              # Label y-axis
title("Reflection and Transmission Coefficient",fontsize=18)
                                             # Title of graph
grid(True)                                   # Show grid
axis([0.0,90,-1,1])                          # Plot region
legend(fontsize=20,loc='lower right')      # Legend and font size
plt.tick_params(labelsize=20)      # Axis scales and font size
tight_layout()  # Commands to make the graph fit into a frame.
show()                             # Display graph
```

The results obtained using this program are shown in Figure 1.2. The transmission coefficients at an incident angle of 0° is equal for both polarizations: $t_{12}^{\rm p} = t_{12}^{\rm s} = 0.8$. This is because there is no distinction between the polarizations. As the angle increases, the transmission coefficient decreases monotonically and is zero at 90°. The reflection coefficient $r_{12}^{\rm p}$ is 0.2 and $r_{12}^{\rm s} = -0.2$ at 0°. The values with opposite signs between p- and s-polarization are due to the definitions of the positive direction of the electric field for the incident and reflected light in p-polarization. $r_{12}^{\rm p}$ crosses zero. Before and after the crossing, the sign of the electric field of the reflected light is reversed.

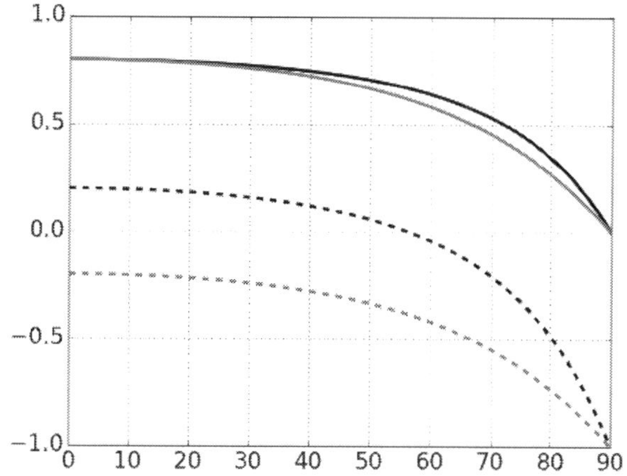

FIGURE 1.2
Transmission and reflection coefficients ($n_1 = 1$, $n_2 = 1.5$).

Next, using Program 1.2, we consider the reflectance R and transmittance T calculated for each polarization when the light is incident from Medium 1. The results are shown in Figure 1.3. The direction of polarization is shown as superscripts, T_{12}^{p}, T_{12}^{s}, R_{12}^{p}, and R_{12}^{s}. At the angle of incidence of 0°, both R_{12}^{p} and R_{12}^{s} are 0.04. The reflectivity for p-polarized light R_{12}^{p} decreases until the

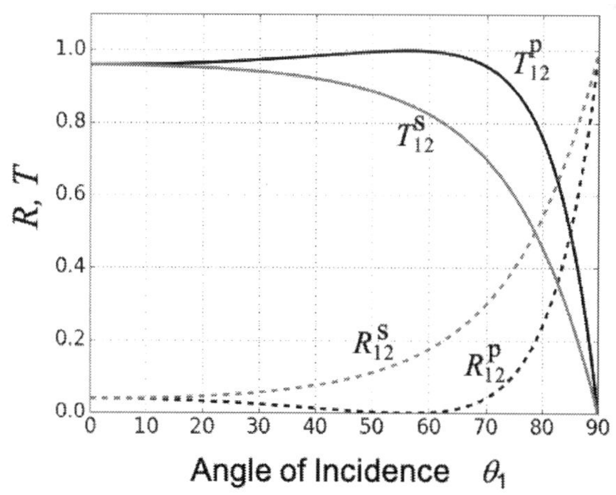

FIGURE 1.3
Reflectivity R and transmittance T ($n_1 = 1.0$, $n_2 = 1.5$). The Brewster angle θ_{B} is 56.3°.

Brewster angle, θ_{B}, where it becomes zero. After that, as the angle of incidence increases, T_{12}^{s} elevates sharply and reaches unity at an angle of incidence of 90°. On the other hand, the reflectivity for s-polarized light R_{12}^{s} increases monotonically to unity at an angle of incidence of 90°. Note that $R_{12}^{\mathrm{p}} \leq R_{12}^{\mathrm{s}}$ at any angle of incidence. s-polarized light is more reflective than p-polarized light.

Brewster angle θ_{B} is obtained by finding the angle at which r_{p} in Eq. (1.13) is zero. Therefore,

$$\tan\theta_{\mathrm{B}} = \frac{n_2}{n_1}. \tag{1.18}$$

At the Brewster angle, the reflectance for p-polarized light is zero, i.e., the transmittance is one. The output window of a high-power laser is designed to be at this angle to prevent damage to the laser crystal. As for the transmittance, both T_{12}^{p} and T_{12}^{s} are 0.96 at normal incidence. As the angle of incidence increases, T_{12}^{s} decreases monotonically for s-polarized light and becomes zero at the angle of incidence of 90°. On the other hand, T_{12}^{p} increases up to the Brewster angle θ_{B}, where it is unity. Then, as the angle of incidence increases, T_{12}^{s} rapidly decreases to zero at an angle of incidence of 90°.

Program 1.2

```
import scipy as sp
import matplotlib as mpl
import matplotlib.pyplot as plt
from numpy import pi,sin,cos,tan,arcsin,linspace,arrange
from matplotlib.pyplot import plot,show,xlabel,ylabel,title,
    legend,grid, axis,tight_layout

n1 = 1       # Refractive index of medium 1
n2 = 1.5     # Refractive index of medium 2

t1Deg = linspace(0, 90, 90)  # Generation of an array of incident
      angles t1. (deg.)
t1  = t1Deg /180*pi          # Convert the angle of incidence
    into radians.
t2 = arcsin((n1/n2)*sin(t1))  # Find the refraction angle t2.

tp = 2*n1*cos(t1)/(n2*cos(t1)+n1*cos(t2))
               # tp: transmission coefficient for p-pol
rp = (n2*cos(t1)-n1*cos(t2))/(n2*cos(t1)+n1*cos(t2))
               # rp: reflection coefficient for p-pol
ts = 2*n1*cos(t1)/(n1*cos(t1)+n2*cos(t2))
               # ts: transmission coefficient for s-pol
rs = (n1*cos(t1)-n2*cos(t2))/(n1*cos(t1)+n2*cos(t2))
               # rs: tp: reflection coefficient for s-pol

Rp = rp**2                             # Tp: Transmittance for p-pol
Tp = tp**2*(n2*cos(t2))/(n1*cos(t1)) # Rp: Reflectance for p-pol
Rs = rs**2                             # Ts: Transmittance for s-pol
Ts = ts**2*(n2*cos(t2))/(n1*cos(t1)) # Rs: Reflectance for s-pol

plt.figure(figsize=(8,6))              # figure size
plot(t1Deg,Rp, label=r"$R_{12}^{\rm{p}}$",linewidth = 3.0, color
    ='black', linestyle='dashed')    # Plot Rp
```

```
plot(t1Deg,Tp, label=r"$T_{12}^{\rm{p}}$",linewidth = 3.0, color='black')  # Plot Tp
plot(t1Deg,Rs, label=r"$R_{12}^{\rm{s}}$",linewidth = 3.0, color='gray', linestyle='dashed')    # Plot Rs
plot(t1Deg,Ts, label=r"$T_{12}^{\rm{s}}$",linewidth = 3.0, color='gray')          # Plot Ts
xlabel(r"$\theta_1$ (deg.)",fontsize=20)    # Label x-axis
ylabel(r"$R, T$",fontsize=20)    # Label y-axis
title("Reflectivity and Transmittance",fontsize=18)    # Graph title
grid(True)  # Show grid.
axis([0.0,90,0,1.1])  # Plot range
legend(fontsize=20,loc='lower left')      # Show legend and set font size
plt.tick_params(labelsize=20)              # Axis scales
tight_layout()  # Commands to make the graph fit into a frame.
show()           # Show graph.
```

Next, consider the case where the refractive index of the incident medium is larger than that of the transmission medium. Snell's law (Eq. (1.2)) gives $\sin\theta_2$ as

$$\sin\theta_2 = \frac{n_1}{n_2}\sin\theta_1. \tag{1.19}$$

Total reflection occurs when the angle of incidence is greater than the critical angle, θ_c, which is given by $\sin\theta_c = n_2/n_1$, $\sin\theta_2 > 1$. This problem is mathematically solved by making θ_2 a complex number. However, this requires some ingenuity when programming in Python. $\cos\theta_2$ is a purely imaginary number. This is because it is

$$\cos\theta_2 = \pm\sqrt{1-\sin\theta_2} = \pm i\sqrt{\left(\frac{n_1}{n_2}\right)^2\sin\theta_1 - 1}. \tag{1.20}$$

Here, $i = \sqrt{-1}$. Therefore, in Program 1.3, $\sin\theta_1$, $\sin\theta_2$, $\cos\theta_1$, and $\cos\theta_2$ are written as complex variables s1, s2, c1, and c2, respectively. The reflection coefficients, rs and rp, are also complex numbers, so when converting them into reflectivity and transmittance, the square of their absolute value must be taken, as described in Lines 24 and 25. In Program 1.3, the reflectance is obtained using the abs function.

Program 1.3

```
1 import scipy as sp
2 import matplotlib as mpl
3 import matplotlib.pyplot as plt
4 from scipy import pi,sin,cos,tan,arcsin,linspace,arrange,sqrt,zeros
5 from matplotlib.pyplot import plot,show,xlabel,ylabel,title,legend,grid, axis
```

```

n1 = 1.5  # refractive index of medium 1
n2 = 1.0  # refractive index of medium 2
ep1 = n1**2  # dielectric constant of medium 1
ep2 = n2**2  # dielectric constant of medium 2

t1Deg = linspace(0, 90, 90)  # Generation of an array of incident angles t1. (deg.)
t1 = t1Deg /180*pi   # Convert the angle of incidence into radians.
s1 = sin(t1)          # sin(t1)
c1 = cos(t1)          # cos(t1)
s2 = n1/n2*s1         # sin(t1)
c2 = sqrt(1-s2**2)    # cos(t2)
n1z = n1*c1          # n1z=k1z/k0
n2z = n2*c2          # n2z=k1z/k0

rs = (n1z-n2z)/(n1z+n2z)  # Reflection coefficient for s-pol
rp = (ep2*n1z-ep1*n2z)/(ep2*n1z+ep1*n2z)  # Reflection coefficient for p-pol

RsAbs = abs(rs)**2  # Reflectiveigy for s-pol
RpAbs = abs(rp)**2  # Reflectiveigy for p-pol

plot(t1Deg,RpAbs, label=r"$R_{12}^{\rm{p}}$") # Plot Rp
plot(t1Deg,RsAbs, label=r"$R_{12}^{\rm{s}}$") # Plot Rs
xlabel(r"$\theta_1$ (deg.)",fontsize=20)   # Label x-axis
ylabel(r"$R, T$",fontsize=20)              # Label y-axis
title("Reflectivity",fontsize=20)          # Graph title
grid(True)              # Show grid
axis([0.0,90,0,1.1])    # Plot range
legend(fontsize=20,loc='lower right')   # Show legend and set font size
plt.tick_params(labelsize=20)          # Axis scales
tight_layout()  # Commands to make the graph fit into a frame.
show()          # Show graph.
```

Figure 1.4 shows the results obtained using this program. The Brewster angle for p-polarized light exists even when the light is incident from a higher refractive index side. The reflectance is always higher for s-polarized light than for p-polarized light. As the incidence angle increases, the reflectance reaches unity after the critical angle, indicating that all light energy is reflected. This state is called total reflection. All light energy is reflected in Medium 1, but there is an extinction wave (called "evanescent wave") in Medium 2. It is an electromagnetic wave decaying with the distance from the interface. The following equation can express evanescent waves.

$$E_2^+ = t_{12}E_1^+ \exp\left(i\left(\frac{n_1}{n_2}k_2 \sin\theta_1 x\right)\right) \times \exp\left(-k_2\sqrt{\left(\frac{n_1}{n_2}\right)^2 - \sin^2\theta_1} \cdot z\right)\exp(-i\omega t) \tag{1.21}$$

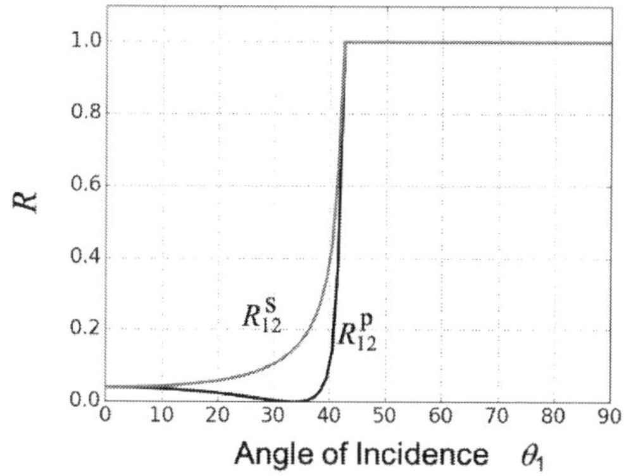

FIGURE 1.4
Reflectance as a function of incident angle when the refractive index of Medium 1 is greater than that of Medium 2 ($n_1 = 1.5$, $n_2 = 1$).

In the x-direction, there is an oscillating wave with a wavenumber of $(\frac{n_1}{n_2})k_2$, and its amplitude decays in the z-direction with distance from the interface. The distance at which the amplitude intensity is $1/e$ is called the penetration depth z_{d}. From Eq. (1.21), z_{d} is given by

$$z_{\mathrm{d}} = \frac{\lambda_0}{2\pi\sqrt{n_1^2 \sin^2\theta_1 - n_2^2}}. \tag{1.22}$$

The penetration length is generally in the order of the wavelength of light. It increases rapidly as the angle of incidence approaches the critical angle θ_{c} and diverges to infinity at the critical angle.

1.3 Reflection and transmission of thin films

The previous section describes the calculation of reflection and transmission at the interface of two isotropic media. While this can be applied to reflection on water surfaces or thick glass plates, the object is often a membrane or slab. It becomes necessary to determine the reflectance and transmittance of a multilayer. In this section, we first consider a simple three-layer problem and then describe the transfer matrix method, which allows the calculation of the reflectance and transmittance of any multilayers.

Consider the light incident on a thin film, as shown in Figure 1.5. For example, this is the case for a soap bubble where Mediums 1 and 3 are air

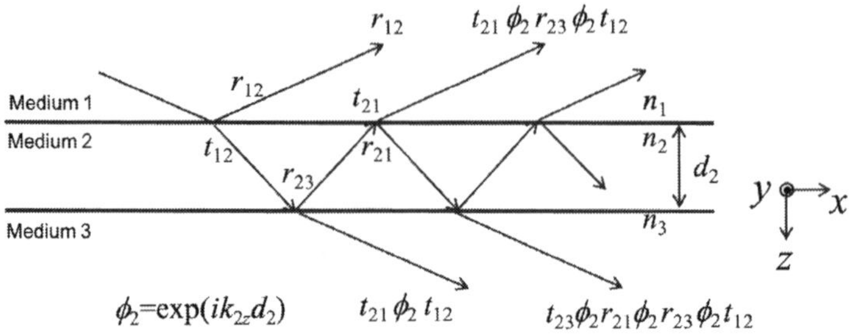

FIGURE 1.5
Multiple reflection in a thin film.

and Medium 2 is a soap film. Let the refractive indices in each layer be n_1, n_2, and n_3, and the reflection coefficient r_{13} and transmission coefficient t_{13}, the total reflectivity R_{13}, and the total transmittance T_{13} of this multilayer film. As shown in the figure, the reflection and transmission coefficients can be calculated as an infinite series sum of the reflected and transmitted optical electric fields. The results are

$$r_{13} = \frac{r_{12} + r_{23}\exp(2k_{2z}d_2 i)}{1 + r_{23}r_{12}\exp(2k_{2z}d_2 i)} \tag{1.23}$$

$$t_{13} = \frac{t_{12}t_{23}\exp(k_{2z}d_2 i)}{1 + r_{23}r_{12}\exp(2k_{2z}d_2 i)}. \tag{1.24}$$

The r_{ij} and t_{ij} are the reflection and transmission coefficients when light is incident from medium i to medium j. Also, k_{2z} is the z-directional component of the wavenumber vector and is described using the refraction angle θ_2 and vacuum wavelength λ_0 in layer 2 as follows:

$$k_{2z} = \frac{2\pi}{\lambda_0} n_2 \cos\theta_2 \tag{1.25}$$

Program 1.4 shows an example program to calculate the transmission coefficient or transmittance at different angles of incidence. In Line 17, the angle of incidence, t1Deg, is defined as an array; in Lines 18–24, the trigonometric functions are defined as variables, such as s1= $\sin\theta_1$ and c1= $\cos\theta_1$. Corresponding to the arrays defined in t1Deg, t1, s1–s3, and c1–c3 are also arrays. The $1j$ used in the argument of exp in Lines 35 and 36 is the way to give imaginary units in Python.

Program 1.4

```
import scipy as sp
import matplotlib as mpl
import matplotlib.pyplot as plt
from scipy import pi,sin,cos,tan,exp,arcsin,linspace,arrange,sqrt,zeros
from matplotlib.pyplot import plot,show,xlabel,ylabel,title,legend,grid,axis,tight_layout

n1=1.0   # refractive index of medium 1
n2=1.5   # refractive index of medium 2
n3=1.0   # refractive index of medium 3
ep1=n1**2   # dielectric constant of medium 1
ep2=n2**2   # dielectric constant of medium 2
ep3=n3**2   # dielectric constant of medium 3
d2=100   # Thickness of medium 2 d2 (nm)
WL=500   # Vacuum wavelength WL (nm)
k0=2*pi/WL  # Vacuum wavenumber

t1Deg = linspace(0, 90, 90)  # Generation of an array of incident angles t1. (deg.)
t1 = t1Deg /180*pi   # Convert the angle of incidence into radians.
s1 = sin(t1)         # sin(t1)
c1 = cos(t1)         # cos(t1)
s2 = n1/n2*s1        # sin(t1)
c2 = sqrt(1-s2**2)   # cos(t2)
s3 = n1/n3*s1        # sin(t1)
c3 = sqrt(1-s3**2)   # cos(t3)

n1z=n1*c1          # n1z=k1z/k0
n2z=n2*c2          # n2z=k1z/k0
n3z=n3*c3          # n2z=k1z/k0

rs12=(n1z-n2z)/(n1z+n2z)  # Reflection coefficient for s-pol rs12
rp12=(ep2*n1z-ep1*n2z)/(ep2*n1z+ep1*n2z)  # Reflection coefficient for p-pol rp12
rs23=(n2z-n3z)/(n2z+n3z)          # Reflection coefficient for s-pol rs23
rp23=(ep3*n2z-ep2*n3z)/(ep3*n2z+ep2*n3z)  # # Reflection coefficient for p-pol rp23

rs=(rs12+rs23*exp(2*1j*n2z*k0*d2))/(1+rs23*rs12*exp(2*1j*n2z*k0*d2))
rp=(rp12+rp23*exp(2*1j*n2z*k0*d2))/(1+rp23*rp12*exp(2*1j*n2z*k0*d2))

RsAbs=abs(rs)**2  # Reflectivity for s-pol
RpAbs=abs(rp)**2  # Reflectivity for s-pol

plt.figure(figsize=(8,6))    # figure size
plot(t1Deg,RpAbs, label="Rp",linewidth = 3.0, color='black') # Plot Rp
plot(t1Deg,RsAbs, label="Rs",linewidth = 3.0, color='gray') # Plot Rs
xlabel(r"$\theta_1$ (deg.)",fontsize=20)    # Label x-axis
```

```
45 ylabel(r"Reflectivity",fontsize=20)                # Label y-axis
46 title("Reflectivity",fontsize=20)              # Graph title
47 grid(True)  # Show grid
48 axis([0.0,90,-1,1])       # Plot Range
49 legend(fontsize=20,loc='lower right')    #  Show legend and set
       font size
50 plt.tick_params(labelsize=20)            # Axis scales
51 tight_layout()  # Commands to make the graph fit into a frame.
52 show()          # Show graph.
```

The angle of incidence depends on the reflectance and transmittance obtained is shown in Figure 1.6(a). As the refractive indices are all real numbers, there exists an angle at which the reflectance of p-polarized light is zero (Brewster angle). On the other hand, for s-polarized light, the reflectance increases monotonically as the incidence angle increases.

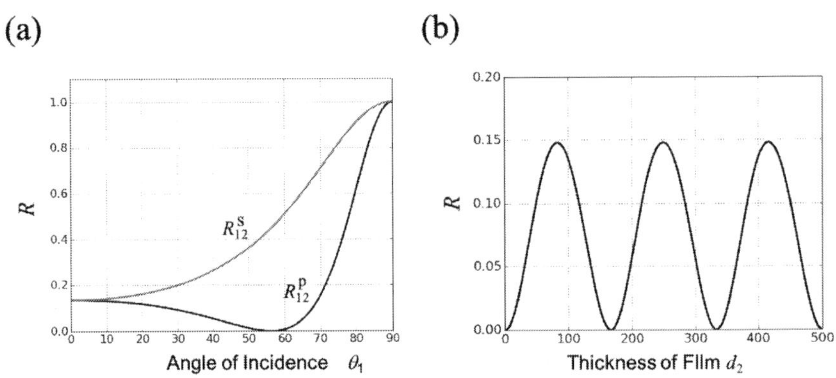

FIGURE 1.6
(a) Reflectivity R from a free-standing film 100 nm-thick for s- and p-polarized light as a function of angle of incidence θ_1. The refractive index of the film is 1.5. (b) Reflectivity R from the free-standing film as a function of the thickness at a wavelength of 500 nm.

Next, using Program 1.5, we calculate the thickness dependence of the reflectance when $\lambda = 500$ nm light is incident perpendicular to this thin film. The incident angle t1Deg defined in Line 18 is set to a constant zero, and s1–s3 and c1–c3 defined in Lines 20–25 are constants. The results are shown in Figure 1.6(b). The reflectance increases as the film thickness increases and reaches a maximum value. After that, the reflectance decreases as the film thickness increases and reaches zero. Thus, it can be seen that the reflectance oscillates with the film thickness.

Program 1.5

```
import scipy as sp
import matplotlib as mpl
import matplotlib.pyplot as plt
from scipy import pi,sin,cos,tan,arcsin,linspace,sqrt,exp
from matplotlib.pyplot import plot,show,xlabel,ylabel,title,
    legend,grid,axis,tight_layout

n1=1.0   # refractive index of medium 1
n2=1.5   # refractive index of medium 2
n3=1.0   # refractive index of medium 3
ep1=n1**2   # dielectric constant of medium 1
ep2=n2**2   # dielectric constant of medium 2
ep3=n3**2   # dielectric constant of medium 3
WL=500      # Vacuum wavelength WL (nm
k0=2*pi/WL  # Vacuum wavenumber

d2=linspace(0, 500, 501) # Thickness of medium 2

t1Deg = 0  # Angle of incidence
t1 = t1Deg /180*pi   # Convert the angle of incidence into
    radians.
s1 = sin(t1)         # sin(t1)
c1 = cos(t1)         # cos(t1)
s2 = n1/n2*s1        # sin(t1)
c2 = sqrt(1-s2**2)   # cos(t2)
s3 = n1/n3*s1        # sin(t1)
c3 = sqrt(1-s3**2)   # cos(t3)

n1z=n1*c1          # n1z=k1z/k0
n2z=n2*c2          # n2z=k1z/k0
n3z=n3*c3          # n2z=k1z/k0

rs12=(n1z-n2z)/(n1z+n2z)  # Reflection coefficient for s-pol rs12
rp12=(ep2*n1z-ep1*n2z)/(ep2*n1z+ep1*n2z)  # Reflection
    coefficient for p-pol rp12
rs23=(n2z-n3z)/(n2z+n3z)  # Reflection coefficient for s-pol rs23
rp23=(ep3*n2z-ep2*n3z)/(ep3*n2z+ep2*n3z)  # Reflection
    coefficient for s-pol rp23

rs=(rs12+rs23*exp(2*1j*n2z*k0*d2))/(1+rs23*rs12*exp(2*1j*n2z*k0*
    d2))
rp=(rp12+rp23*exp(2*1j*n2z*k0*d2))/(1+rp23*rp12*exp(2*1j*n2z*k0*
    d2))

RsAbs=abs(rs)**2  # Reflectivity for s-pol
RpAbs=abs(rp)**2  # Reflectivity for s-pol

plot(d2,RpAbs, label="$R_p$",linewidth = 3.0, color='black')#
    Plot Rp
xlabel(r"$d_2$ (nm)",fontsize=20)   # Label x-axis
ylabel("Reflectivity",fontsize=20)  # Label y-axis
title("Reflectivity",fontsize=20)   # Graph title
grid(True)                          # Show grid
axis([0.0,500,0,0.2])               # Plot Range
plt.tick_params(labelsize=20)       # Axis scales
tight_layout()  # Commands to make the graph fit into a frame.
show()          # Show graph.
```

1.4 Transfer matrix

The transfer matrix method [4] is a simple method for calculating the optical response of multilayers. Here, we consider the thin film structure of an isotropic medium with f layers as shown in Figure 1.7. Let n_1, n_2, $\cdots$ n_f be the refractive indices in each layer medium, and find the reflection coefficient r_{1f} and transmission coefficient t_{1f} and the reflectance R_{1f} and transmission T_{1f} of this multilayer. Polarization is expressed as R^{p}_{1f} as a superscript, and if the expression is the same for both polarizations, the polarization notation is suppressed. Thicknesses of the first and f-th layers are not considered, but the thicknesses of the second through $(f-1)$ layers are considered, as d_2, d_3, $\cdots d_{f-1}$.

The transmission and reflection coefficients of this multilayer structure can be obtained using the transfer matrix $\boldsymbol{G}$, which is calculated as follows:

$$\boldsymbol{G} = \boldsymbol{M}_{f(f-1)}\boldsymbol{\Phi}_{f-1}\cdots\boldsymbol{M}_{32}\boldsymbol{\Phi}_2\boldsymbol{M}_{21}. \tag{1.26}$$

Here, $\boldsymbol{M}_{ij}$ is a matrix representing the boundary condition between i-layer and j-layer, and $\boldsymbol{\Phi}_i$ is a matrix representing the phase change when light propagates through layer i. While $\boldsymbol{M}_{ij}$ varies with polarization, $\boldsymbol{\Phi}_i$ is independent

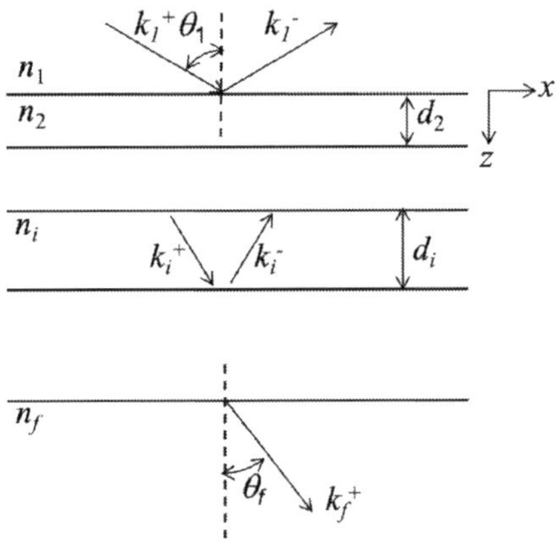

FIGURE 1.7
Geometry for the transfer matrix method.

of polarization.

$$\boldsymbol{M}_{ij} = \frac{1}{2n_i n_j k_{iz}} \begin{pmatrix} n_i^2 k_{jz} + n_j^2 k_{iz} & n_i^2 k_{jz} - n_j^2 k_{iz} \\ n_i^2 k_{jz} - n_j^2 k_{iz} & n_i^2 k_{jz} + n_j^2 k_{iz} \end{pmatrix} \quad \text{(p-polarization)} \tag{1.27}$$

$$\boldsymbol{M}_{ij} = \frac{1}{2k_{iz}} \begin{pmatrix} k_{iz} + k_{jz} & k_{iz} - k_{jz} \\ k_{iz} - k_{jz} & k_{iz} + k_{jz} \end{pmatrix} \quad \text{(s-polarization)} \tag{1.28}$$

$$\boldsymbol{\Phi}_i = \begin{pmatrix} \exp(ik_{iz}d_i) & 0 \\ 0 & \exp(-ik_{iz}d_i) \end{pmatrix} \quad \text{(both polarizations)} \tag{1.29}$$

Here, k_{iz} is the wavenumber vector component in the z-direction. With the angle of incidence in layer 1, θ_1, and the refraction angle in the other layer i, θ_i, they are described for the light of vacuum wavelength λ_0 as follows:

$$k_{iz} = \frac{2\pi}{\lambda_0} n_i \cos\theta_i \tag{1.30}$$

Using the component G_{ij} of the 2×2 matrix $\boldsymbol{G}$, the reflection coefficient r_{1f} and transmission coefficient t_{1f} of the entire layer are written as

$$r_{1f} = -\frac{G_{21}}{G_{22}} \tag{1.31}$$

$$t_{1f} = G_{11} + r_{1f} G_{12} = G_{11} - G_{12}\frac{G_{21}}{G_{22}}. \tag{1.32}$$

The reflectance R_{1f} and transmittance T_{1f} are given as follows:

$$R_{1f} = r_{1f} r_{1f}^* \tag{1.33}$$

$$T_{1f} = \frac{k_{fz}}{k_{1z}} t_{1f} t_{1f}^*, \tag{1.34}$$

where r_{1r}^* and r_{1r}^* are the complex conjugates of r_{1r} and t_{1r}, respectively.

Program 1.6 calculates the reflectance of a thin film using the transfer matrix method. Lines 7, 10, and 15 define functions to create $\boldsymbol{M}_{ij}$ and $\boldsymbol{\Phi}_i$ using the matrix command. Lines 46 through 52 prepare an array containing each angle's reflection and transmission coefficients. Lines 59 through 64 calculate the reflection coefficient of a thin film. In Lines 66 and 67, we multiply $\boldsymbol{M}_{ij}$ and $\boldsymbol{\Phi}_i$ to create the transfer matrix $\boldsymbol{G}$. From this, the total reflection and transmission coefficients are obtained in Lines 69–75, and the total reflectance and transmittance in Lines 78 and 79. The results obtained are the same as those obtained in Program 1.5.

Program 1.6

```
import scipy as sp
import matplotlib as mpl
import matplotlib.pyplot as plt
from scipy import pi,sin,cos,tan,arcsin,exp,linspace,arrange,sqrt,zeros,array,matrix,asmatrix
from matplotlib.pyplot import plot,show,xlabel,ylabel,title,legend,grid, axis

def mMATs(n1z,n2z):
    return (1/(2*n1z))*matrix([[n1z+n2z,n1z-n2z],[n1z-n2z,n1z+n2z]])
                                  # Definition of Mij Matrix for s-pol
def mMATp(n1z,n2z,n1,n2):
    return (1/(2*n1*n2*n1z))*\backslash
            matrix([[n1**2*n2z+n2**2*n1z,n1**2*n2z-n2**2*n1z],\backslash
                   [n1**2*n2z-n2**2*n1z,n1**2*n2z+n2**2*n1z]])
                                  # Definition of Mij Matrix for s-pol
def matFAI(n1z,d1,k0):
    return matrix([[exp(1j*n1z*k0*d1), 0],[0,exp(-1j*n1z*k0*d1)]])
                                 # Definition of Phi Matrix for s-pol

n1=1.0             # refractive index of medium 1
n2=1.5             # refractive index of medium 2
n3=1.0             # refractive index of medium 3
ep1=n1**2          # dielectric constant of medium 1
ep2=n2**2          # dielectric constant of medium 2
ep3=n3**2          # dielectric constant of medium 3
d2=100             # Thickness of medium 2 d2 (nm)
WL=500             # Vacuum wavelength WL (nm)
k0=2*pi/WL         # Vacuum wavenumber

t1start=0          # Start angle
t1end=89            # End angle
t1points=90         # Number of Plots

t1Deg = linspace(t1start,t1end,t1points)  # Generation of an array of incident angles t1. (deg.)
t1 = t1Deg /180*pi   # Convert the angle of incidence into radians.
s1 = sin(t1)         # sin(t1)
c1 = cos(t1)         # cos(t1)
s2 = n1/n2*s1        # sin(t1)
c2 = sqrt(1-s2**2)   # cos(t2)
s3 = n1/n3*s1        # sin(t1)
c3 = sqrt(1-s3**2)   # cos(t3)

n1z=n1*c1          # n1z=k1z/k0
n2z=n2*c2          # n2z=k1z/k0
n3z=n3*c3          # n2z=k1z/k0

mMats21=zeros((t1points,2,2),dtype=complex)         # M21 matrix initialization for s-pol
```

```
mMats32=zeros((t1points,2,2),dtype=complex)          # M32 matrix initialization for s-pol
mMatp21=zeros((t1points,2,2),dtype=complex)          # M21 matrix initialization for p-pol
mMatp32=zeros((t1points,2,2),dtype=complex)          # M32 matrix initialization for p-pol
matFAI2=zeros((t1points,2,2),dtype=complex)          # Phi2 matrix initialization
matTs=zeros((t1points,2,2),dtype=complex)            # Transfer matrix  initialization for s-pol
matTp=zeros((t1points,2,2),dtype=complex)            # Transfer matrix  initialization for p-pol
rs=zeros((t1points),dtype=complex)                   # rs initialization
ts=zeros((t1points),dtype=complex)                   # ts initialization
rp=zeros((t1points),dtype=complex)                   # rp initialization
tp=zeros((t1points),dtype=complex)                   # tp initialization

for i in range(t1points):
    mMats21[i]=mMATs(n2z[i],n1z[i])       # M21 generation
    mMats32[i]=mMATs(n3z[i],n2z[i])       # M32 generation
    mMatp21[i]=mMATp(n2z[i],n1z[i],n2,n1)      # M21 generation
    mMatp32[i]=mMATp(n3z[i],n2z[i],n3,n2)      # M32 generation

    matFAI2[i]=matFAI(n2z[i],d2,k0)       # Phi2 generation

    matTs[i]=mMats32[i]@matFAI2[i]@mMats21[i]  # Generation of transfere matrix for s-pol
    matTp[i]=mMatp32[i]@matFAI2[i]@mMatp21[i] # Generation of transfere matrix for p-pol

    rs[i]=-matTs[i,1,0]/matTs[i,1,1]
                 # reflection coefficient for s-pol
    ts[i]=matTs[i,0,0]-matTs[i,0,1]*matTs[i,1,0]/matTs[i,1,1]
                 # transmission coefficient for s-pol
    rp[i]=-matTp[i,1,0]/matTp[i,1,1]
                 # reflection coefficient for p-pol
    tp[i]=matTp[i,0,0]-matTp[i,0,1]*matTp[i,1,0]/matTp[i,1,1]
                 # transmission coefficient for p-pol

RsAbs=abs(rs)**2         # Reflectivity for s-pol
RpAbs=abs(rp)**2         # Reflectivity for p-pol

plot(t1Deg,RpAbs, label="Rp")          # Plot Rp
plot(t1Deg,RsAbs, label="Rs")          # Plot Rs
xlabel(r"$\theta_1$ (deg.)",fontsize=20)    # Label x-axis
ylabel(r"$r, t$",fontsize=20)               # Label y-axis
title("Reflectivity",fontsize=20)           # Plot range
grid(True)                                  # Show grid
legend(fontsize=16)              # Show legend and set font size
plt.tick_params(labelsize=20)           # Axis scales
tight_layout()  # Commands to make the graph fit into a frame.
show()          # Show graph.
```

In Python 2, we cannot use "@" for the multiplication of $\boldsymbol{M}_{ij}$ and $\boldsymbol{\Phi}_i$ in part to create the propagation matrix $\boldsymbol{G}$ in Lines 66 and 67. This part must be written with nested dot commands, as in Program 1.7.

Program 1.7

```
tmatrixs = mMATs(n3z,n2z).dot(matFAI(n2z,d2,k0).dot(mMATs(n2z,n1z
    )))
tmatrixp = mMATp(n3z,n2z,n3,n2).dot(matFAI(n2z,d2,k0).dot(mMATp(
    n2z,n1z,n2,n1)))
```

Finally, as an application example, we calculate the surface plasmon resonance spectrum using the total reflection attenuation method. Surface plasmon resonance is free-electron waves in a thin metal film interacting with light at the surface under a certain condition. The resonance appears as an optical absorption or an enhancement of the electric field intensity near the surface. For example, when the reflectance of the p-polarized light incident through a prism is measured, as in Figure 1.8(a), the reflectance drops to a minimum at the angle of incidence on resonance. This angle is called the resonance angle. The resonance angle changes when a dielectric layer is adsorbed on a metal surface or when the ambient medium's refractive index changes, so it is used as an optical sensor for refractive index or biological substances such as proteins and DNA.

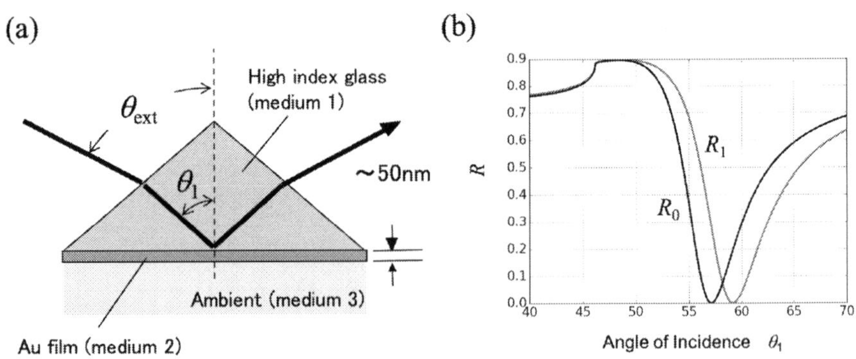

FIGURE 1.8
Surface Plasmon Resonance (a) optical geometry and (b) calculated results.

The results of the calculations using Program 8.1 in the Appendix are shown in Figure 1.8(b). Since water (refractive index 1.33) is assumed as the ambient medium, we used a high refractive index glass (refractive index 1.86) for the prism. Gold was used as the metallic film, with a thickness of 47 nm. The angle of incidence t1DegOut to the right angle prism is defined as an array. The internal angle θ_1 (s1 in the program) caused by refraction at the slope when light is incident on the prism is different. There is a relationship $\theta_1 = 45° + \sin(\theta_{\text{ext}} - 45°)/n_1$ between the angle of incidence θ_{ext} (external angle) into the right angle prism with refractive index n_1 in air and the angle of incidence θ_1 inside the prism. The results show that a resonance angle shift of 2.1°, which stems from a 10-nm-thick film with a refractive index of 1.5 on the surface of the gold film. Since the accuracy of the angle measurement is 1/1000∼1/100°, the adsorption and desorption of minute substances corresponding to a film thickness of 0.1 Å or less can be measured.

1.5 Transfer matrix method for anisotropic media

1.5.1 Eigen propagation modes and boundary conditions

An optically anisotropic medium is a medium whose refractive index differs depending on the polarization direction. Most optical crystals, liquid crystals, and stretched polymer films are anisotropic. Biaxial media have three different refractive indices, and uniaxial media have two different refractive indices. Here, we consider a uniaxial medium in which the optical axis coincides with the direction normal to the surface of the layer, as shown in Figure 1.9. In this case, there is an extraordinary principal refractive index n_e, which is the refractive index for polarized light in the direction of the optical axis, and an ordinary refractive index n_o, which is the refractive index for polarized light[5]. Light propagating in a uniaxial medium is divided into ordinary and extraordinary light, with polarization directions differing by 90°. The refractive indices of ordinary and extraordinary light are n_o and $n_e(\theta_{2e})$, respectively, where the angle between the direction of light and the optical axis is θ. The extraordinary light refractive index has a value between the extraordinary principal refractive index and the ordinary light refractive index. The exception is when light propagates along the optical axis, in which case only ordinary light exists, and the ordinary refractive index applies in all polarization directions.

We introduce methods for calculating reflection and transmission in multilayers of anisotropic media. The optical configuration is shown in Figure 1.9. The z-axis is defined as normal to the surface, and the direction of light propagation is positive. Light is incident from isotropic Medium 1 at an angle of incidence θ_1, passes through a thin film of anisotropic Medium 2 (film

[5] Note that the extraordinary *principal* refractive index, n_e, differs from the extraordinary refractive index, $n_e(\theta_{2e})$. The former is an optical constant, while the latter is a function of the direction of light propagation (θ_{2e}).

thickness d_2), and is transmitted to isotropic Medium 3 at a refraction angle θ_3. Let the refractive indices of Medium 1 and Medium 3 be n_1 and n_3, respectively, and $\eta_i = k_{iz}/k_0$ ($i = 1$ or 3), where k_{iz} is the z-component of the wavevector in Medium i. We also have the extraordinary light principal refractive index n_e and the ordinary light refractive index n_o of Medium 2. Let θ_{2e} and θ_{2o} be the corresponding refraction angles.

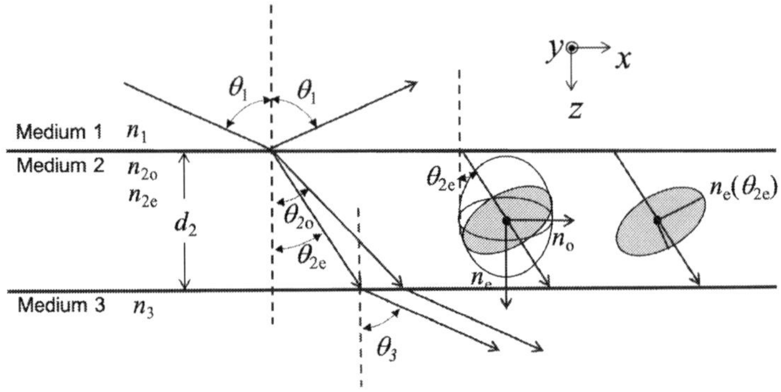

FIGURE 1.9
Reflection and transmission of an anisotropic film.

Find the relationship between the wavenumber vector $\boldsymbol{k}_2$ of light propagating through Medium 2 and the intrinsic polarization. The eigenequation of the light is written with $\boldsymbol{E}_2$ for the light electric field of Medium 2 and k_0 as the wavenumber in a vacuum as

$$\left(k_2^2 - \boldsymbol{k}_2\boldsymbol{k}_2 - k_0^2\hat{\epsilon}\right)\boldsymbol{E}_2 = 0. \tag{1.35}$$

$\hat{\epsilon}$ is the dielectric constant tensor of Medium 2. With a refractive index n_e for extraordinary light and that for ordinary light n_o, it is

$$\hat{\epsilon} = \begin{pmatrix} n_o^2 & 0 & 0 \\ 0 & n_o^2 & 0 \\ 0 & 0 & n_e^2 \end{pmatrix}. \tag{1.36}$$

$\boldsymbol{E}_2 = (E_{2x}, E_{2y}, E_{2z})$ are calculated as follows:

$$\begin{pmatrix} \eta_2^2 - n_o^2 & 0 & -\kappa\eta_2 \\ 0 & \kappa^2 + \eta_2^2 - n_o^2 & 0 \\ -\kappa\eta_2^2 & 0 & \kappa^2 - n_e^2 \end{pmatrix} \begin{pmatrix} E_{2x} \\ E_{2y} \\ E_{2z} \end{pmatrix} = 0 \tag{1.37}$$

Here, κ corresponds to the x-directional component k_x of the wavenumber vector, defined as $\kappa = k_x/k_0$, and is equal in each layer. Also, η_2 is the quantity corresponding to the z component k_{2z} of the wavenumber vector in Medium

2, defined by $\eta_2 = k_{2z}/k_0$. For this expression to make sense, the determinant must equal zero, leading to the following relationship:

$$(\kappa^2 + \eta_2^2 - n_o^2)(n_e^2 n_o^2 - \eta_2^2 n_o^2 - \kappa^2 n_e^2) = 0 \tag{1.38}$$

Since κ, n_o, n_o, and n_e is known, η_2 is unknown. The absolute value of η_2 corresponding to normal light is written as η_{2o}, and that corresponding to extraordinary light as η_{2e}. Each has multiple solutions, with positive and negative corresponding to the direction of light propagation – positive to light propagating forward (in the positive direction of the z-axis), and negative to light propagating backward. They are as follows:

$$\eta_2 = \pm\eta_{2o}, \text{ where } \eta_{2o} = \sqrt{n_o^2 - \kappa^2} \tag{1.39}$$

$$\eta_2 = \pm\eta_{2e}, \text{ where } \eta_{2e} = \left(\frac{n_o}{n_e}\right)\sqrt{n_e^2 - \kappa^2} \tag{1.40}$$

To distinguish the four eigen propagation modes, we number them as $\eta_2^{(1)} = \eta_{2e}$, $\eta_2^{(2)} = -\eta_{2e}$, $\eta_2^{(3)} = \eta_{2o}$, and $\eta_2^{(4)} = -\eta_{2o}$. $\eta_2^{(1)}$ corresponds to extraordinary light propagating forward in Medium 2, $\eta_2^{(2)}$ to extraordinary light propagating backward, $\eta_2^{(3)}$ to ordinary light propagating forward, and $\eta_2^{(4)}$ to ordinary light propagating backward.

In the case of a uniaxial medium with an optical axis normal to the surface, p- and s-polarized light do not affect each other and can be treated independently. First, the boundary conditions between Medium 1 and Medium 2 in p-polarized light are described because the tangential components of the electric and magnetic fields must be continuous,

$$\frac{\eta_1}{n_1}E_1^+ - \frac{\eta_1}{n_1}E_1^- = \cos\theta_2' E_2^+ - \cos\theta_2' E_2^- \tag{1.41}$$

$$n_1 E_1^+ + n_1 E_1^- = \frac{n_{2o}^2}{n_{2e}}\cos\theta_2' E_2^+ + \frac{n_{2o}^2}{n_{2e}}\cos\theta_2' E_2^-. \tag{1.42}$$

Here, θ_2' is the angle between the surface normal and the Pointing vector. In an anisotropic medium, the wavenumber vector and the Pointing vector do not have the same direction, which differs from the refraction angle θ_{2o} and θ_{2e}. The cosine and sine of θ_2' are represented by n_e and n_o as follows:

$$\cos\theta_2' = \frac{n_e\sqrt{n_e^2 - \kappa^2)}}{\sqrt{n_e^4 + \kappa^2(n_o^2 - n_e^2)}} = \frac{n_e^2\eta_{2e}}{n_o\sqrt{n_e^4 + \kappa^2(n_o^2 - n_e^2)}} \tag{1.43}$$

$$\sin\theta_2' = \frac{n_o\kappa}{\sqrt{n_e^4 + \kappa^2(n_o^2 - n_e^2)}} \tag{1.44}$$

The boundary conditions between Medium 2 and Medium 3 are

$$\cos\theta_2'\phi_{2e}^+ E_2^+ - \cos\theta_2'\phi_{2e}^- E_2^- = \frac{\eta_3}{n_3}E_3^+ \tag{1.45}$$

$$\frac{n_{2o}^2}{n_{2e}}\cos\theta_2'\phi_{2e}^+ E_2^+ + \frac{n_{2o}^2}{n_{2e}}\cos\theta_2'\phi_{2e}^- E_2^- = n_3 E_3^+. \tag{1.46}$$

Here, ϕ_2 is the phase difference during propagation and $\phi_{2e}^{\pm} = \exp(\pm i\eta_{2e}d_2)$. Although there are five unknowns, these equations can be solved because the ratio of E_1^- to E_1^+ is the reflection coefficient $r = E_1^-/E_1^+$ and the ratio of E_3^+ to E_1^+ is the transmission coefficient $t = E_3^+/E_1^+$.

Next, we describe the boundary conditions for Mediums 1 and 2 for s-polarized light.

$$E_1^+ + E_1^- = E_2^+ + E_2^- \tag{1.47}$$
$$\eta_1 E_1^+ - \eta_1 E_1^- = \eta_{2o} E_2^+ - \eta_{2o} E_2^- \tag{1.48}$$

The boundary conditions between Medium 2 and Medium 3 are as follows:

$$\phi_{2o}^+ E_2^+ + \phi_{2o}^- E_2^- = E_3^+ \tag{1.49}$$
$$\eta_{2o}\phi_{2o}^+ E_2^+ - \eta_{2o}\phi_{2o}^- E_2^- = \eta_3 E_3^+. \tag{1.50}$$

Here, $\phi_{2o}^{\pm} = \exp(\pm i\eta_{2o}d_2)$. The transmission coefficient and transmittance for s-polarized light can be obtained from these equations.

1.5.2 Transfer matrix method for anisotropic medium

Solving a series of equations with the boundary conditions is easy to understand in its physical meaning, but it is not practical for calculations dealing with multilayers. It is complicated when the optical axis is not normal to the surface or in the plane of incidence. Here, we introduce a calculation method using the transfer matrix that solves these problems, proposed by Bethune [5].

First, the polarization unit vector $\boldsymbol{u}$ corresponding to each eigenvector is obtained. In the case of uniaxial media where the optical axis is along the surface normal, $\boldsymbol{u}$ can be easily obtained, as shown below, but in other cases, some calculation is required. For extraordinary light, from the three conditions, the electric field vector is in the xz plane, is orthogonal to the wavenumber vector, and is a unit vector. We have

$$\boldsymbol{u}^{(1)} = \begin{pmatrix} -\cos\theta' \\ 0 \\ \sin\theta' \end{pmatrix} \qquad \boldsymbol{u}^{(2)} = \begin{pmatrix} \cos\theta' \\ 0 \\ \sin\theta' \end{pmatrix}. \tag{1.51}$$

Here, θ' is the angle between the Pointing vector and the z-axis. The direction of polarization for ordinary light is

$$\boldsymbol{u}^{(3)} = \boldsymbol{u}^{(4)} = \begin{pmatrix} 0 \\ 1 \\ 0 \end{pmatrix}. \tag{1.52}$$

The boundary condition is that the tangential components of the electric and magnetic fields are continuous. As shown in Section 1.4, for p- and s-polarized light in medium i, the electric fields of the forward and backward propagating light are summarized, and a vector $\boldsymbol{E}_i$ is defined. The polarization and the direction of light propagation are expressed as superscripts, respectively.

$$\boldsymbol{E}_i = \begin{pmatrix} E_i^{\mathrm{p}+} \\ E_i^{\mathrm{p}-} \\ E_i^{\mathrm{s}+} \\ E_i^{\mathrm{s}-} \end{pmatrix} \tag{1.53}$$

On the other hand, to incorporate the continuity condition at the boundary, it is sufficient to consider the following four components of the electric and magnetic field components, called the Berreman vector $\boldsymbol{\psi}$.

$$\boldsymbol{\psi}_i = \begin{pmatrix} E_x \\ B_y \\ E_y \\ -B_x \end{pmatrix} \tag{1.54}$$

The relationship between $\boldsymbol{\psi}_i$ and $\boldsymbol{E}_i$ is written as

$$\boldsymbol{\psi}_i = \boldsymbol{\Pi}_i \boldsymbol{E}_i. \tag{1.55}$$

Here, $\boldsymbol{\Pi}_i$ can be written using η and u of Medium 2 from Eq. (1.5) as follows:

$$\boldsymbol{\Pi}_i = \begin{pmatrix} u_x^{(1)} & u_x^{(2)} & u_x^{(3)} & u_x^{(4)} \\ \eta^{(1)}u_x^{(1)} - \kappa u_z^{(1)} & \eta^{(2)}u_x^{(2)} - \kappa u_z^{(2)} & \eta^{(3)}u_x^{(3)} - \kappa u_z^{(3)} & \eta^{(4)}u_x^{(4)} - \kappa u_z^{(4)} \\ u_y^{(1)} & u_y^{(2)} & u_y^{(3)} & u_y^{(4)} \\ \eta^{(1)}u_y^{(1)} & \eta^{(2)}u_y^{(2)} & \eta^{(3)}u_y^{(3)} & \eta^{(4)}u_y^{(4)} \end{pmatrix} \tag{1.56}$$

If Medium 2 is uniaxial and the optical axis is along the surface normal,

$$\boldsymbol{\Pi}_2 = \begin{pmatrix} \cos\theta_2' & -\cos\theta_2' & 0 & 0 \\ \frac{n_{2\mathrm{o}}^2}{n_{2\mathrm{e}}}\cos\theta_2' & \frac{n_{2\mathrm{o}}^2}{n_{2\mathrm{e}}}\cos\theta_2' & 0 & 0 \\ 0 & 0 & 1 & 1 \\ 0 & 0 & \eta_{2\mathrm{o}} & -\eta_{2\mathrm{o}} \end{pmatrix}. \tag{1.57}$$

It consists of two independent 2×2 matrices, indicating that p- and s-polarized light can be treated independently, although a diagonal component generally arises, and p- and s-polarized light interact. In other words, when p-polarized light is incident, the reflected or transmitted light involves an s-polarized component.

Medium 1 and Medium 3 are isotropic media, then $\boldsymbol{\Pi}_i$ $(i = 1 \text{ or } 3)$ is written as follows:

$$\boldsymbol{\Pi}_i = \begin{pmatrix} \frac{\eta_i}{n_i} & -\frac{\eta_i}{n_i} & 0 & 0 \\ n_i & n_i & 0 & 0 \\ 0 & 0 & 1 & 1 \\ 0 & 0 & \eta_i & -\eta_i \end{pmatrix} \tag{1.58}$$

Now, let $\boldsymbol{\psi}_i$ be the Berreman vector in medium i,

$$\boldsymbol{\Pi}_2\boldsymbol{\psi}_2 = \boldsymbol{\Pi}_1\boldsymbol{\psi}_1 \tag{1.59}$$

$$\boldsymbol{\Pi}_3\boldsymbol{\psi}_3 = \boldsymbol{\Pi}_2\Phi_2\boldsymbol{\psi}_2. \tag{1.60}$$

The Φ_2 gives the phase difference of the light propagating through Medium 2 and is expressed as follows:

$$\Phi_2 = \begin{pmatrix} \phi_{2\mathrm{e}}^{+} & 0 & 0 & 0 \\ 0 & \phi_{2\mathrm{e}}^{-} & 0 & 0 \\ 0 & 0 & \phi_{2\mathrm{o}}^{+} & 0 \\ 0 & 0 & 0 & \phi_{2\mathrm{o}}^{-} \end{pmatrix} \tag{1.61}$$

Here, $\phi_{2\mathrm{e}}^{\pm} = \exp(\pm i\eta_{2\mathrm{e}}d_2)$ and $\phi_{2\mathrm{o}}^{\pm} = \exp(\pm i\eta_{2\mathrm{o}}d_2)$.

From Eqs. (1.59) and (1.60), we have

$$\boldsymbol{\Pi}_3\boldsymbol{\psi}_3 = (\boldsymbol{\Pi}_3^{-1}\boldsymbol{\Pi}_2)\Phi_2(\boldsymbol{\Pi}_2^{-1}\boldsymbol{\Pi}_1)\boldsymbol{\psi}_1 = \mathbf{M}_{32}\Phi_2\mathbf{M}_{21}, \tag{1.62}$$

where $\mathbf{M}_{ji} = \boldsymbol{\Pi}_j^{-1}\boldsymbol{\Pi}_i$. Although the inverse matrix of $\boldsymbol{\Pi}$ does not exist in isotropic media or in uniaxial media when the surface normal and the optical axis coincide, the optical model considered here allows the use of the effective inverse matrix shown below.

$$\boldsymbol{\Pi}_2^{-1} = \frac{1}{2}\begin{pmatrix} \frac{1}{\cos\theta_2'} & \frac{\eta_{2\mathrm{e}}}{n_{2\mathrm{o}}^2\cos\theta_2'} & 0 & 0 \\ -\frac{1}{\cos\theta_2'} & \frac{\eta_{2\mathrm{e}}}{n_{2\mathrm{o}}^2\cos\theta_2'} & 0 & 0 \\ 0 & 0 & 1 & \frac{1}{\eta_{2\mathrm{o}}} \\ 0 & 0 & 1 & -\frac{1}{\eta_{2\mathrm{o}}} \end{pmatrix} \tag{1.63}$$

$$\boldsymbol{\Pi}_i^{-1} = \frac{1}{2}\begin{pmatrix} \frac{n_i}{\eta_i} & \frac{1}{n_i} & 0 & 0 \\ -\frac{n_i}{\eta_i} & \frac{1}{n_i} & 0 & 0 \\ 0 & 0 & 1 & \frac{1}{\eta_i} \\ 0 & 0 & 1 & -\frac{1}{\eta_i} \end{pmatrix} \qquad (i = 1 \text{ or } 3) \tag{1.64}$$

1.5.3 Hyperbolic metamaterials

Here, we consider reflectance and transmittance in hyperbolic metamaterials (HMMs) as an example of calculating an effective anisotropic medium. The multilayers composed of thin films thinner than the wavelength of light, as shown in Figure 1.10(a), is an effective anisotropic media with different eigenwavenumber vectors in the surface normal and in-plane directions. In particular, when alternating layers of metal and dielectric are used, the iso-wavenumber surface becomes an HMM, which is a hyperbolic surface, and the wavenumber in the surface normal direction can be increased, and other peculiar optical properties are exhibited [6]. This can be understood by the effective medium approximation (EMA), in which the multilayer is regarded as an effective medium, as shown in Figure 1.10(b). Consider the case where the HMM comprises two types of mediums, A and B. Let the dielectric constants of each be $\epsilon_{\rm A}$ and $\epsilon_{\rm B}$. The thicknesses of the thin films are equal and are $d_{\rm A}$ and $d_{\rm B}$, respectively. The effective dielectric constant depends on the direction of polarization and can be written as follows, where the dielectric constants in the z and in-plane directions are ϵ_z and $\epsilon_{\parallel}$, respectively.

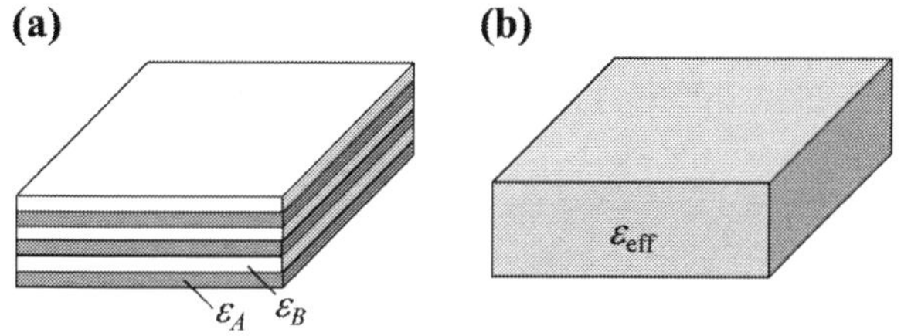

FIGURE 1.10
Mutilayer(a) and effective medium(b).

$$\epsilon_z = \frac{\epsilon_{\rm A}\epsilon_{\rm B}}{q\epsilon_{\rm B} + (1-q)\epsilon_{\rm A}} \tag{1.65}$$

$$\epsilon_{\parallel} = q\epsilon_{\rm A} + (1-q)\epsilon_{\rm B} \tag{1.66}$$

Here, q is the volume fraction of medium A and $q = d_{\rm A}/(d_{\rm A} + d_{\rm B})$. Therefore, the z-directional component perpendicular to the surface differs from the in-plane surface component, and the HMM can be regarded as a uniaxial anisotropic medium with the optical axis perpendicular to the surface.

Generally, anisotropic media with optical axes in the x-, y-, and z-directions have a diagonalized permittivity tensor. If the components in the x-, y-, and z-directions are ϵ_{xx}, ϵ_{yy}, and ϵ_{zz} respectively, we can consider the dielectric constant ellipsoid shown by the following equation:

$$\frac{x^2}{\epsilon_{xx}} + \frac{y^2}{\epsilon_{yy}} + \frac{z^2}{\epsilon_{zz}} = 1 \tag{1.67}$$

The dispersion relation of light propagating through this medium can be obtained from Eq. (1.38). As $\epsilon_{xx} = \epsilon_{yy} = \epsilon_{\parallel}$ and $\epsilon_z = \epsilon_{zz}$, and by multiplying both sides by the wavenumber in the vacuum, for the ordinary light, the dispersion relation is

$$\frac{k_x^2}{\epsilon_{\parallel}} + \frac{k_y^2}{\epsilon_{\parallel}} + \frac{k_z^2}{\epsilon_{\parallel}} = \left(\frac{\omega}{c}\right)^2, \tag{1.68}$$

and for the extraordinary light,

$$\frac{k_x^2}{\epsilon_z} + \frac{k_y^2}{\epsilon_z} + \frac{k_z^2}{\epsilon_{\parallel}} = \left(\frac{\omega}{c}\right)^2. \tag{1.69}$$

Since it is clear that the dispersion relation for ordinary light is a sphere of radius $\sqrt{\epsilon_{\|}}\left(\frac{\omega}{c}\right)$, we now deal with the dispersion relation for extraordinary light. Considering three cases according to the signs of $\epsilon_{\|}$ and ϵ_z is necessary. Program 1.8 calculates the dispersion relation between the values taken by the wavenumber vector components when the signs of $\epsilon_{\|}$ and ϵ_z are positive. Eq. (1.68) can be easily understood if it is expressed in terms of the polar angle θ and azimuthal angle ϕ regarding the mediating variables. With this, we can write

$$x = \sqrt{\epsilon_z}\sin\theta\cos\phi \tag{1.70}$$

$$y = \sqrt{\epsilon_z}\sin\theta\sin\phi \tag{1.71}$$

$$z = \sqrt{\epsilon_{\|}}\cos\theta. \tag{1.72}$$

Program 1.8

```
1  import scipy as sp
2  from scipy import pi,sin,cos,tan, meshgrid,arrange
3  import pylab as pylab
4  import mpl_toolkits.mplot3d.axes3d as pylab3
5
6  u=arrange(0,2*pi,0.1)    # Creation of mesh Phi
7  v=arrange(0,1*pi,0.1)    # Creation of mesh Theta
8
9  epz = 4          # Dielectric constant in z-direction
10 epx = 9          # Dielectric constant in x-direction
11
12 uu,vv=meshgrid(u,v)          # Creation of mesh
13
14 x=epz*cos(uu)*sin(vv)    # Dielectric constant in x-direction
15 y=epz*sin(uu)*sin(vv)    #  Dielectric constant in y-direction
16 z=epx*cos(vv)            # Dielectric constant in z-direction
17
18 fig=pylab.figure()
19 ax = pylab3.Axes3D(fig,aspect=1)    # Declaration of the creation
       of 3D diagrams.
20 ax.plot_wireframe(x,y,z)            # Wireframe plotting.
21 ax.set_xlabel('X')    # x-direction label
22 ax.set_ylabel('Y')    # y-direction label
23 ax.set_zlabel('Z')    # z-direction label
24
25 ax.set_xlim3d(-10, 10)    # x-directional plotting range
26 ax.set_ylim3d(-10, 10)    # y-directional plotting range
27 ax.set_zlim3d(-10, 10)    # z-directional plotting range
28
29 pylab.show( )    # Display graph
```

The mesh is created with the grid mesh command in Line 12 before plotting and storing in the lists uu and vv. Plotting is made with the plot_wireframe command. The calculation result is a rotating ellipsoid as shown in Figure 1.11(a).

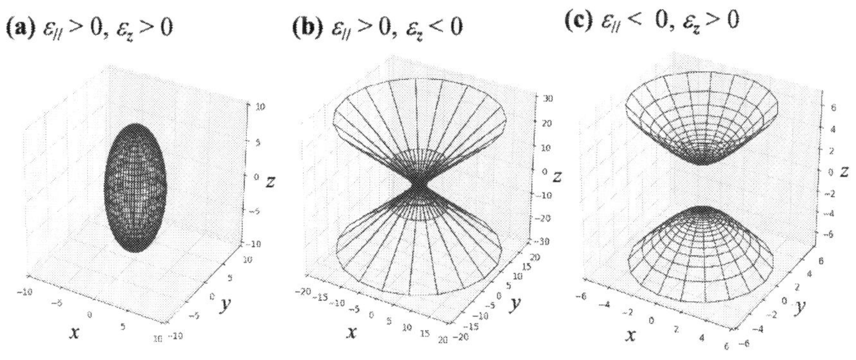

FIGURE 1.11
Dispersion relation of the multilayer film. (a) Both $\epsilon_{||}$ and ϵ_z are positive, (b) positive $\epsilon_{||}$ and negative ϵ_z, and (c) negative $\epsilon_{||}$ and positive ϵ_z.

On the other hand, if $\epsilon_{||}$ is positive and ϵ_z is negative, the dispersion relation becomes hyperbolic as shown in Figure 1.11(b). Eq. (1.68) in terms of the parameters is expressed as

$$x = \sqrt{\epsilon_z} \sec\theta \cos\phi \quad (1.73)$$
$$y = \sqrt{\epsilon_z} \sec\theta \sin\phi \quad (1.74)$$
$$z = \sqrt{\epsilon_{||}} \tan\theta. \quad (1.75)$$

When $\epsilon_{||}$ is negative and ϵ_z is positive, it is

$$x = \sqrt{\epsilon_z} \tan\theta \cos\phi \quad (1.76)$$
$$y = \sqrt{\epsilon_z} \tan\theta \sin\phi \quad (1.77)$$
$$z = \sqrt{\epsilon_{||}} \sec\theta. \quad (1.78)$$

Program 1.9 illustrates this relation, in which epx and epz are both absolute dielectric permittivity values, the positive and negative of which are chosen by describing the parameters. The fact that the dispersion relation is hyperbolic indicates the possibility of obtaining large wavenumber vector components.

Program 1.9

```
import scipy as sp
import matplotlib.pyplot as plt
from matplotlib.pyplot import plot,show,grid,axis,figure
from scipy import pi,sin,cos,tan,arcsin,meshgrid,linspace,sqrt
import mpl_toolkits.mplot3d.axes3d as p3d

def sec(x):
    return 1/cos(x)  #  Define function sec
```

```

u=linspace(0, 2*pi, 20)  #  Creation of mesh Theta
v=linspace(0, 2*pi, 20)  #  Creation of mesh Phi

epz = 3  #  Dielectric constant in z-direction (negative value)
epx = 5  #  Dielectric constant in x-direction (positive value)

uu,vv=meshgrid(u,v)  #  Create mesh

x=sqrt(epz)*sec(uu)*cos(vv)  #  Dielectric constant in x-
    direction
y=sqrt(epz)*sec(uu)*sin(vv)  #  Dielectric constant in y-
    direction
z=sqrt(epx)*tan(uu)  #  Dielectric constant in z-direction

fig=figure()
ax = p3d.Axes3D(fig,aspect=1)  #  Declaration of the creation of
    3D diagrams.
ax.plot_wireframe(x,y,z)  #  Wireframe plotting.
ax.set_xlabel('X')  #  x-direction label
ax.set_ylabel('Y')  #  y-direction label
ax.set_zlabel('Z')  #  z-direction label

ax.set_xlim3d(-20, 20)  #  x-directional plotting range
ax.set_ylim3d(-20, 20)  #  y-directional plotting range
ax.set_zlim3d(-30, 30)  #  z-directional plotting range

show()  #  Display graph
```

Finally, the results of the calculation for negative $\epsilon_{||}$ and positive ϵ_z are shown in Figure 1.11(c). Program 1.10 shows epx and epz, which are absolute dielectric permittivity values. For the convenience of mesh grid fabrication, the positive and negative portions of the z-axis are plotted separately and finally shown as a single surface. This case is characterized by the existence of a gap in k_z.

Program 1.10

```
import scipy as sp
import matplotlib.pyplot as plt
from matplotlib.pyplot import plot,show,grid, axis,subplot,figure
from scipy import pi,sin,cos,tan,arcsin,meshgrid,linspace,sqrt
import mpl_toolkits.mplot3d.axes3d as p3d

def sec(x):
    return 1/cos(x)  #  Define function sec

u=linspace(0, 0.4*pi, 20)  #  Creation of mesh Theta
v=linspace(0, 2*pi, 20)  #  Creation of mesh Phi

```

```
epz = 3  #  Dielectric constant in z-direction (positive value)
epx = 5  #  Dielectric constant in x-direction (negative value)

uu,vv=meshgrid(u,v)        #  Create mesh

x1=sqrt(epz)*tan(uu)*cos(vv)  #  Dielectric constant in x-direction
y1=sqrt(epz)*tan(uu)*sin(vv)  #  Dielectric constant in y-direction
z1=sqrt(epx)*sec(uu)  #  Dielectric constant in z-direction

x2=sqrt(epz)*tan(uu)*cos(vv)  #  Dielectric constant in x-direction
y2=sqrt(epz)*tan(uu)*sin(vv)  #  Dielectric constant in y-direction
z2=-sqrt(epx)*sec(uu)  #  Dielectric constant in z-directio

fig=figure()
ax = p3d.Axes3D(fig,aspect=1)  #  Wireframe Plotting
ax.plot_wireframe(x1,y1,z1)       #  Wireframe Plotting
ax.plot_wireframe(x2,y2,z2)  #  Wireframe Plotting

ax.set_xlabel('X')  #  x-direction label
ax.set_ylabel('Y')  #  y-direction label
ax.set_zlabel('Z')  #  z-direction label

ax.set_xlim3d(-6, 6)       #  x-direction plotting range
ax.set_ylim3d(-6, 6)       #  y-direction plotting range
ax.set_zlim3d(-7, 7)       #  z-direction plotting range

show()  #  Show graph
```

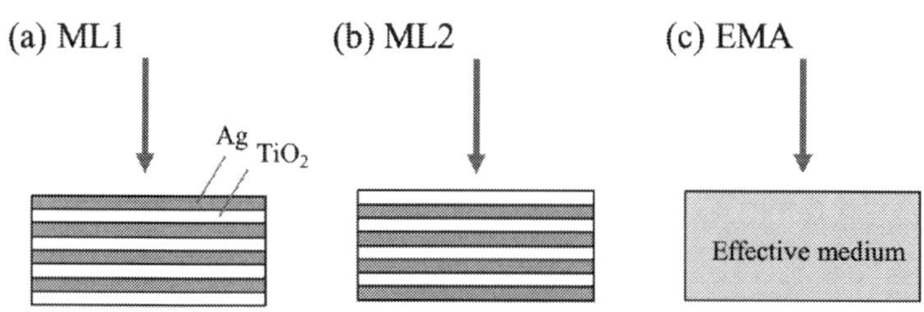

FIGURE 1.12
Silver and titanium dioxide multilayer films (a) and (b). Their effective medium (c).

For a multilayer structure consisting of eight layers of dielectric (TiO_2) and metal (Ag) with a thickness of 10 nm, as shown in Figures 1.12(a) and 1.12(b), we calculate the wavelength dependence of reflectivity and transmittance. They are compared with the wavelength dependence of reflectivity and transmittance of an 80-nm thick film of an effective medium shown in Figure 1.12(c). The difference in the structures between (a) and (b) is whether the top layer is metal (Ag) or dielectric (TiO_2). Calculations for multilayer structures can be performed using the method introduced in Section 1.4. On the other hand, since the effective medium is a single layer but an anisotropic medium, the calculation introduced in Section 1.5 was adopted. In addition, the dielectric constants ϵ_{TiO2} of the dielectric (TiO_2) and ϵ_{Ag} of the metal (Ag) are assumed to follow the following functions [7].

$$\epsilon_{\mathrm{TiO2}} = 5.193 + \frac{0.244}{(\lambda/1000)^2 - 0.0803} \quad (1.79)$$

$$\epsilon_{\mathrm{Ag}} = 3.691 - \frac{9.1522}{(1242/\lambda)^2 + i0.021 * (1242/\lambda)} \quad (1.80)$$

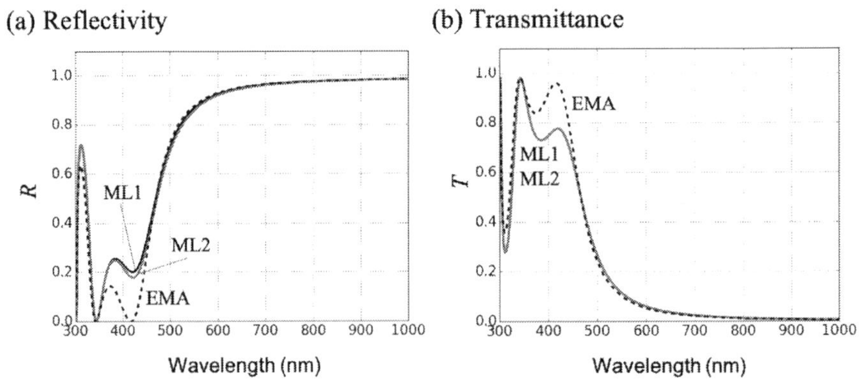

FIGURE 1.13
Calculated reflectance and transmittance of the multilayer at normal incidence; ML1 and ML2 refer to the structures in Figures 1.12(a) and (b), respectively; EMA is the effective medium shown in Figure 1.12(c).

Using Program 8.2 shown in the Appendix, the reflectivity and transmittance are shown for Figure 1.13 at an incidence angle of 0° and for Figure 1.14 at an incidence angle of 45°. The reflectivity differs slightly between ML1 and ML2, but the transmittance agrees. The difference in reflectance is due to a slight difference in absorption by Ag, and the agreement in transmittance can be explained by the optical reciprocity theorem in transmission. The values differ on the short wavelength side, but the characteristics are similar. When the incident angle is 45°, the EMA and ML models almost coincide on the wavelength side longer than 450 nm at 0° incident angle.

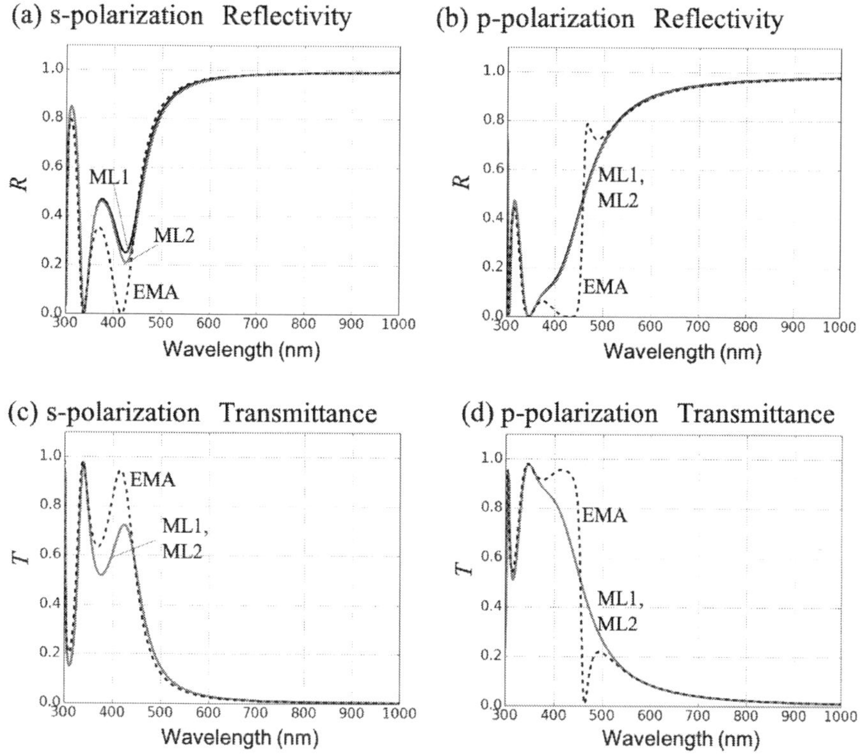

FIGURE 1.14
Calculated reflectance and transmittance of the multilayer at an angle of incidence of 45°. (a) Reflection spectrum of s-polarized light, (b) reflection spectrum of p-polarized light, (c) transmission spectrum of s-polarized light, and (d) reflection spectrum of p-polarized light. ML1 and ML2 refer to the structures in Figures 1.12(a) and (b), respectively. EMA is the effective medium shown in Figure 1.12(c).

A plot of the effective dielectric constant $\epsilon_{\parallel}$ and ϵ_z of the effective medium using Program 1.11 is shown in Figure 1.15. The imaginary part of $\epsilon_{\parallel}$ is extremely small, but the real part is positive on the short wavelength side and negative on the long wavelength side after 450 nm. In other words, the effective medium behaves as a dielectric at short wavelengths and as a metal at long wavelengths. Also, $\epsilon_{\parallel} = 0$ is realized at 450 nm, indicating that the real part of the refractive index is close to zero. On the other hand, the sign of the real part of ϵ_z is opposite to $\epsilon_{\parallel}$.

Program 1.11

```
import scipy as sp
import matplotlib as mpl
import matplotlib.pyplot as plt
from scipy import pi,arrange,sqrt,zeros,array
from matplotlib.pyplot import plot,show,xlabel,ylabel,title,
    legend,grid,axis

def func_nAg(WLs):
    ep=3.691-9.1522**2/((1240/WLs)**2+1j*0.021*(1240/WLs))
    index=sqrt(ep)
    return index  #  Dielectric function of silver

def func_nTiO2(WLs):
    ep=5.193 + 0.244/((WLs/1000)**2-0.0803)
    index=sqrt(ep)
    return index   #  Dielectric function of TiO2

WLmin = 300  #  Wavelength (shortest)  (nm)
WLmax = 1000  #  Wavelength (longest)  (nm)
WLperiod = 1  #  Wavelength period  (nm)
WLx = arrange(WLmin, WLmax+1, WLperiod)  #  Array of wavelengths
NumWLx = int((WLmax-WLmin)/WLperiod)+1  #  Number of wavelengths
k0=2*pi/WLx  #  Wavenumber

nTiO2=zeros((NumWLx),dtype=complex)
                          # TiO2 refractive index initialization
nAg=zeros((NumWLx),dtype=complex)
                          #  Ag refractive index initialization

for i in range(NumWLx):
    nTiO2[i]=func_nTiO2(WLx[i])
               #  Generation of refractive index of TiO2
    nAg[i]=func_nAg(WLx[i])
               # Generation of refractive index of Ag

epx=0.5*(nTiO2**2 + nAg**2)
                       #  Dielectric constant by EMA x-direction
epz=2*(nTiO2**2)*(nAg**2)/((nTiO2**2)+(nAg**2))
                       #  Delectric constant by EMA z-direction

plot(WLx,epx.real, label=r"Re$(\epsilon_{\rm \parallel})$")
                              #  Plot x (real)
plot(WLx,epx.imag, label=r"Im$(\epsilon_{\rm \parallel})$")
                              #  Plot x  (imaginary)
xlabel(r"Wavelength(nm)",fontsize=20)  #  x-axis labelong
ylabel(r"$ \epsilon_{\rm \parallel}$",fontsize=20)
                                 #  y-axis labeling
title("",fontsize=20)            #  Graph title
grid(True)                       #  Show grid
axis([300,1000,-30,30])          #  Plot range
legend(fontsize=16)              #  Show legend
plt.tick_params(labelsize=20)    #  Axis scales
show()                           #  Show graph

plot(WLx,epz.real, label=r"Re$(\epsilon_{\rm z})$")
```

```
                                                   #  Plot  z  (real)
plot(WLx,epz.imag,  label=r"Im$(\epsilon_{\rm z})$")
                                                   #  Plot  z   (imaginary)
xlabel(r"Wavelength(nm)",fontsize=20)                    #  Label x-axis
ylabel(r"$ \epsilon_{\rm z}$",fontsize=20)               #  Label y-axis
title("",fontsize=20)                                    #  Graph title
grid(True)                                               #  Show grid
axis([300,1000,-100,100])                                #  Plot range
legend(fontsize=16)                                      #  Show legend
plt.tick_params(labelsize=18)                            #  Axis scales
show()                                                   #  Show graph
```

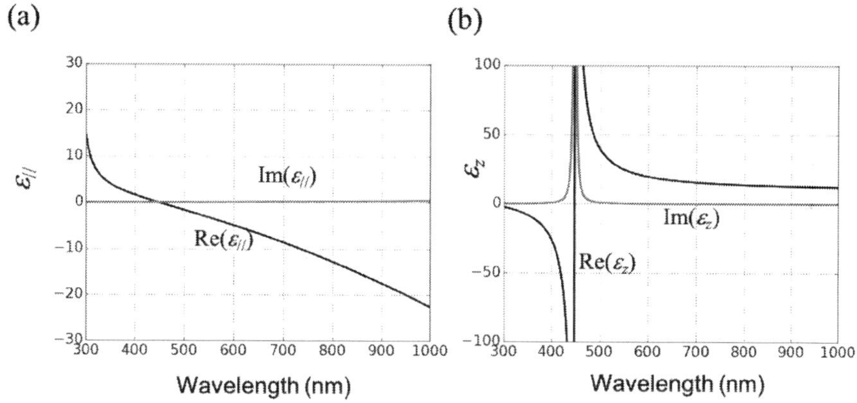

FIGURE 1.15
Dielectric constant of the effective medium (a) $\epsilon_{\|}$ and (b) ϵ_z. Both the real and imaginary parts are plotted.

2

Electromagnetic Analysis of Spheres

When light is irradiated to a small substance, scattering and absorption occur. Analytical solutions are available for small spheres and cylinders. This chapter uses Python to describe calculation programs for scattering and absorption by small spheres. If the sphere is much smaller than the wavelength of light, the long wavelength approximation can be adopted, and calculation is simple. Otherwise, calculation based on the Mie theory is necessary, which is complicated. This chapter deals with both cases.

2.1 Theory

2.1.1 Long wavelength approximation

The long wavelength approximation (quasi-static approximation) can be applied when the diameter of the sphere is sufficiently small compared to the wavelength of the light (approximately 1/7th of the wavelength or less). When the light electric field can be considered static, it is much easier to handle than calculations incorporating the retardation. Consider the absorption and scattering spectra of a nanosphere, as shown in Figure 2.1(a), where the refractive index of the surrounding medium and that of the sphere are n_1 and n_2, respectively. The following equation gives the polarizability α of the sphere with a radius R [8, 9, 10]:

$$\alpha = 4\pi n_1^2 R^3 \frac{n_1^2 - n_2^2}{2n_1^2 + n_2^2} \tag{2.1}$$

In the case of metallic spheres, n_2 is complex and wavelength-dependent. At a wavelength where $2n_1^2 + n_2^2$ is minimum, the polarizability α is maximum. This phenomenon is called localized surface plasmon resonance. If only the real part is considered, it is at a wavelength where $n_2^2 = -2n_1^2$. On the other hand, note that the resonance wavelength is determined solely by the sphere's refractive index n_2 and is independent of the sphere's radius. The scattering cross-section C_{sca}, extinction cross-section C_{ext} and absorption cross-section

DOI: 10.1201/9781003357674-2

C_{abs} are given by

$$\begin{aligned} C_{\text{sca}} &= \frac{k^4}{6\pi}|\alpha|^2 \\ C_{\text{abs}} &= k\ \text{Im}(\alpha) \\ C_{\text{ext}} &= C_{\text{sca}} + C_{\text{abs}}. \end{aligned} \tag{2.2}$$

Here, k is the wavenumber of light in the surrounding medium. The scattering efficiency Q_{sca}, extinction efficiency Q_{ext} and absorption efficiency Q_{abs} are obtained by normalizing the cross-section by the cross-sectional area.

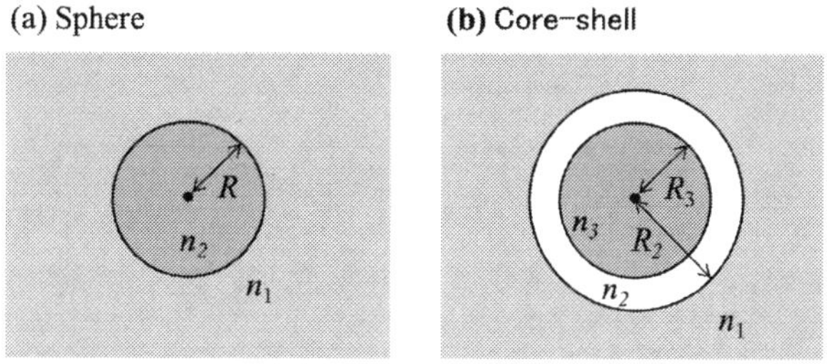

FIGURE 2.1
(a) Sphere and (b) core-shell structure.

2.1.2 Calculation of optical response of sphere with retardation

If the sphere is large, calculations incorporating retardation is necessary. We define the relative refractive index m, as $m = n_2/n_1$. Let λ and k_0 be the wavelength and wavenumber of light in vacuum, respectively, and ω be the angular frequency. The incident light electric field vector $\boldsymbol{E}_i$ is expanded into a spherical wave using the vector spherical harmonic functions

$$\boldsymbol{E}_i = \sum_{n=1}^{\infty} E_n(\mathbf{M}^{(1)}_{o1n} - i\mathbf{N}^{(1)}_{e1n}). \tag{2.3}$$

Here, $E_n = i^n \dfrac{2n+1}{n(n+1)} E_0$. Also, the magnetic field $\boldsymbol{H}_i$ is

$$\boldsymbol{H}_i = -\frac{k_0}{\omega\mu_0} \sum_{n=1}^{\infty} E_n(\mathbf{M}^{(1)}_{e1n} + i\mathbf{N}^{(1)}_{o1n}). \tag{2.4}$$

The electric field $\boldsymbol{E}_2$ and magnetic field $\boldsymbol{H}_2$ inside the sphere are obtained by using the coefficients c_n and d_n as

$$\begin{aligned}\boldsymbol{E}_2 &= \sum_{n=1}^{\infty} E_n(c_n\mathbf{M}_{o1n}^{(1)} - id_n\mathbf{N}_{e1n}^{(1)})\\ \boldsymbol{H}_2 &= -\frac{mk_0}{\omega\mu_0}\sum_{n=1}^{\infty} E_n(d_n\mathbf{M}_{e1n}^{(1)} + ic_n\mathbf{N}_{o1n}^{(1)}).\end{aligned} \tag{2.5}$$

Furthermore, the scattering fields $\boldsymbol{E}_\text{s}$ and $\boldsymbol{H}_\text{s}$ are expressed with coefficients a_n and b_n as follows:

$$\begin{aligned}\boldsymbol{E}_\text{s} &= \sum_{n=1}^{\infty} E_n(ia_n\mathbf{N}_{e1n}^{(3)} - ib_n\mathbf{M}_{o1n}^{(3)})\\ \boldsymbol{H}_\text{s} &= \frac{k_0}{\omega\mu_0}\sum_{n=1}^{\infty} E_n(ib_n\mathbf{N}_{o1n}^{(3)} + ia_n\mathbf{M}_{e1n}^{(3)})\end{aligned} \tag{2.6}$$

The coefficients a_n, b_n, c_n, and d_n are from the boundary conditions at the sphere surface. They are written as follows:

$$\begin{aligned}a_n &= \frac{m\psi_n(mx)\psi_n'(x) - \psi_n(x)\psi_n'(mx)}{m\psi_n(mx)\xi_n'(x) - \xi_n(x)\psi_n'(mx)}\\ b_n &= \frac{\psi_n(mx)\psi_n'(x) - m\psi_n(x)\psi_n'(mx)}{\psi_n(mx)\xi_n'(x) - m\xi_n(x)\psi_n'(mx)}\\ c_n &= \frac{m\psi_n(x)\xi_n'(x) - m\xi_n(x)\psi_n'(x)}{\psi_n(mx)\xi_n'(x) - m\xi_n(x)\psi_n'(mx)}\\ d_n &= \frac{m\psi_n(x)\xi_n'(x) - m\xi_n(x)\psi_n'(x)}{m\psi_n(mx)\xi_n'(x) - \xi_n(x)\psi_n'(mx)}\end{aligned} \tag{2.7}$$

The $'$ denotes the derivative due to the variables in parentheses. x is called the size parameter and $x = k_0R$. $\psi_n(\rho)$ and $\xi_n(\rho)$ are the Ruccati-Bessel functions. The spherical Bessel function $j_n(\rho)$ and the spherical Hankel function $h_n(\rho)$ are used to obtain them.

$$\begin{aligned}\psi_n(\rho) &= \rho j_n(\rho)\\ \xi_n(\rho) &= \rho h_n(\rho)\end{aligned} \tag{2.8}$$

Since we set the time dependence to $e^{-i\omega t}$, the spherical Hankel function of the first kind is used. The scattering cross-section C_sca and extinction cross section C_ext are expressed as

$$\begin{aligned}C_\text{sca} &= \frac{2\pi}{k_0^2}\sum_{n=1}^{\infty}(2n+1)(|a_n|^2 + |b_n|^2)\\ C_\text{ext} &= \frac{2\pi}{k_0^2}\sum_{n=1}^{\infty}(2n+1)\text{Re}(a_n + b_n).\end{aligned} \tag{2.9}$$

2.1.3 Core-shell structure (long-wavelength approximation)

Consider a core-shell structure with a core of refractive index n_3 and a shell of refractive index n_2 covering it in an ambient medium of refractive index n_1, as in Figure 2.1(b). First, we consider the case where the core-shell is sufficiently small compared to the wavelength to apply the long-wavelength approximation.

The spherical coordinate system (ρ,θ,ϕ) is used in the calculation. To simplify the calculation, ρ is normalized by the core radius R_3. Then $r = 1$ corresponds to the core's surface, and the shell's surface is $r = R_2/R_3 = s$. The potential at the medium i that arises when a unit electric field is applied in the z-direction is ψ_i, using the Legendre function $P_j(t)$.

$$\begin{aligned}
\psi_1 &= rt + \sum_{j=1}^{\infty} B_{1j} r^{-(j+1)} P_j(t) \\
\psi_2 &= \sum_{j=1}^{\infty} \Big(A_{2j} r^j P_j(t) + B_{2j} r^{-(j+1)} P_j(t) \Big) \\
\psi_3 &= \sum_{j=1}^{\infty} A_{3j} r^j P_j(t) \qquad (2.10)
\end{aligned}$$

where $t = \cos\theta$. Also, A_{ij} and B_{ij} are the coefficients of order j in medium i. The boundary conditions at the interface $r = 1$ between the core and shell and at the interface $r = s$ between the shell and the surrounding medium lead to the following four equations[1]:

$$\begin{aligned}
A_{31} - A_{21} - B_{21} &= 0 \\
\epsilon_3 A_{31} - \epsilon_2 A_{21} + 2\epsilon_2 B_{21} &= 0 \\
s^3 A_{21} + B_{21} - B_{11} - s^3 &= 0 \\
\epsilon_2 s^3 A_{21} - 2\epsilon_2 B_{21} + 2\epsilon_1 B_{11} &= \epsilon_1 s^3, \qquad (2.11)
\end{aligned}$$

where the dielectric constant ϵ_i in medium i is $\epsilon_i = n_i^2$. The following results from solving this system of equations [11].

$$\begin{aligned}
A_{21} &= \frac{s^3}{\Delta}(3\epsilon_1(2\epsilon_2 + \epsilon_3)) \\
B_{21} &= \frac{s^3}{\Delta}(3\epsilon_1(\epsilon_2 - \epsilon_3)) \\
A_{31} &= \frac{s^3}{\Delta}(9\epsilon_1\epsilon_2) \\
B_{11} &= \frac{s^3}{\Delta}(\epsilon_2(\epsilon_1(1 + 2s^3) - \epsilon_3(2 + s^3)) + (2\epsilon_2^2 - \epsilon_1\epsilon_3)(1 - s^3)). \quad (2.12)
\end{aligned}$$

[1]The potential is continuous at the boundary and the derivative of the potential multiplied by the dielectric permittivity is continuous.

Here,

$$\Delta = \epsilon_2(2\epsilon_1(1+2s^3) + \epsilon_3(2+s^3)) - 2(\epsilon_2^2 + \epsilon_1\epsilon_3)(1-s^3) \tag{2.13}$$

The polarizability α is obtained by the following equation:

$$\alpha = -4\pi\epsilon_1 R_2^3 B_{11} \tag{2.14}$$

The scattering cross-section C_{sca}, absorption cross-section C_{abs}, and extinction cross-section C_{ext} can be obtained using Eq. (2.2).

2.1.4 Core-shell structure (considering retardation)

Next, in the case of large core-shell spheres, retardation must be considered [8]. The treatment is similar to the Mie scattering case described in Section 2.1.2. Consider a core-shell structure with a core of radius R_3 and refractive index n_3 and a shell of radius R_2 and refractive index n_2 covering it in an ambient medium of refractive index n_1, as in Figure 2.1(b). The relative refractive indices are $m_3 = n_3/n_1$ and $m_2 = n_2/n_1$ for the core and shell, respectively, and two size parameters are defined as $x - k_0R_3$ and $y - k_0R_2$.

Let $\boldsymbol{E}_3$ and $\boldsymbol{H}_3$ denote the electric and magnetic fields inside the sphere, $\boldsymbol{E}_2$ and $\boldsymbol{H}_2$ the electric and magnetic fields in the shell, $\boldsymbol{E}_{\text{i}}$ and $\boldsymbol{H}_{\text{i}}$ the incident field electric and magnetic fields, and $\boldsymbol{E}_{\text{s}}$ and $\boldsymbol{H}_{\text{s}}$ the scattering field electric and magnetic fields. The incident and scattered fields are

$$\begin{aligned}
\boldsymbol{E}_{\text{i}} &= \sum_{n=1}^{\infty} E_n(\mathbf{M}_{o1n}^{(1)} - i\mathbf{N}_{e1n}^{(1)}) \\
\boldsymbol{H}_{\text{i}} &= -\frac{k_0}{\omega\mu_0}\sum_{n=1}^{\infty} E_n(\mathbf{M}_{e1n}^{(1)} + i\mathbf{N}_{o1n}^{(1)}) \\
\boldsymbol{E}_{\text{s}} &= \sum_{n=1}^{\infty} E_n(ia_n\mathbf{N}_{e1n}^{(3)} - ib_n\mathbf{M}_{o1n}^{(3)}) \\
\boldsymbol{H}_{\text{s}} &= \frac{k_0}{\omega\mu_0}\sum_{n=1}^{\infty} E_n(ib_n\mathbf{N}_{o1n}^{(3)} + ia_n\mathbf{M}_{e1n}^{(3)}).
\end{aligned} \tag{2.15}$$

The electric and magnetic fields in the core (Medium 3) are described as follows:

$$\begin{aligned}
\boldsymbol{E}_3 &= \sum_{n=1}^{\infty} E_n(c_n\mathbf{M}_{o1n}^{(1)} - id_n\mathbf{N}_{e1n}^{(1)}) \\
\boldsymbol{H}_3 &= -\frac{m_3k_0}{\omega\mu_0}\sum_{n=1}^{\infty} E_n(d_n\mathbf{M}_{e1n}^{(1)} + ic_n\mathbf{N}_{o1n}^{(1)})
\end{aligned} \tag{2.16}$$

On the other hand, in the shell, since it is the sum of the waves travelling to the inside of the sphere and the waves travelling to the outside, it is written as

$$\boldsymbol{E}_2 = \sum_{n=1}^{\infty} E_n (f_n \mathbf{M}^{(1)}_{o1n} - i g_n \mathbf{N}^{(1)}_{e1n} + v_n \mathbf{M}^{(3)}_{o1n} - i w_n \mathbf{N}^{(3)}_{e1n})$$

$$\boldsymbol{H}_2 = -\frac{m_2 k_0}{\omega \mu_0} \sum_{n=1}^{\infty} E_n (g_n \mathbf{M}^{(1)}_{e1n} + i f_n \mathbf{N}^{(1)}_{o1n} + w_n \mathbf{M}^{(3)}_{e1n} + i v_n \mathbf{N}^{(3)}_{o1n}). \quad (2.17)$$

Solving these with the boundary conditions at $\rho = R_2$ and $\rho = R_3$ yields

$$a_n = \frac{\psi_n(y)(\psi'_n(m_2 y) - A_n \chi'_n(m_2 y)) - m_2 \psi'_n(y)(\psi_n(m_2 y) - A_n \chi_n(m_2 y))}{\xi_n(y)(\psi'_n(m_2 y) - A_n \chi'_n(m_2 y)) - m_2 \xi'_n(y)(\psi_n(m_2 y) - A_n \chi_n(m_2 y))}$$
$$b_n = \frac{m_2 \psi_n(y)(\psi'_n(m_2 y) - B_n \chi'_n(m_2 y) - \psi'_n(y)(\psi_n(m_2 y) - B_n \chi_n(m_2 y))}{m_2 \xi_n(y)(\psi'_n(m_2 y) - B_n \chi'_n(m_2 y) - \xi'_n(y)(\psi_n(m_2 y) - B_n \chi_n(m_2 y))} \quad (2.18)$$

where A_n and B_n are

$$A_n = \frac{m_2 \psi_n(m_2 x)\psi'_n(m_3 x) - m_1 \psi'_n(m_2 x)\psi(m_3 x)}{m_2 \chi_n(m_2 x)\psi'(m_3 x) - m_1 \chi'_n(m_2 x)\psi(m_3 x)}$$
$$B_n = \frac{m_2 \psi_n(m_2 x)\psi'_n(m_2 x) - m_1 \psi_n(m_2 x)\psi'(m_3 x)}{m_2 \chi'_n(m_2 x)\psi(m_3 x) - m_1 \psi'_n(m_3 x)\chi(m_2 x)}. \quad (2.19)$$

Here, the Riccati Bessel function $\chi(\rho)$ is $\chi(\rho) = -\rho y_n(\rho)$ using the spherical Bessel function of the second kind $y_n(\rho)$.

2.2 Programing

2.2.1 Long-wavelength approximation

Here, we calculate the scattering spectra of scattering, absorption, and extinction spectra. The cross-sections of scattering (C_{sca}), absorption (C_{abs}), and extinction (C_{ext}) of a metal sphere that are small compared to the wavelength are calculated. Data on the refractive indices of the metals (gold and silver) at various wavelengths are needed. Refractive indices of metals are wavelength-dependent, which is discretely given in papers as [12]. Therefore, interpolation is used to obtain continuous spectra. The simplest linear interpolation, interpolate.interp1d, is used, although various other types of interpolation, such as spline interpolation, are available. These are made into a module, RI.py, and placed in the same directory (folder). By importing this module (Program 2.1), one can use the refractive index and dielectric constant in the program.

Program 2.1

```
from scipy import array,interpolate, arrange,zeros

RIAu=array([
    [292.4, 1.49, 1.878], [300.9, 1.53, 1.889], [310.7, 1.53, 1.893],
    [320.4, 1.54, 1.898], [331.5, 1.48, 1.883], [342.5, 1.48, 1.871],
    [354.2, 1.50, 1.866], [367.9, 1.48, 1.895], [381.5, 1.46, 1.933],
    [397.4, 1.47, 1.952], [413.3, 1.46, 1.958], [430.5, 1.45, 1.948],
    [450.9, 1.38, 1.914], [471.4, 1.31, 1.849], [495.9, 1.04, 1.833],
    [520.9, 0.62, 2.081], [548.6, 0.43, 2.455], [582.1, 0.29, 2.863],
    [616.8, 0.21, 3.272], [659.5, 0.14, 3.697], [704.5, 0.13, 4.103],
    [756.0, 0.14, 4.542], [821.1, 0.16, 5.083], [892.0, 0.17, 5.663],
    [984.0, 0.22, 6.350], [1088.0, 0.27, 7.150]])

RIAg=array([
    [292.4, 1.39, 1.161],[300.9, 1.34, 0.964],[310.7, 1.13, 0.616],
    [320.4, 0.81, 0.392],[331.5, 0.17, 0.829],[342.5, 0.14, 1.142],
    [354.2, 0.10, 1.419],[367.9, 0.07, 1.657],[381.5, 0.05, 1.864],
    [397.4, 0.05, 2.070],[413.3, 0.05, 2.275],[430.5, 0.04, 2.462],
    [450.9, 0.04, 2.657],[471.4, 0.05, 2.869],[495.9, 0.05, 3.093],
    [520.9, 0.05, 3.324],[548.6, 0.06, 3.586],[582.1, 0.05, 3.858],
    [616.8, 0.06, 4.152],[659.5, 0.05, 4.483],[704.5, 0.04, 4.838],
    [756.0, 0.03, 5.242],[821.1, 0.04, 5.727],[892.0, 0.04, 6.312],
    [984.0, 0.04, 6.992],[1088.0, 0.04, 7.795]])

NumWL = 26
WL=zeros(NumWL, dtype=int)
RIAuRe=zeros(NumWL, dtype=float)
RIAuIm=zeros(NumWL, dtype=float)
RIAgRe=zeros(NumWL, dtype=float)
RIAgIm=zeros(NumWL, dtype=float)

WLmin = 300
WLmax = 1000
WLperiod = 1
WLx = arrange(WLmin, WLmax+1, WLperiod)
  # Interpolated wavelengths 300-1000nm 1nm intervals
NumWLx = int((WLmax+1-WLmin)/WLperiod)  # number of wavelengths interpolated

```

```
for i in range(NumWL):
    WL[i]=RIAu[i,0]    #  zeroth of 2D array is wavelength
    RIAuRe[i]=RIAu[i,1]  # 1st of 2D array is real part (Au)
    RIAuIm[i]=RIAu[i,2]  # 2nd of 2D array is imaginary part (Au)
    RIAgRe[i]=RIAg[i,1]  # 1st of 2D array is real part (Ag)
    RIAgIm[i]=RIAg[i,2]  # 2nd of 2D array is imaginary part (Ag)

fRIAuReInt2 = interpolate.splrep(WL,RIAuRe,s=0)
   # Interpolation (Au real part)
RIAuReInt2 = interpolate.splev(WLx,fRIAuReInt2,der=0)
   # Interpolation (Au, real part)

fRIAuImInt2 = interpolate.splrep(WL,RIAuIm,s=0)
   # Interpolation (Au, imaginary part)
RIAuImInt2 = interpolate.splev(WLx,fRIAuImInt2,der=0)
   # Interpolation (Au, imaginary part)

fRIAgReInt2 = interpolate.splrep(WL,RIAgRe,s=0)
   #  Interpolation (Ag, real part)
RIAgReInt2 = interpolate.splev(WLx,fRIAgReInt2,der=0)
   # Interpolation (Ag, real part)

fRIAgImInt2 = interpolate.splrep(WL,RIAgIm,s=0)
   # Interpolation (Ag, imaginary part)
RIAgImInt2 = interpolate.splev(WLx,fRIAgImInt2,der=0)
   # Interpolation (Ag, imaginary part)

RIAu=zeros(NumWLx, dtype=complex)
epAu=zeros(NumWLx, dtype=complex)
RIAg=zeros(NumWLx, dtype=complex)
epAg=zeros(NumWLx, dtype=complex)

RIAu=RIAuReInt2+1j*RIAuImInt2  #   RIAu: Refractive index of Au
RIAg=RIAgReInt2+1j*RIAgImInt2  #   RIAg: Refractive index of Ag
epAu=RIAu**2      #   epAu: Dielectric constant of Au
epAg=RIAg**2      #   epAu: Dielectric constant of Ag
```

First, the calculations are performed for silver nanospheres. In Program 2.2, after loading the refractive index and dielectric constant of silver from RI.py in Line 6, the program finds the polarizability of the nanosphere. Figure 2.2(a) is the refractive index spectrum of silver, and Figure 2.2(b) is the dielectric-constant spectrum of silver. Silver has a small imaginary part of the dielectric constant, indicating a small loss in the visible light region. Figure 2.2(c) shows the spectra of the scattering (C_{sca}) and absorption (C_{abs}) cross-sections of silver nanospheres (R = 25 nm). Figure 2.2(d) shows the spectra of the scattering efficiency (Q_{sca}) and absorption efficiency (Q_{abs}) of the nanosphere. The imaginary part of the polarizability peaks at a wavelength of 360 nm, where the real part of the silver dielectric constant becomes -2. This peak stems from the localized surface plasmon resonance of silver nanospheres. Using a similar program, the wavelength dependence of the optical constant versus the change in polarizability for gold nanospheres is plotted in Figure 2.3. Localized surface plasmon resonance occurs at approximately 510 nm. Compared to silver, the imaginary part of the dielectric constant is larger, so the width of the polarizability peak is broader, and its absolute value is smaller.

Program 2.2

```
import scipy as sp
import matplotlib as mpl
import matplotlib.pyplot as plt
from matplotlib.pyplot import plot,show,xlabel,ylabel,title,
    legend,grid,axis,rcParams,tight_layout
from scipy import real,imag,pi
from RI import WLx, epAg, epAu, RIAu, RIAg

n1 = 1                  #   refractive index of ambient
n2 = RIAg               #   refractive index of sphere
r=25                    #   radius of sphere
k = 2 * pi / WLx        #   array of wavenumber
alpha = 4 * pi * (r**3) * (n1**2) * (n2**2 - n1**2) / (n2**2 + 2
    * n1**2)
                                          #  Calculation of polarizability
Csca = k**4 / (6 * pi) * abs(alpha)**2
                                          # scattering cross-section
Cabs = k * imag(alpha)                    # absorption cross-section
Qsca = Csca / ((r**2) * pi)               # scattering efficiency
Qabs = Cabs / ((r**2) * pi)               # absorption efficiency

plt.figure(figsize=(8,6))
plot(WLx,real(RIAg), label="real",linewidth = 3.0, color='black')
plot(WLx,imag(RIAg), label="imaginary",linewidth = 3.0, color='
    gray')
xlabel("wavelength (nm)",fontsize=22)
ylabel("refractive index",fontsize=22)
title("Refractive index of Ag",fontsize=22)
grid(True)
axis([300,700,0,5])
plt.tick_params(labelsize=20)
legend(fontsize=20,loc='lower right')
tight_layout()
show()
```

2.2.2 Calculation of sphere with retardation

In Line 5 of Program 2.3, we read the data of the refractive indices of metals from RI.py. In Line 8, we use Bessel and Hankel functions, so we read them in and prepare their derivatives. In Lines 10 and 12, we define Riccati's Bessel functions (ψ and ξ) and their derivatives. We need to take the sum when finding C_{sca} and C_{abs}. We take the sum for n starting at Line 39. Note that n starts at zero.

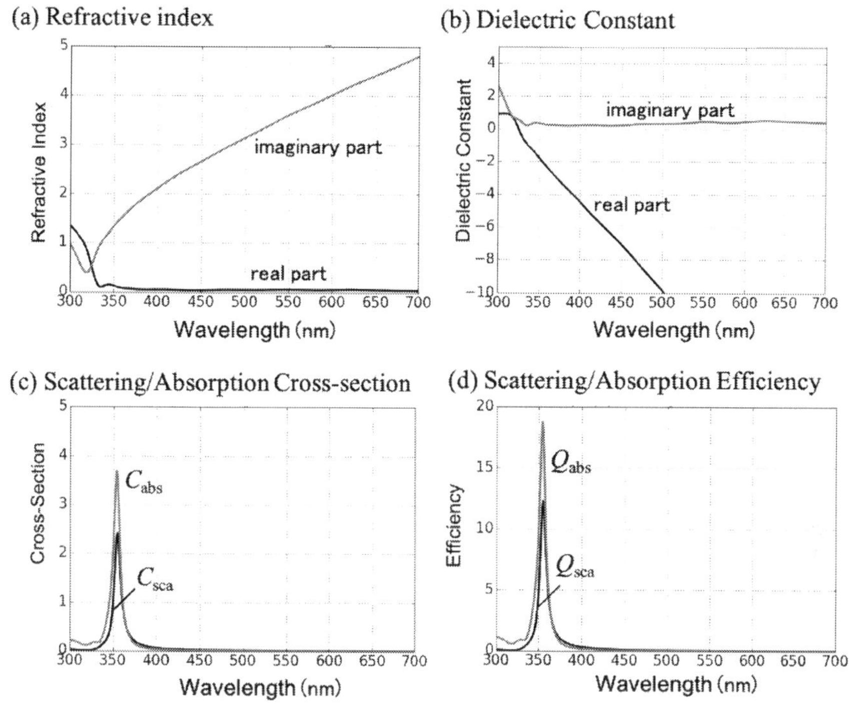

FIGURE 2.2
(a) Refractive index of silver, (b) dielectric constant of silver, (c) scattering cross section (C_{sca}) and absorption cross section (C_{abs}) of silver spheres (R = 25 nm), and its (d) scattering efficiency (Q_{sca}) and absorption efficiency (Q_{abs}) spectra.

Program 2.3

```
import scipy as sp
import scipy.special
import matplotlib as mpl
import matplotlib.pyplot as plt
from RI import WLx, NumWLx, epAu, RIAu
from scipy import pi,arrange,zeros,array,real,imag
from matplotlib.pyplot import plot,show,xlabel,ylabel,title,
    legend,grid, axis
from scipy.special import spherical_jn,spherical_yn

def psi(n,z):      # Riccati-Bessel Function of 1st kind
    return z*spherical_jn(n,z)
def psiDz(n,z):    # Deliberative of Riccati-Bessel of 1st kind
    return spherical_jn(n,z)+z*spherical_jn(n,z,1)
def xi(n,z):       # Riccati-Bessel function of 3rd kind
```

```
    return z*(spherical_jn(n,z)+1j*spherical_yn(n,z))
def xiDz(n,z):    # Deliberative of Riccati-Bessel function of 3 rd kind
    return (spherical_jn(n,z)+1j*spherical_yn(n,z))  #  spherical Bessel function
               +z*(spherical_jn(n,z,1)+1j*spherical_yn(n,z,1))
def a(n,m,x):
    return (m*psi(n,m*x)*psiDz(n,x)-psi(n,x)*psiDz(n,m*x))/  #an
               (m*psi(n,m*x)*xiDz(n,x)-xi(n,x)*psiDz(n,m*x))
def b(n,m,x):
    return (psi(n,m*x)*psiDz(n,x)-m*psi(n,x)*psiDz(n,m*x))/  #bn
               (psi(n,m*x)*xiDz(n,x)-m*xi(n,x)*psiDz(n,m*x))

n1 = 1.0          # refractive index of ambient
n2 = RIAu         # refractive index of sphere
r = 100           # radius of sphere
qq = 50           # order of Bessel function

Csca = zeros(NumWLx, dtype=complex)
Cext = zeros(NumWLx, dtype=complex)
Cabs = zeros(NumWLx, dtype=complex)

k0 = 2*pi/WLx   # vacuum wavenumber
x = k0*n1*r     # size parameter
m = n2/n1       # relative refractive index

for n in range(qq):                 # Sum of 0-qq order
    Csca = Csca + (2*pi/k0**2)*(2*(n+1)+1)*(abs(a(n+1,m,x)**2)+abs(b(n+1,m,x)**2))
    Cext = Cext + (2*pi/k0**2)*(2*(n+1)+1)*(real(a(n+1,m,x)+b(n+1,m,x)))
    Cabs = Cext - Csca

Qsca = Csca / ((r**2) * pi)      #   scattering efficiency
Qabs = Cabs / ((r**2) * pi)      #   absorption efficiency

plot(WLx,Qsca, label=r"$Q_{\rm sca}$",linewidth = 3.0, color='black')
plot(WLx,Qabs, label=r"$Q_{\rm abs}$",linewidth = 3.0, color='gray')

xlabel("wavelength (nm)",fontsize=22)
ylabel("efficiency",fontsize=22)
title(r"$Q_{{\rm sca}}, Q_{{\rm abs}} of Au sphere",fontsize=22)
grid(True)
axis([400,800,0,5])
legend(fontsize=20,loc='lower right')
plt.tick_params(labelsize=18)
show()
```

The calculated scattering ($C_{\rm sca}$), absorption ($C_{\rm abs}$), and extinction ($C_{\rm ext}$) cross-sections for gold nanospheres with radii of 10, 25, 50, and 100 nm are shown in Figure 2.4. For small sphere radii, the scattering is negligible, and

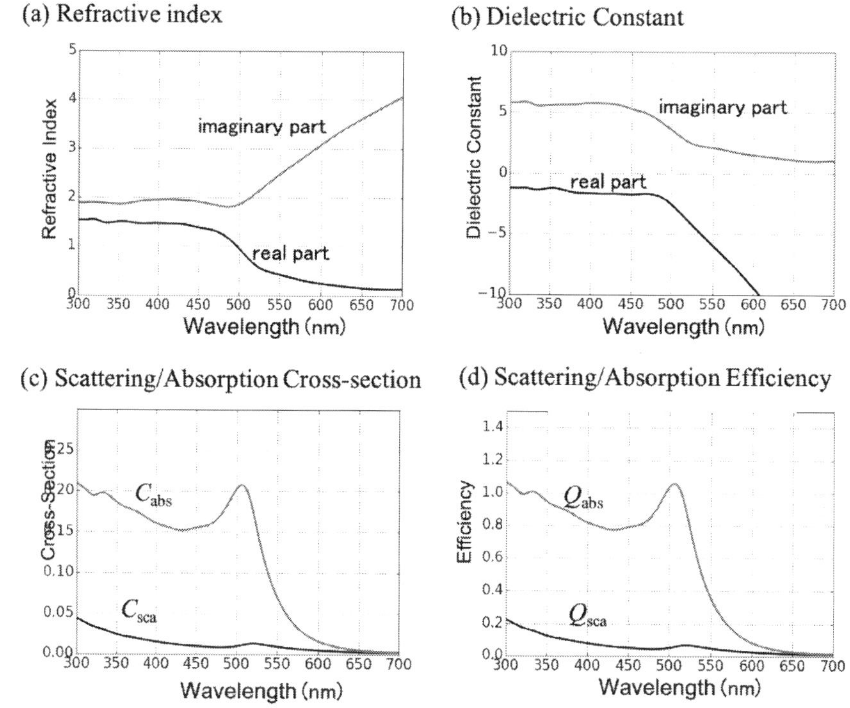

FIGURE 2.3
(a) Refractive index spectrum of gold, (b) dielectric-constant spectrum of gold, (c) scattering cross section (C_{sca}) and absorption cross section (C_{abs}) of gold nanospheres ($R = 25$ nm), and its (d) scattering efficiency (Q_{sca}) and absorption efficiency (Q_{abs}) spectra.

the absorption and extinction cross-sections are almost equal. The peak wavelength originating from the localized surface plasmon is also close to the result under the long wavelength approximation. As the radius increases, the peak wavelength shifts to the long wavelength side and the peak width becomes broader. This is because the multipole effect becomes non-negligible. The scattering cross-section then becomes larger, especially on the long wavelength side. On the other hand, the spectral shape of the absorption cross-section remains unchanged with changing size.

2.2.3 Core-shell structure

Program 2.4 shows the calculation program of scattering efficiency (Q_{sca}) and absorption efficiency spectrum (Q_{abs}) of the core-shell structure under the

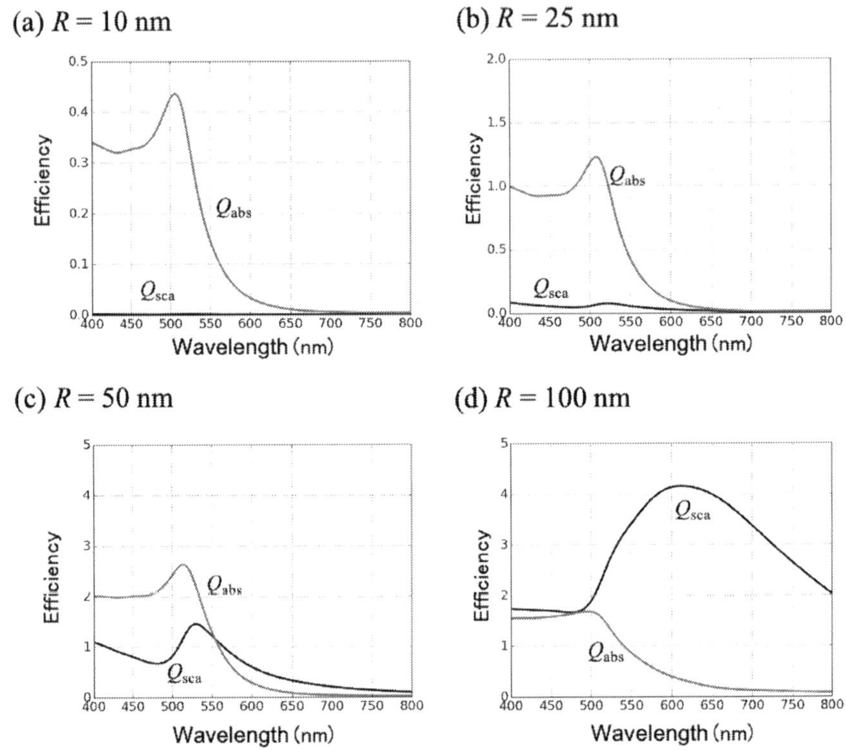

FIGURE 2.4
Calculated scattering (Q_{sca}) and absorption (Q_{abs}) efficiencies for gold nanospheres of various sizes (a)R = 10 nm, (b)R = 25 nm, (c)R = 50 nm, and (d)R = 100 nm.

assumption that retardation is absent. A sphere with a refractive index of 1.5 (radius 25 nm) is considered as a core, and a thin gold film is used as the shell. First, the data of the refractive index of the metal is loaded, and the calculation is performed using Eqs. (2.12)–(2.14).

The results of the calculations performed using Program 2.4 are shown in Figure 2.5. The shell thickness is described by $s = R_2/R_3$ with the structure shown in Figure 2.1(b). While the spectral peak of the gold nanosphere, which is sufficiently small compared to the wavelength, is around 510 nm, it can be seen that for the same size nanosphere, a large extinction (mainly absorption) efficiency can be obtained over a wide range up to the near-infrared region by choosing the thickness of the gold shell. In addition, as can be seen when compared with the gold sphere of the same size shown in Figure 2.4(b), the extinction, absorption, and scattering efficiencies are all about ten times greater.

Program 2.4

```
import scipy as sp
import matplotlib as mpl
import matplotlib.pyplot as plt
from matplotlib.pyplot import plot,show,xlabel,ylabel,title,
    legend,grid,axis,rcParams,tight_layout
from scipy import real,imag,pi
from RI import WLx, epAg, epAu, RIAu, RIAg

r3 = 25      # radius of core
s = 1.1      # adius of shell / radius of core
r2 = r3*s    # radius of shell
n1 = 1       #   refractive index of ambient
n2 = 1.5    #   refractive index of core
n3 = RIAu     #   refractive index of shell
k = 2 * pi / WLx   #  array of wavenumber

delta = (n2**2) * (2 * (n1**2) * (1 + 2 * (s**3)) + (n3**2) * (2
    + s**3)) - 2 * ((n2**2)**2 + (n1**2) * (n3**2)) * (1 - s**3)
b11 = (s**3 / delta) * ((n2**2) * ((n1**2) * (1 + 2*s**3) - (n3
    **2) * (2 + s**3)) + (2 * (n2**2)**2 - (n1**2) * (n3**2)) * (1
     - s**3))

alpha = -4 * pi * r2**3 * (n1**2) * b11  # polarizability
Csca = k**4 / (6 * pi) * abs(alpha)**2   # scattering cross-
    section
Cabs = k * imag(alpha)          #  absorption cross-section
Qsca = Csca / ((r2**2) * pi)   # scattering efficiency
Qabs = Cabs / ((r2**2) * pi)   # absorption efficiency

plt.figure(figsize=(8,6))
plot(WLx,Qsca, label=r"$Q_{{\rm sca}}$",linewidth = 3.0, color='
    black')
plot(WLx,Qabs, label=r"$Q_{{\rm abs}}$",linewidth = 3.0, color='
    gray')
xlabel("wavelength (nm)",fontsize=22)
ylabel("efficiency",fontsize=22)
title(r"$Q_{{\rm sca}}, Q_{{\rm abs}}$ of Au ($R=25$ nm)",
    fontsize=22)
grid(True)
axis([300,1000,0,15])
plt.tick_params(labelsize=20)
legend(fontsize=20,loc='lower left')
tight_layout()
show()
```

Next, Program 2.5 shows the calculation considering the retardation of the core-shell structure. It can be applied to spheres of large size. At the beginning, we define Riccati's Bessel functions ($\psi_n(\rho)$, $\xi_n(\rho)$, $\chi_n(\rho)$). Since there are two interfaces, we define two size parameters x and y. The scattering coefficients are then obtained using Eqs. (2.18) and (2.19). The calculation results are shown in Figure 2.6. The results under the long-wavelength approximation

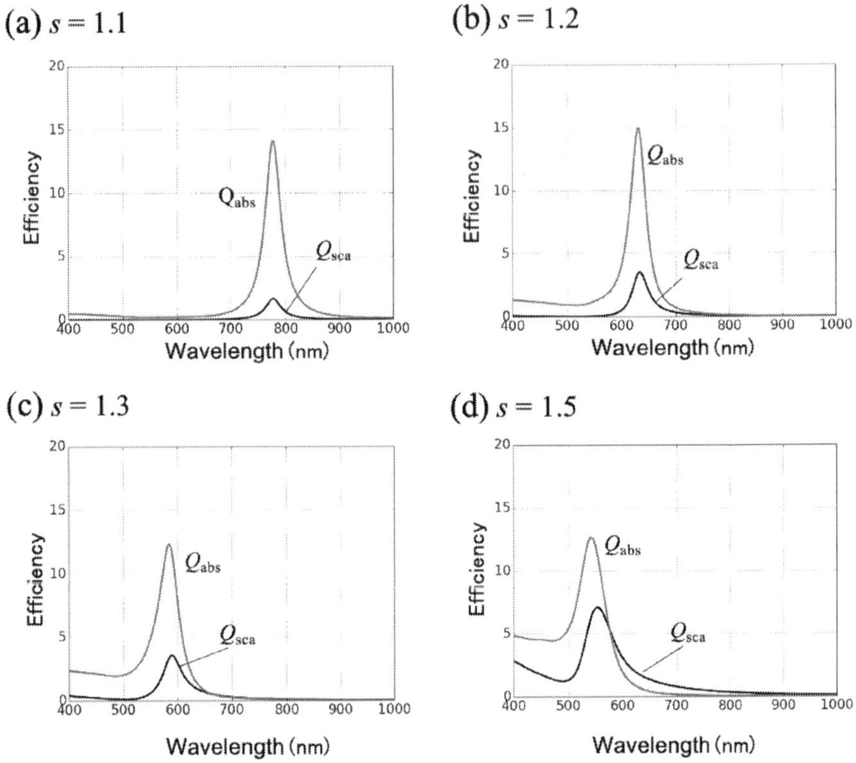

FIGURE 2.5
Scattering efficiency ($Q_{\rm sca}$), absorption efficiency ($Q_{\rm abs}$), and extinction efficiency ($Q_{\rm ext}$) of the core-shell structure under the long wavelength approximation. A dielectric core (radius 25 nm) with n=1.5 is coated with a gold shell. The thickness of the gold film is described by the ratio s to the radius of the dielectric core. (a) $s = 1.1$, (b) $s = 1.2$, (c) $s = 1.3$, and (d) $s = 1.5$.

shown in Figure 2.5(a), where the radius of the inner shell is 25 nm and $s = 1.1$, can be compared with those of Figure 2.6(b), where the retardation is considered. The positions of the peaks are almost the same, but the intensities are different. This is due to the retardation effect at a radius of 25 nm. As the sphere is smaller, the two become closer. Compared the spectrum of Figure 2.5(a) to the calculation results for gold nanospheres with the retardation effect shown in Figure 2.4, each cross-section of the core-shell structure is more than one order of magnitude larger, and the peak is shifted to the long-wavelength side. The widths of the peaks are also narrower.

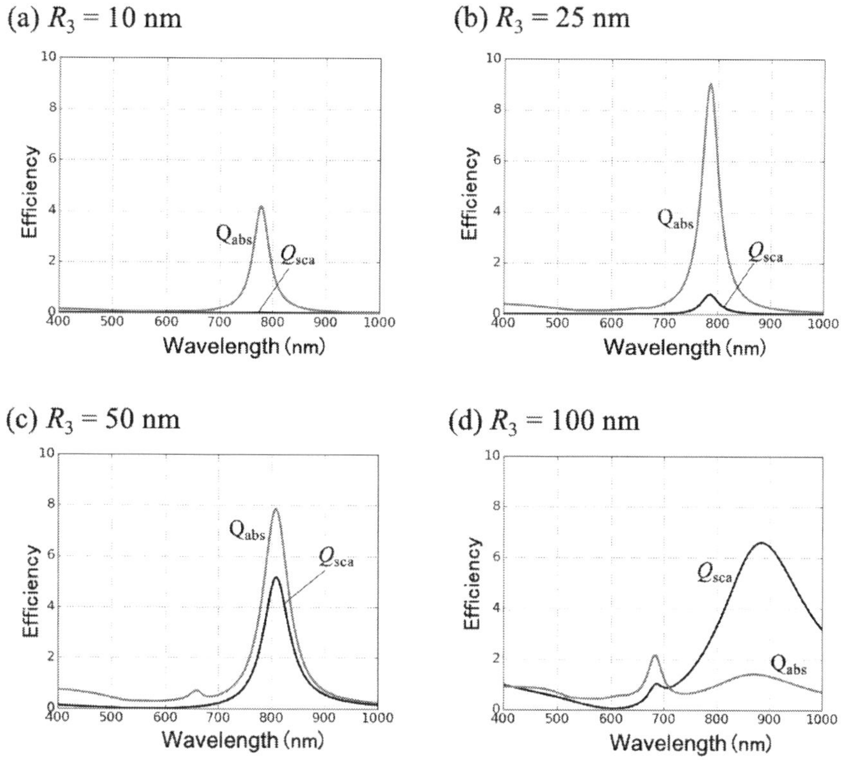

FIGURE 2.6
Scattering efficiency (Q_{sca}), absorption efficiency (Q_{abs}) and extinction efficiency (Q_{ext}) of the core-shell structure considering retardation at various radii of the core, R_3. The refractive index of the core is 1.5, and the core is coated with an Au film. The thickness ratio is $s = 1.1$. (a) $R_3 = 10$ nm, (b) $R_3 = 25$ nm, (c) $R_3 = 50$ nm, and (d) $R_3 = 100$ nm.

Program 2.5

```
import scipy as sp
import scipy.special
import matplotlib as mpl
import matplotlib.pyplot as plt
from RI import WLx, NumWLx, epAu, RIAu
from scipy import pi,arrange,zeros,array,real,imag
from matplotlib.pyplot import plot,show,xlabel,ylabel,title, legend,grid, axis
from scipy.special import spherical_jn,spherical_yn

def psi(n,z):                          # Riccati-Bessel function of first kind
```

```
    return z*spherical_jn(n,z)
def psiDz(n,z):                      # Derivative of Riccati-Bessel function of first kind
    return spherical_jn(n,z)+z*spherical_jn(n,z,1)
def xi(n,z):                         # Riccati-Bessel function of third kind
    return z*(spherical_jn(n,z)+1j*spherical_yn(n,z))
def xiDz(n,z):                       # Derivative of Riccati-Bessel function of third kind
    return (spherical_jn(n,z)+1j*spherical_yn(n,z)) \
           +z*(spherical_jn(n,z,1)+1j*spherical_yn(n,z,1))
def chi(n,z):
    return -z*spherical_yn(n,z)
def chiDz(n,z):
    return -spherical_yn(n,z)-z*spherical_yn(n,z,1)

def aa(n,m1,m2,x):
    return (m2*psi(n,m2*x)*psiDz(n,m1*x)-m1*psiDz(n,m2*x)*psi(n,m1*x)) \
           /(m2*chi(n,m2*x)*psiDz(n,m1*x)-m1*chiDz(n,m2*x)*psi(n,m1*x))
def bb(n,m1,m2,x):
    return (m2*psi(n,m1*x)*psiDz(n,m2*x)-m1*psi(n,m2*x)*psiDz(n,m1*x)) \
           /(m2*chiDz(n,m2*x)*psi(n,m1*x)-m1*psiDz(n,m1*x)*chi(n,m2*x))

def a(n,m1,m2,x,y):
    return (psi(n,y)*(psiDz(n,m2*y)-aa(n,m1,m2,x)*chiDz(n,m2*y)) \
            -m2*psiDz(n,y)*(psi(n,m2*y)-aa(n,m1,m2,x)*chi(n,m2*y))) \
            /(xi(n,y)*(psiDz(n,m2*y)-aa(n,m1,m2,x)*chiDz(n,m2*y)) \
            -m2*xiDz(n,y)*(psi(n,m2*y)-aa(n,m1,m2,x)*chi(n,m2*y)))
def b(n,m1,m2,x,y):
    return (m2*psi(n,y)*(psiDz(n,m2*y)-bb(n,m1,m2,x)*chiDz(n,m2*y)) \
            -psiDz(n,y)*(psi(n,m2*y)-bb(n,m1,m2,x)*chi(n,m2*y))) \
            /(m2*xi(n,y)*(psiDz(n,m2*y)-bb(n,m1,m2,x)*chiDz(n,m2*y)) \
              -xiDz(n,y)*(psi(n,m2*y)-bb(n,m1,m2,x)*chi(n,m2*y)))

r3 = 100   # core radius
s = 1.1   # shell radius/core radius
r2 = r3*s   # shell radius
qq = 20    # order of Bessel function

k0 = 2*pi/WLx   # vacuum wavenumber
n1 = 1  # ambient refractive index
n2 = RIAu  # core refractive index
n3 = 1.5  # shell refractive index

x = k0 * n1 * r3   # size parameter(core)
y = k0 * n1 * r2   # size parameter(shell)
```

```
m2 = n2 / n1   # relative refractive index(shell)
m3 = n3 / n1   # relative refractive index(core)

Csca = zeros(NumWLx, dtype=complex)
Cext = zeros(NumWLx, dtype=complex)
Cabs = zeros(NumWLx, dtype=complex)

for n in range(qq):
    Csca = Csca + (2*pi / k0**2) * \
           (2 * (n+1)+1) * (abs(a(n+1,m3,m2,x,y))**2 + abs(b(n+1,
               m3,m2,x,y)**2))
    Cext = Cext + (2*pi / k0**2) * \
           (2 * (n+1)+1) * (real(a(n+1,m3,m2,x,y) + b(n+1,m3,m2,x
               ,y)))
    Cabs = Cext - Csca

Qsca = Csca / ((r2**2) * pi)    #   scattering efficiency
Qabs = Cabs / ((r2**2) * pi)  #   absorption efficiency

plot(WLx,abs(Qsca), label=r"$Q_{\rm sca}$",linewidth = 3.0, color
    ='black')
plot(WLx,abs(Qabs), label=r"$Q_{\rm abs}$",linewidth = 3.0, color
    ='gray')

xlabel("wavelength (nm)",fontsize=22)
ylabel("efficiency",fontsize=22)
title(r"$Q_{{\rm sca}}, Q_{{\rm abs}}, Q_{{\rm ext}}$ of Au
    sphere",fontsize=22)
grid(True)
axis([400,1000,0,10])
legend(fontsize=20,loc='lower left')
plt.tick_params(labelsize=18)
show()
```

3

Electromagnetic Analysis of Cylinders

Similarly to spheres, analytical solutions can be obtained for cylinder structures with infinitely long lengths. In this case, the long wavelength approximation can be applied if the diameter of the cylinder is sufficiently small compared to the wavelength. Then, rigorous calculations of the scattering, absorption, and extinction by cylinders are also given in this chapter.

3.1 Introduction

As shown in Eq. (2.1), the polarizability α could be described in the case of a sphere, but it cannot be expressed in this form for a cylinder. Instead, we describe the magnitude E of the electric field of scattered light at a position r away from the central axis of the cylinder. The optical geometry is shown in Figure 3.1. The magnitude E of the electric field of scattered light is described by using the incident light electric field E_0 with polarization perpendicular to the axis (TE polarization), the refractive index n_1 of the cylinder (radius R) and n_2 of the surrounding medium as follows:

$$E = 4\pi n_2^2 \left(\frac{R}{r}\right)^2 \frac{n_1^2 - n_2^2}{n_1^2 + n_2^2} E_0. \tag{3.1}$$

In the case of a metallic cylinder, the refractive index n_1 has wavelength dependence (wavelength dispersion), and the scattered electric field E is maximized at the wavelength where $n_1^2 + n_2^2$ of the molecule is minimum, indicating that resonance may occur[1]. In the case of a sphere, the resonance condition is achieved at the wavelength where $n_1^2 + 2n_2^2$ is at a minimum, but the condition is slightly different in the case of a cylinder. To the extent that the electrostatic approximation holds, it is similarly independent of the size of the cylinder. Since the calculations are similar to those for spheres, we present the solution incorporating the retardation effect here.

[1]Note that in this chapter, the order of the media numbers is reversed from the sphere case in Chapter 2.

DOI: 10.1201/9781003357674-3

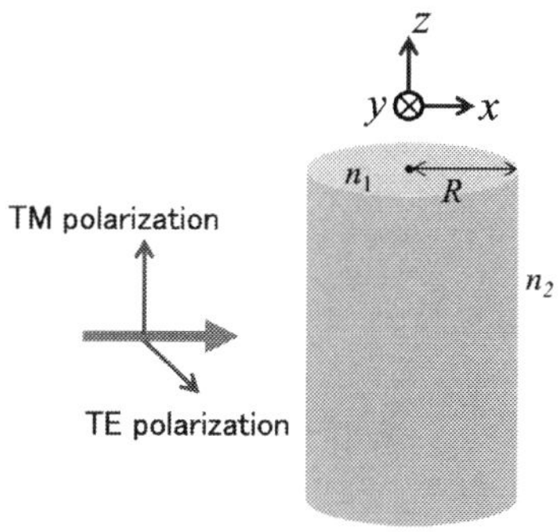

FIGURE 3.1
Geometry of a cylinder structure.

3.2 Theory

3.2.1 Cylinder

Consider the scattering and absorption by a cylinder structure of radius R (refractive index n_1), as shown in Figure 3.1. The refractive index of the ambient medium is n_2. The potentials generated by the application of an optical electric field $\boldsymbol{E}$ propagating in the positive direction of the x-axis can be expressed using the cylinder coordinate system as follows for ϕ_1 outside the cylinder and ϕ_2 inside the cylinder, respectively [8, 9],

$$\phi_1 = \sum_{n=-\infty}^{\infty} F_n(b_n J_n(kr)) \tag{3.2}$$

$$\phi_2 = \sum_{n=-\infty}^{\infty} F_n(J_n(kr) - a_n H_n(kr)), \tag{3.3}$$

where $F_n = Ee^{in\theta + i\omega t}(-1)^n$ and $E = |\boldsymbol{E}|$. Also, k is the wavenumber in each medium, $k = mk_0$, using the wavenumber k_0 in vacuum and the refractive index m of the medium.

The continuity conditions at the surface of the cylinder $r = R$ depend on the polarization. For polarization parallel to the cylinder axis (TM polarization),

$$m_1\phi_1 = m_2\phi_2 \tag{3.4}$$

$$m_1\frac{\partial \phi_1}{\partial r} = m_2\frac{\partial \phi_2}{\partial r}, \tag{3.5}$$

and the continuity conditions for polarization perpendicular to the cylinder axis (TE polarization) are

$$m_1^2\phi_1 = m_2^2\phi_2 \tag{3.6}$$
$$\frac{\partial\phi_1}{\partial r} = \frac{\partial\phi_2}{\partial r}. \tag{3.7}$$

In the case of TM polarization, we have two equations

$$m_1 J_n(m_1x)b_n = m_2 J_n(m_2x) - m_2 H_n(m_2x)a_n \tag{3.8}$$
$$m_1^2 J_n'(m_1x)b_n = m_2^2 J_n'(m_2x) - m_2^2 H_n'(m_2x)a_n. \tag{3.9}$$

Here, x is called the size parameter and $x = k_0R$. Then, the coefficient a_n on scattered light is as follows:

$$a_n = \frac{m_1 J_n'(m_1x)J_n(m_2x) - m_2 J_n(m_1x)J_n'(m_2x)}{m_1 J_n'(m_1x)H_n(m_2x) - m_2 H_n'(m_2x)J_n(m_1x)} \tag{3.10}$$

In the case of TE polarization, we have two equations

$$m_1^2 J_n(m_1x)b_n = m_2^2 J_n(m_2x) - m_2^2 H_n(m_2x)a_n \tag{3.11}$$
$$m_1 J_n'(m_1x)b_n = m_2 J_n'(m_2x) - m_2 H_n'(m_2x)a_n. \tag{3.12}$$

Then, the coefficient a_n on scattered light is as follows:

$$a_n = \frac{m_2 J_n(m_2x)J_n'(m_1x) - m_1 J_n(m_1x)J_n'(m_2x)}{m_2 J_n'(m_1x)H_n(m_2x) - m_1 H_n'(m_2x)J_n(m_1x)} \tag{3.13}$$

The scattering cross-section Q_{sca}, extinction cross-section Q_{ext}, and absorption cross-section Q_{abs} are calculated using b_n as follows:

$$Q_{\text{sca}} = \frac{2}{x}\sum_{n=-\infty}^{\infty} |a_n|^2 \tag{3.14}$$
$$Q_{\text{ext}} = \frac{2}{x}\sum_{n=-\infty}^{\infty} \text{Re}(a_n) \tag{3.15}$$
$$Q_{\text{abs}} = Q_{\text{ext}} - Q_{\text{sca}} \tag{3.16}$$

3.2.2 Core-shell cylinder

Next, we discuss calculating the optical response of a core-shell cylinder structure incorporating retardation. The structure is shown in Figure 3.2. The radius of the core is R_1, and that of the shell is R_2. The thickness of the shell is $R_2 - R_1$. The medium is numbered from the inside. The potential $\phi_1 - \phi_3$ in each medium generated by the application of a photoelectric field $\boldsymbol{E}$ propagating in the positive direction of the x-axis is written using the coefficient $a_n - d_n$ as follows [13].

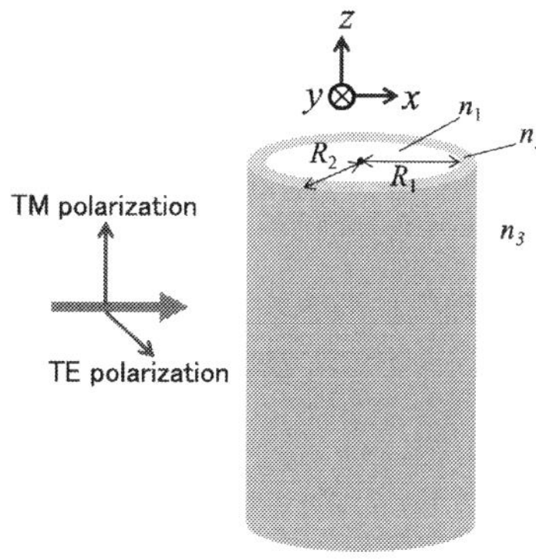

FIGURE 3.2
Geometry of a core-shell cylinder structure.

From the boundary conditions, the following four equations can be obtained for TM polarization,

$$
\begin{aligned}
m_1 J_n(m_1x_1)a_n &= m_2 J_n(m_2x_1)b_n - m_2 H_n(m_2x_1)c_n \quad (3.17)\\
m_1^2 J'_n(m_1x_1)a_n &= m_2^2 J'_n(m_2x_1)b_n - m_2^2 H'_n(m_2x_1)c_n \quad (3.18)\\
m_2 J_n(m_2x_2)b_n - m_2 H_n(m_2x_2)c_n &= m_3 J_n(m_3x_2) - m_3 H_n(m_3x_2)d_n \quad (3.19)\\
m_2^2 J'_n(m_2x_2)b_n - m_2^2 H'_n(m_2x_2)c_n &- m_3^2 J'_n(m_3x_2) - m_3^2 H'_n(m_3x_2)d_n, \quad (3.20)
\end{aligned}
$$

where $x_1 = k_0R_1$ and $x_2 = k_0R_2$.

In the case of TE polarization, the following is also given:

$$
\begin{aligned}
m_1^2 J_n(m_1x_1)a_n &= m_2^2 J_n(m_2x_1)b_n - m_2^2 H_n(m_2x_1)c_n \quad (3.21)\\
m_1 J'_n(m_1x_1)a_n &= m_2 J'_n(m_2x_1)b_n - m_2 H'_n(m_2x_1)c_n \quad (3.22)\\
m_2^2 J_n(m_2x_2)b_n - m_2^2 H_n(m_2x_2)c_n &= m_3^2 J_n(m_3x_2) - m_3^2 H_n(m_3x_2)d_n \quad (3.23)\\
m_2 J'_n(m_2x_2)b_n - m_2 H'_n(m_2x_2)c_n &= m_3 J'_n(m_3x_2) - m_3 H'_n(m_3x_2)d_n. \quad (3.24)
\end{aligned}
$$

Use matrices to solve these equations; Python has commands for solving simultaneous equations so that you can use these commands. The scattering cross-section Q_{sca}, extinction cross-section Q_{ext}, and absorption cross-section Q_{abs} are calculated using d_n as follows:

$$Q_{\text{sca}} = \frac{2}{x_2}\sum_{n=-\infty}^{\infty} |d_n|^2 \quad (3.25)$$

$$Q_{\text{ext}} = \frac{2}{x_2}\sum_{n=-\infty}^{\infty} \text{Re}(d_n) \quad (3.26)$$

$$Q_{\text{abs}} = Q_{\text{ext}} - Q_{\text{sca}} \quad (3.27)$$

3.3 Programing

3.3.1 Cylinder

Here, silver cylinders are considered to confirm that the rigorous calculation, including retardation, give the resonance wavelength of around 330 nm, as predicted under the long-wavelength approximation.

In Lines 11 and 14, we describe the functions that give the scattering coefficients a_n and b_n for TE and TM polarization, respectively. F_n in Line 17 is not used in this calculation but is necessary for calculating the angular dependence of the scattered light intensity.

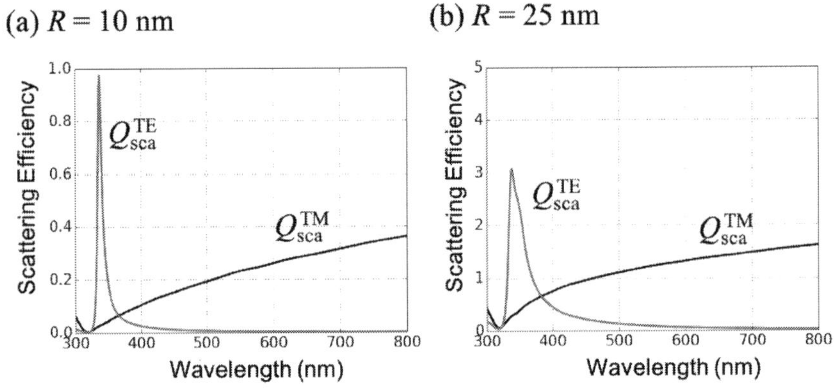

FIGURE 3.3
Calculated scattering efficiency of cylinder Q_{sca} (a) $R = 10$ nm and (b) $R = 50$ nm.

The calculation results of this program are shown in Figure 3.3. Here, the radius of the silver cylinder is set to 10 nm. As shown in Figure 2.2(b), the real part of the dielectric constant of silver has a value of -1 around 335 nm. The value of the real part of the silver dielectric constant decreases monotonically with wavelength thereafter. On the other hand, the imaginary part has an almost constant value of $0 - 0.1$, indicating little loss. The scattering efficiencies for TE and TM polarization are plotted in Figures 3.3(a) and 3.3(b), respectively, showing that the TE polarization has a peak at 335 nm due to the localized plasmon resonance as expected, while the TM polarization only monotonically increases and shows no resonance. When the radius of the silver cylinder is 50 nm, the peak shifts to around 350 nm and the peak width becomes broader. This is due to the retardation effect, which is non-negligible as the radius of the cylinder increases.

Program 3.1

```
import scipy as sp
import matplotlib as mpl
import matplotlib.pyplot as plt
from matplotlib.pyplot import plot,show,xlabel,ylabel,title,legend,grid,axis,tight_layout
from scipy import pi,sqrt,zeros,array,real,imag
from scipy.special import jv,jvp,hankel1,h1vp
from RI import WLx, NumWLx, epAg, epAu, RIAu, RIAg

def h1v(n,x):
    return hankel1(n,x)                                        # Hankel function
def a(n,x,mA,mB):
    return (mB*jv(n,mB*x)*jvp(n,mA*x)-mA*jv(n,mA*x)*jvp(n,mB*x))/     # an
    (mB*jv(n,mB*x)*h1vp(n,mA*x)-mA*h1v(n,mA*x)*jvp(n,mB*x))
def b(n,x,mA,mB):
    return (mA*jv(n,mB*x)*jvp(n,mA*x)-mB*jv(n,mA*x)*jvp(n,mB*x))/     # bn
    (mA*jv(n,mB*x)*h1vp(n,mA*x)-mB*h1v(n,mA*x)*jvp(n,mB*x))
def fn(n,phi):
    return (1/k0)*(cos(n*phi)+1j*sin(n*phi))*(pow(-1j,n))    # fn

rr = 25    # radius of cylinder
qq = 20    # order of Bessel function

k0 = 2 * pi / WLx    # vacuum wavenumber
m1 = RIAg            # refractive index of cylinder
m2 = 1.0             # refractive index of ambient
x = k0 * m2 * rr     # size parameter

Qsca_tm = zeros(NumWLx, dtype=float)
Qext_tm = zeros(NumWLx, dtype=float)
Qsca_te = zeros(NumWLx, dtype=float)
Qext_te = zeros(NumWLx, dtype=float)

for n in range(-qq,qq):
    Qsca_tm = Qsca_tm + (2/x) * abs(b(n,x,m2,m1))**2    # TM scattering efficiency
    Qext_tm = Qext_tm + (2/x) * real(b(n,x,m2,m1))      # TM extinction efficiency
    Qsca_te = Qsca_te + (2/x) * abs(a(n,x,m2,m1))**2    # TE scattering efficiency
    Qext_te = Qext_te + (2/x) * real(a(n,x,m2,m1))      # TE extinction efficiency

Qabs_tm = Qext_tm - Qsca_tm
Qabs_te = Qext_te - Qsca_te

plt.figure(figsize=(8,6))
plot(WLx,Qsca_tm, label=r"$Q_{\rm sca}(\rm TM)$",linewidth = 3.0, color='black')
plot(WLx,Qsca_te, label=r"$Q_{\rm sca}(\rm TE)$",linewidth = 3.0, color='gray')
xlabel("wavelength (nm)",fontsize=22)
```

```
ylabel("efficiency",fontsize=22)   1
title("Qsca",fontsize=22)
grid(True)
axis([300,800,0,5])
legend(fontsize=20,loc='lower right')
plt.tick_params(labelsize=20)
tight_layout()
show()
```

3.3.2 Core-shell cylinder

Next, the calculation program for the core-shell cylinder structure is described: the scattering coefficient a_n was given in the program for the cylinder in Section 3.3.1. In the case of the core-shell cylinder structure, the scattering coefficient d_n is as follows. First, in the case of TM polarization, the scattering coefficient $d_n = p_n/q_n$ is

$$
\begin{aligned}
p_n = & \, m_1 m_2^3 m_3^2 J_n'(m_2x_1)J_n'(m_3x_2)H_n(m_2x_2)J_n(m_1x_1) \\
& - m_1^2 m_2^2 m_3^2 J_n'(m_1x_1)J_n'(m_3x_2)H_n(m_2x_2)J_n(m_2x_1) \\
& + m_1^2 m_2^2 m_3^2 J_n'(m_1x_1)J_n'(m_3x_2)H_n(m_2x_1)J_n(m_2x_2) \\
& - m_1 m_2^3 m_3^2 J_n'(m_2x_1)J_n'(m_3x_2)J_n(m_1x_1)J_n(m_2x_2) \\
& - m_1^2 m_2^3 m_3 J_n'(m_1x_1)J_n'(m_2x_2)H_n(m_2x_1)J_n(m_3x_2) \\
& - m_1 m_2^4 m_3 J_n'(m_2x_2)J_n'(m_2x_1)J_n(m_1x_1)J_n(m_3x_2) \\
& + m_1 m_2^4 m_3 J_n'(m_2x_1)J_n', m_2x_2)J_n(m_1x_1)J_n(m_3x_2) \\
& + m_1^2 m_2^3 m_3 J_n', m_2x_2)J_n'(m_1x_1)J_n(m_2x_1)J_n(m_3x_2) \\
q_n = & -m_1^2 m_2^3 m_3 J_n'(m_1x_1)J_n'(m_2x_2)H_n(m_2x_1)H_n(m_3x_2) \\
& + m_1 m_2^3 m_3^2 J_n'(m_3x_2)J_n'(m_2x_1)H_n(m_2x_2)J_n(m_1x_1) \\
& - m_1 m_2^4 m_3 J_n'(m_2x_2)J_n'(m_2x_1)H_n(m_3x_2J_n(m_1x_1) \\
& + m_1 m_2^4 m_3 J_n'(m_2x_1)J_n'(m_2x_2)H_n(m_3x_2)J_n(m_1x_1) \\
& - m_1^2 m_2^2 m_3^2 J_n'(m_3x_2)J_n'(m_1x_1)H_n(m_2x_2)J_n(m_2x_1) \\
& + m_1^2 m_2^3 m_3 J_n'(m_2x_2)J_n'(m_1x_1)H_n(m_3x_2)J_n(m_2x_1) \\
& + m_1^2 m_2^2 m_3^2 J_n'(m_3x_2)J_n'(m_1x_1)H_n(m_2x_1)J_n(m_2x_2) \\
& - m_1 m_2^3 m_3^2 J_n'(m_2x_1)J_n'(m_3x_2)J_n(m_1x_1)J_n(m_2x_2).
\end{aligned}
\tag{3.28}
$$

In the case of TM polarization, the scattering coefficient $d_n = p_n/q_n$ is

$$
\begin{aligned}
p_n = & \, m_1^2 m_2^3 m_3 J_n'(n, m_2x_1)J_n'(n, m_3x_2)H_n(m_2x_2)J_n(m_1x_1) \\
& - m_1 m_2^4 m_3 J_n'(n, m_1x_1)J_n'(n, m_3x_2)H_n(m_2x_2)J_n(m_2x_1) \\
& + m_1 m_2^4 m_3 J_n'(n, m_1x_1)J_n'(n, m_3x_2)H_n(m_2x_1)J_n(m_2x_2) \\
& - m_1^2 m_2^3 m_3 H_n'(n, m_2x_1)J_n'(n, m_3x_2)J_n(m_1x_1)J_n(m_2x_2)
\end{aligned}
$$

$$
\begin{aligned}
&- m_1 m_2^3 m_3^2 J'_n(n, m_1x_1) J'_n(n, m_2x_2) H_n(m_2x_1) J_n(m_3x_2) \\
&- m_1^2 m_2^2 m_3^2 H'_n(n, m_2x_2) J'_n(n, m_2x_1) J_n(m_1x_1) J_n(m_3x_2) \\
&+ m_1^2 m_2^2 m_3^2 H'_n(n, m_2x_1) J'_n(n, m_2x_2) J_n(m_1x_1) J_n(m_3x_2) \\
&+ m_1 m_2^3 m_3^2 H'_n(n, m_2x_2) J'_n(n, m_1x_1) J_n(m_2x_1) J_n(m_3x_2) \\
q_n = &-m_1 m_2^3 m_3^2 J'_n(n, m_1x_1) J'_n(n, m_2x_2) H_n(m_2x_1) H_n(m_3x_2) \\
&+ m_1^2 m_2^3 m_3 H'_n(n, m_3x_2) J'_n(n, m_2x_1) H_n(m_2x_2) J_n(m_1x_1) \\
&- m_1^2 m_2^2 m_3^2 H'_n(n, m_2x_2) J'_n(n, m_2x_1) H_n(m_3x_2) J_n(m_1x_1) \\
&+ m_1^2 m_2^2 m_3^2 H'_n(n, m_2x_1) J'_n(n, m_2x_2) H_n(m_3x_2) J_n(m_1x_1) \\
&- m_1 m_2^4 m_3 H'_n(n, m_3x_2) J'_n(n, m_1x_1) H_n(m_2x_2) J_n(m_2x_1) \\
&+ m_1 m_2^3 m_3^2 H'_n(n, m_2x_2) J'_n(n, m_1x_1) H_n(m_3x_2) J_n(m_2x_1) \\
&+ m_1 m_2^4 m_3 H'_n(n, m_3x_2) J'_n(n, m_1x_1) H_n(m_2x_1) J_n(m_2x_2) \\
&- m_1^2 m_2^3 m_3 H'_n(n, m_2x_1) H'_n(n, m_3x_2) J_n(m_1x_1) J_n(m_2x_2). \qquad (3.29)
\end{aligned}
$$

If these are written as functions and used in calculations, calculations can be performed quickly. However, writing these functions in a program can be troublesome.

Python provides commands for solving simultaneous equations; if calculation speed is not a priority, it is easier to use these commands. In other words, the coefficients of Eqs. (3.17)–(3.20) and (3.21)–(3.24) are written as an array. The program is shown below. sp.array is used in Lines 21 and 36 to write the left-hand side of the matA of coefficients. Similarly, Lines 27 and 42 define the right-hand side as the longitudinal vector matF, and Lines 33 or 48 use the sp.linalg.solve command to find the solution of the system of equations as the longitudinal vector matX. This program finds the solution of the simultaneous equations for each wavelength and order n, while it is not fast, it is readable.

Program 3.2

```
1  import scipy as sp
2  import scipy.special
3  import matplotlib as mpl
4  import matplotlib.pyplot as plt
5  from scipy import pi,arrange,sqrt,zeros,array,matrix,asmatrix,
       real,imag
6  from matplotlib.pyplot import plot,show,xlabel,ylabel,title,
       legend,grid, axis,tight_layout
7  from scipy.special import jv,jvp,hankel1,h1vp
8  from RI import WLx, NumWLx, epAg, epAu, RIAu, RIAg
9
10 def h1v(n,x):
11     return hankel1(n,x)
12 def a(n,x,mA,mB):
```

```
    return (mB*jv(n,mB*x)*jvp(n,mA*x)-mA*jv(n,mA*x)*jvp(n,mB*x))/ \
            (mB*jv(n,mB*x)*h1vp(n,mA*x)-mA*h1v(n,mA*x)*jvp(n,mB*x))
def b(n,x,mA,mB):
    return (mA*jv(n,mB*x)*jvp(n,mA*x)-mB*jv(n,mA*x)*jvp(n,mB*x))/ \
            (mA*jv(n,mB*x)*h1vp(n,mA*x)-mB*h1v(n,mA*x)*jvp(n,mB*x))
def fn(n,phi):
    return (1/k0)*(cos(n*phi)+1j*sin(n*phi))*(pow(-1j,n))

def matA_tm(n,m1,m2,m3,x1,x2):
    return array([[    m1*jv(n,m1*x1),     -m2*jv(n,m2*x1), m2*h1v(n,m2*x1),         0],
                  [m1**2*jvp(n,m1*x1), -m2**2*jvp(n,m2*x1), m2**2*h1vp(n,m2*x1),         0],
                  [                 0,     m2*jv(n,m2*x2), -m2*h1v(n,m2*x2),     m3*h1v(n,m3*x2)],
                  [                 0,  m2**2*jvp(n,m2*x2),-m2**2*h1vp(n,m2*x2), m3**2*h1vp(n,m3*x2)]])

def matF_tm(n,m3,x2):
    return array([[                 0],
                  [                 0],
                  [     m3*jv(n,m3*x2)],
                  [m3**2*jvp(n,m3*x2)]])

def matX_tm(n,m1,m2,m3,x1,x2):
    return sp.linalg.solve(matA_tm(n,m1,m2,m3,x1,x2), matF_tm(n,m3,x2))

def matA_te(n,m1,m2,m3,x1,x2):
    return array([[ m1**2*jv(n,m1*x1),  -m2**2*jv(n,m2*x1),  m2**2*h1v(n,m2*x1),         0],
                  [  m1*jvp(n,m1*x1),     -m2*jvp(n,m2*x1), m2*h1vp(n,m2*x1),         0],
                  [                 0,   m2**2*jv(n,m2*x2), -m2**2*h1v(n,m2*x2), m3**2*h1v(n,m3*x2)],
                  [                 0,     m2*jvp(n,m2*x2),   -m2*h1vp(n,m2*x2),   m3*h1vp(n,m3*x2)]])

def matF_te(n,m3,x2):
    return array([[                 0],
                 [                 0],
                 [m3**2*jv(n,m3*x2)],
                 [  m3*jvp(n,m3*x2)]])

def matX_te(n,m1,m2,m3,x1,x2):
    return sp.linalg.solve(matA_te(n,m1,m2,m3,x1,x2), matF_te(n,m3,x2))

r1=50  # radius of core
r2=55  # radius of shell
qq=5   # order of Bessel function

Qsca_tm=zeros(NumWLx, dtype=float)
```

```
Qext_tm=zeros(NumWLx, dtype=float)
Qabs_tm=zeros(NumWLx, dtype=float)
Qsca_te=zeros(NumWLx, dtype=float)
Qext_te=zeros(NumWLx, dtype=float)
Qabs_te=zeros(NumWLx, dtype=float)

k0 = 2 * pi / WLx   # vacuum wavenumber
m1 = 1.5        # refractive index of shell
m2 = RIAg       # refractive index of core
m3 = 1.0    # refractive index of ambient
x1 = k0*r1  # size parameter(core)
x2 = k0*r2  # size parameter(shell)

for i in range(NumWLx):
    for n in range(-qq,qq):
        Qsca_tm[i] = Qsca_tm[i] + (2/x2[i])*abs(matX_tm(n,m1,m2[i],m3,x1[i],x2[i])[3,0])**2
        Qext_tm[i] = Qext_tm[i] + (2/x2[i])*real(matX_tm(n,m1,m2[i],m3,x1[i],x2[i])[3,0])
        Qsca_te[i] = Qsca_te[i] + (2/x2[i])*abs(matX_te(n,m1,m2[i],m3,x1[i],x2[i])[3,0])**2
        Qext_te[i] = Qext_te[i] + (2/x2[i])*real(matX_te(n,m1,m2[i],m3,x1[i],x2[i])[3,0])

Qabs_tm = Qext_tm - Qsca_tm
Qabs_te = Qext_te - Qsca_te

plt.figure(figsize=(8,6))
plot(WLx,Qsca_tm, label=r"$Q_{\rm sca}$",linewidth = 3.0, color=‘black’)
plot(WLx,Qabs_tm, label=r"$Q_{\rm abs}$",linewidth = 3.0, color=‘gray’)
xlabel("wavelength (nm)",fontsize=22)      # x-axis label
ylabel("efficiency",fontsize=22)        # y-axis label
title(r"$Q_{\rm sca}$(TE) and $Q_{\rm abs}$(TE)",fontsize=22)
grid(True)
axis([300,800,0,1])
legend(fontsize=20,loc=‘lower right’)
plt.tick_params(labelsize=20)
tight_layout()
show()

plt.figure(figsize=(8,6))
plot(WLx,Qsca_te, label=r"$Q_{\rm sca}$",linewidth = 3.0,color=‘black’)
plot(WLx,Qabs_te, label=r"$Q_{\rm abs}$",linewidth = 3.0,color=‘gray’)
xlabel("wavelength (nm)",fontsize=22)
ylabel("efficiency",fontsize=22)
title(r"$Q_{\rm sca}$(TE) and $Q_{\rm abs}$(TE)",fontsize=22)
grid(True)
axis([300,1000,0,10])
legend(fontsize=20,loc=‘lower left’)
plt.tick_params(labelsize=20)
tight_layout()
show()
```

The results obtained with this program are shown in Figure 3.4(a). For TM polarization, the scattering efficiency is zero at around 450 nm. Although not completely transparent due to slight absorption, this method can make an object invisible. On the other hand, in the TE polarization shown in Figure 3.4(b), absorption and scattering efficiency peaks can be observed around 880 nm. It is possible to make an optical medium with very large scattering or to propose an optical switching device using this result in combination with nonlinear optical materials.

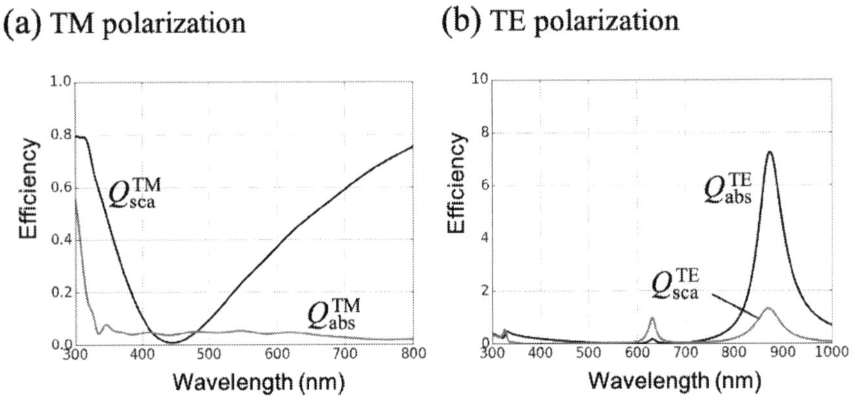

FIGURE 3.4
Calculation results of scattering efficiency Q_{sca} for cylindrical core-shell structure (a) TM polarization and (b) TE polarization.

4

Analytical Calculations for Particles with Other Shapes

This chapter describes calculations of optical response using the long-wavelength approximation for structures other than circles or cylinders, such as nanorods, spheres on substrates, aggregated spheres, and nano-island thin films. They are sometimes used for studies in nanophotonics. The structures discussed in this chapter have analytical solutions to the Maxwell equations. Thus, the optical response can be calculated rigorously.

4.1 Ellipsoid

Ellipsoids are often used as an approximate model when calculating the optical responses of nanostructures such as nanorods and nanodisks, as shown in Figure 4.1. There are two types of rotating ellipsoids: cigar-shaped ($a = b < c$), in which the length in the rotation axis (c-axis) is longer than the radius of the rotating body, and pancake-shaped ($a = b > c$), in which the length in the rotation axis is shorter than the radius of the rotating body. They are treated differently. In both cases, the quasi-static approximation can be applied if the structure is small compared to the optical wavelength. Suppose the refractive index of the rotating ellipsoid is n_1 and the refractive index of the ambient medium is n_2. In this case, the polarizability α_i can be obtained, and the optical response is calculated by finding the depolarization field coefficient L_i about the axis i using the following formula [8]:

$$\alpha_i = 4\pi abc \frac{n_1^2 - n_2^2}{3(n_2^2 + L_i(n_1^2 - n_2^2))} \tag{4.1}$$

Here, the axis of rotation is defined as $\parallel$ and the axis perpendicular to the rotation axis as $\perp$.

4.1.1 Cigar-shaped

In cigar-shaped ellipsoids, the following equation gives the polarizability in the long axis-direction, $\alpha_\parallel$, where $L_\parallel$ is the depolarization field coefficient in

DOI: 10.1201/9781003357674-4

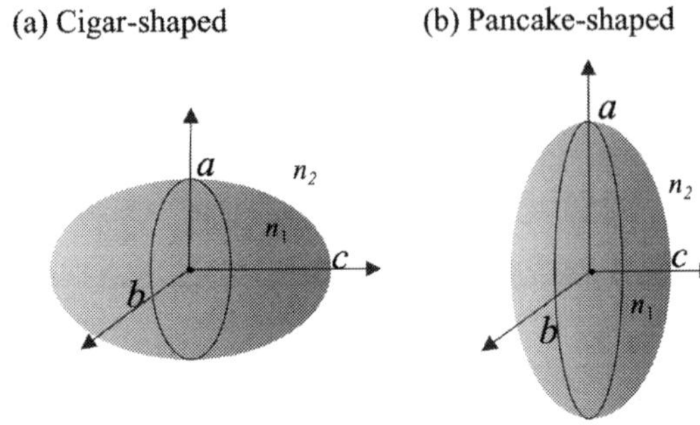

FIGURE 4.1
Optical geometry of an ellipsoid.

the direction of the axis of rotation.

$$L_{\parallel} = \frac{1-e^2}{2e^2}\left(\frac{1}{2e}\ln\frac{1+e}{1-e} - 1\right) \tag{4.2}$$

Here, e is called eccentricity and measures how far the shape is from a sphere. It is

$$e^2 = 1 - \frac{a^2}{c^2} \tag{4.3}$$

for cigar-shaped ellipsoids. The polarizability in the short-axis direction is obtained from the relation $L_{\parallel} + 2L_{\perp} = 1$ to obtain the depolarization field coefficient $L_{\perp}$. With polarizability α, the scattering and absorption cross-sections can be evaluated as in the sphere case.

Program 4.1 calculates the polarizability of a cigar-shaped rotating ellipsoid. The long-axis length c is 50 nm, and the short-axis length (a or b) is 10 nm. The obtained depolarization field coefficients are $L_{\parallel} = 0.058$ and $L_{\perp} = 0.472$. After evaluating polarizability, the scattering cross-section C_{sca} and absorption cross-section C_{abs} are obtained using Eq. (2.2). The scattering efficiency Q_{sca} and absorption efficiency Q_{abs} are obtained by normalizing with the cross-sectional area.

Program 4.1

```
import scipy as sp
import matplotlib as mpl
import matplotlib.pyplot as plt
from matplotlib.pyplot import plot,show,xlabel,ylabel,title,
    legend,grid,axis,rcParams,tight_layout
from scipy import real,imag,pi,sqrt,log
from RI import WLx, epAg, epAu, RIAu, RIAg

```

```
n1 = RIAu  # refractive index of ellipsoid
n2 = 1  # refractive index of ambient
a = b = 10  # length of non-rotation axis(nm)
c = 50  # length of rotation axis(nm)
ee = sqrt(1-(a/c)**2)              # eccentricity
lz = (1-ee**2)/ee**2 * (1/(2*ee) * log((1+ee)/(1-ee))-1)    # depolarization factor in z
lx = (1-lz)/2  # depolarization factor in x
k = 2 * pi / WLx                   # vacuum wavenumber

alphax = 4*pi*a*b*c*(n2**2)*((n1**2)-(n2**2))/(3*((n2**2)+lx*((n1**2)-(n2**2))))   # polarizability in x
alphaz = 4*pi*a*b*c*(n2**2)*((n1**2)-(n2**2))/(3*((n2**2)+lz*((n1**2)-(n2**2))))   # polarizability in z

Csca_x = k**4 / (6 * pi) * abs(alphax)**2       # scattering cross-section in x
Cabs_x = k * imag(alphax)        # absorption cross-section in x
Qsca_x = Csca_x / (a*a*pi)       # scattering efficiency in x
Qabs_x = Cabs_x / (a*a*pi)       # absorption efficiency in x

Csca_z = k**4 / (6 * pi) * abs(alphaz)**2       # scattering cross-section in z
Cabs_z = k * imag(alphaz)        # absorption cross-section in z
Qsca_z = Csca_z / (a*c*pi)       # scattering efficiency in z
Qabs_z = Cabs_z / (a*c*pi)       # absorption efficiency in z

plt.figure(figsize=(8,6))
plot(WLx,Qsca_x, label=r"$Q_{{\rm sca},a}$",linewidth = 3.0, color='black')
plot(WLx,Qabs_x, label=r"$Q_{{\rm abs},a}$",linewidth = 3.0, color='gray')
xlabel("wavelength (nm)",fontsize=22)      # x-axis label
ylabel("efficiency",fontsize=22)          # y-axis label
title("Efficiency $a$-axis",fontsize=22)   # Title of the graph
grid(True)                                 # Show Grid
axis([300,1000,0,1])                       # Plot Range
plt.tick_params(labelsize=20)
legend(fontsize=20,loc='lower right')
tight_layout()
show()

plt.figure(figsize=(8,6))
plot(WLx,Qsca_z, label=r"$Q_{{\rm sca},c}$",linewidth = 3.0, color='black')
plot(WLx,Qabs_z, label=r"$Q_{{\rm abs},c}$",linewidth = 3.0, color='gray')
xlabel("wavelength (nm)",fontsize=22)      # x-axis label
ylabel("efficiency",fontsize=22)          # y-axis label
title("Efficiency $c$-axis",fontsize=22)   # Title of the graph
grid(True)                                 # Show Grid
axis([300,1000,0,50])                       # Plot Range
plt.tick_params(labelsize=20)
legend(fontsize=20,loc='lower left')
tight_layout()
show()
```

The scattering efficiency Q_{sca} and absorption efficiency Q_{abs} are shown in Figure 4.2(a) when light is polarized in the short-axis direction. Figure 4.2(b) shows Q_{sca} and Q_{abs} when light is polarized in the long-axis directions. In the case of light polarized in the short-axis direction, the peak scattering and absorption efficiencies are small, less than 1.0, even at the peak wavelength, which is around 500 nm. It is similar to the surface plasmon resonance of spherical gold particles. On the other hand, when the light is polarized in the long-axis direction, a sharp peak due to surface plasmon is observed at around 700 nm, and the absorption efficiency and scattering efficiency are high, about 35 and 8, respectively. As the rotating ellipsoid's aspect ratio (c/a) increases, the peak shifts to the long wavelength side, and the scattering and absorption efficiencies elevate. These properties are useful for applications of metallic rod structures with a shape similar to the rotating ellipsoid have been studied.

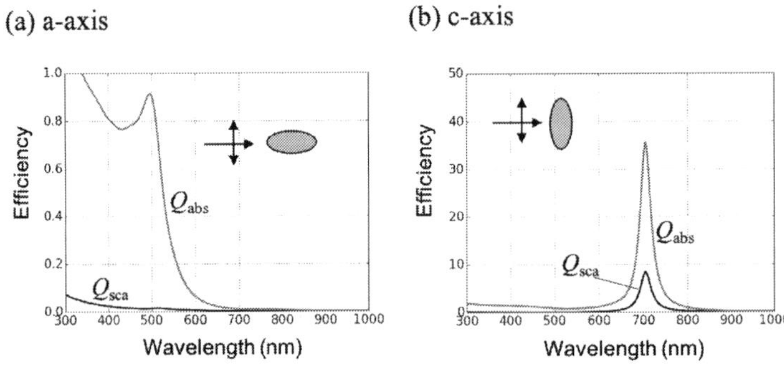

FIGURE 4.2
Calculated scattering efficiency Q_{sca} and absorption efficiency Q_{abs} for a rotating ellipsoid structure of gold at $a = 10$ nm and $c = 50$ nm: (a) when the polarization direction is in the short-axis direction and (b) when the polarization direction is in the long-axis direction.

4.1.2 Pancake-shaped

For a pancake-shaped rotating ellipsoid with a shorter rotation axis, the depolarization field coefficient $L_{\perp}$ in the direction perpendicular to the rotation axis is expressed by the following equation:

$$
\begin{aligned}
L_{\perp} &= \frac{g}{2e^2}\left(\frac{\pi}{2} - \tan^{-1} g\right) - \frac{g^2}{2} \\
g^2 &= \frac{1-e^2}{e^2} \\
e^2 &= 1 - \frac{c^2}{a^2}
\end{aligned}
\tag{4.4}
$$

As in the cigar-shaped case, the depolarization field coefficient $L_{\parallel}$ in the short-axis direction is obtained from the relation $L_{\parallel} + 2L_{\perp} = 1$. The polarizability in each direction can be obtained using Eq. (4.1) with the depolarization field coefficients.

4.1.3 Core-shell ellipsoids

This section deals with the optical response of the core-shell structure of a rotating ellipsoid [8]. Let n_1 be the refractive index of the core, n_2 be the refractive index of the shell, and n_3 be the refractive index of the surrounding medium. Let c_1 be the radius of the core in the rotation axis direction, and $a_1 = b_1$ be the radius of the other axis, and let c_2 and $a_2 = b_2$ be the radius of the axis of rotation and minor axis of the shell, respectively. The depolarization field coefficients can be obtained for both the cigar and pancake shapes from Eq. (4.2) and Eq. (4.4). Let L_i be the depolarization field coefficient for polarization in the i direction determined by the shape. If the thickness of the shell is constant, the depolarization field coefficient of the shell is also equal, and the following equation obtains the polarizability α_i:

$$\alpha_i = \frac{4\pi abc}{3} \frac{L_i(n_2^2 - n_3^2)(n_2^2 + (n_1^2 - n_2^2)(1 - Q))) + Qn_2^2(n_1^2 - n_2^2)}{(n_2^2 + L_i(n_1^2 - n_2^2)(1 - Q)))(n_3^2 + L_i(n_2^2 - n_3^2)) + QL_in_2^2(n_1^2 - n_2^2)} \tag{4.5}$$

Here, Q is the core-to-shell volume ratio, given by $Q = a_1b_1c_1/(a_2b_2c_2)$. In the case of a sphere, $L_i = \frac{1}{3}$, which is reduced to Eq. (2.14) discussed in Chapter 2.

4.2 Sphere above a substrate

Experiments often involve particles on a substrate. Therefore, discussing the optical response of the sphere immobilized on a substrate is sometimes necessary. When the substrate is a dielectric with a relatively low refractive index, such as quartz, the influence of the substrate is small, and the response of the isolated spheres can be discussed. However, when a metal, dielectric with a high refractive index, or semiconductor is used as the substrate, the optical response of the particles is greatly influenced by the substrate. Then, the optical response of spheres on the substrate differs from that of isolated particles. This problem is analytically solved by Wind [14] and is applied to gold spheres on a metallic surface by Okamoto [15].

Consider a sphere of radius R with dielectric constant ϵ_3 immobilized at a distance of gap g on a substrate with dielectric constant ϵ_2 in an ambient medium with dielectric constant ϵ_1 as shown in Figure 4.3. Using the dielectric constant instead of the refractive index makes the description simple. Consider a spherical coordinate system (ρ, θ, ϕ) with the sphere's centre as the origin,

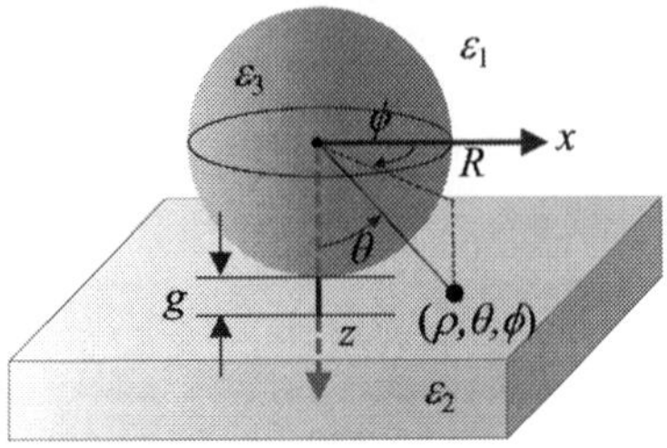

FIGURE 4.3
Optical configuration used to calculate the optical response of a sphere on a substrate. Polar coordinates $(\rho,\ \theta,\ \phi)$ with the sphere's centre as the origin are used.

where $\rho = r/R$. Then, the surface of the sphere is $r = 1$. The potential V_i in the medium i that arises when an electric field E_0 is applied. ψ_i is the potential normalized by $-E_0R$. Define r_0 as $r_0 = 1 + g/R$.

4.2.1 Normal component

The normalized potential ψ in Mediums 1–3 that arises when an electric field E_0 is applied perpendicular to the substrate surface (in the z-direction) is expressed as follows, taking multipoles into account.

$$
\begin{aligned}
\psi_1 &= rt + \sum_{j=1}^{\infty} r^{-(j+1)} P_j^0(t) A_{1j} + V_j^0(r,t) A'_{1j} \\
\psi_2 &= \psi'_2 + \alpha rt + \sum_{j=1}^{\infty} r^{-(j+1)} P_j^0(t) A_{2j} \\
\psi_3 &= \sum_{j=1}^{\infty} r^j P_j^0(t) A_{3j}
\end{aligned}
\tag{4.6}
$$

The $P_j^0(t)$ is the Legendre polynomials, and we set $t = \cos\theta$.

$$A'_{1j} = \frac{\epsilon_1 - \epsilon_2}{\epsilon_1 + \epsilon_2}(-1)^j A_{1j} \tag{4.7}$$

$$A_{2j} = \frac{2\epsilon_1}{\epsilon_1 + \epsilon_2} A_{1j} \tag{4.8}$$

The $V_j^0(r,t)$ in Eq. (4.6) can be written as follows:

$$V_j^m(r,t) = \frac{P_j^m\left(\dfrac{rt - 2r_0}{(r^2 - 4rr_0t + 4r_0^2)^{1/2}}\right)}{(r^2 - 4rr_0t + 4r_0^2)^{(j+1)/2}} \tag{4.9}$$

This represents the contribution of the mirror image in the substrate. The A_{ij} are multipole coefficients of order j in medium i. Although α and ψ'_2 are unknown, the solution solved with appropriate boundary conditions yields

$$\sum_{j=1}^{\infty}\left(\delta_{ij}+\frac{k(\epsilon_2-\epsilon_1)(\epsilon_1-\epsilon_3)}{(\epsilon_2+\epsilon_1)((k+1)\epsilon_1+k\epsilon_3)}\frac{(k+j)!}{k!j!(2r_0)^{k+j+1}}\right)A_{1j}=\frac{\epsilon_1-\epsilon_3}{2\epsilon_1+\epsilon_3}\delta_{k1}. \tag{4.10}$$

Here, δ_{pq} is Kronecker's delta. This equation is a simultaneous equation with an infinite number of undetermined coefficients, but in practice, it is sufficient to consider 10–15 undetermined coefficients. From the obtained coefficients A_{11}, the vertical component of the polarizability α_z is obtained using the following equation:

$$\alpha_z=-4\pi\epsilon_1 R^3 A_{11} \tag{4.11}$$

From above, the scattering cross section $C_{\mathrm{sca},z}$ and absorption cross-section $C_{\mathrm{abs},z}$ for the photoelectric field component perpendicular to the surface are

$$\begin{aligned} C_{\mathrm{sca},z} &= \frac{k^4}{6\pi}|\alpha_z|^2 \\ C_{\mathrm{abs},z} &= k\ \mathrm{Im}(\alpha_z). \end{aligned} \tag{4.12}$$

Here, k is the wavenumber of light. Normalizing it by the area of the great circle of the sphere, the scattering efficiency $Q_{\mathrm{sca},z}$ and absorption efficiency $Q_{\mathrm{abs},z}$.

4.2.2 In-plane component

The potential created when an electric field E_0 is applied horizontally (in the x- or y-direction) to the substrate surface is expressed as follows:

$$\begin{aligned} \psi_1 &= r\sqrt{1-t^2}\cos\phi+\sum_{j=1}^{\infty}r^{-(j+1)}P_j^1(t)B_{1j}\cos\phi+V_j^1(r,t)B'_{1j}\cos\phi \\ \psi_2 &= \psi'_2+\beta r\sqrt{1-t^2}\cos\phi+\sum_{j=1}^{\infty}r^{-(j+1)}P_j^1(t)B_{2j}\cos\phi \\ \psi_3 &= \sum_{j=1}r^jP_j^1(t)\cos\phi B_{3j} \end{aligned} \tag{4.13}$$

Here, $P_j^m(t)$ is the associated Legendre polynomial, and B_{ij} is the multipole coefficient of order j in medium i.

$$B'_{1j} = \frac{\epsilon_1-\epsilon_2}{\epsilon_1+\epsilon_2}(-1)^{j+1}B_{1j} \tag{4.14}$$

$$B_{2j} = \frac{2\epsilon_1}{\epsilon_1+\epsilon_2}B_{1j} \tag{4.15}$$

Including β and ψ_2' and solving for them, as well as the vertical component, yield

$$\sum_{j=1}^{\infty}\left(\delta_{ij}+\frac{k(\epsilon_2-\epsilon_1)(\epsilon_1-\epsilon_3)}{(\epsilon_2+\epsilon_1)((k+1)\epsilon_1+k\epsilon_3)}\frac{(k+j)!}{(k+1)!(j-1)!(2r_0)^{k+j+1}}\right)B_{1j} = \frac{\epsilon_1-\epsilon_3}{2\epsilon_1+\epsilon_3}\delta_{k1}. \quad (4.16)$$

With the obtained coefficient B_{11}, the in-plane component of the polarizability $\alpha_{\|}$ can be obtained, and the scattering efficiency $Q_{\text{sca},\|}$ and absorption efficiency $Q_{\text{abs},\|}$ can be obtained.

4.2.2.1 Programing

Based on the above results, an example of a program to calculate the optical response of a sphere immobilized on a substrate is shown below. The factorial function math.factorial is loaded in advance. However, since this function name is long, it is again defined as kjo in Line 14. Define the coefficients to be calculated in Eqs. (4.10) and (4.16) as perpen and parallel functions, respectively. The number of undetermined coefficients in the simultaneous equation is qq. Here, qq =15, meaning we are solving a 15-element linear system of equations. The matrix of undetermined coefficients is described in Lines 52–70 and solved using linalg.solve in Lines 72 and 73.

Program 4.2

```
1  import numpy as np
2  import scipy as sp
3  import scipy.special
4  import math
5  import cmath
6  import matplotlib as mpl
7  import matplotlib.pyplot as plt
8  from RI import WLx, NumWLx, epAg, epAu, RIAu, RIAg
9
10 from scipy import pi,sin,cos,tan,arcsin,exp,linspace,arrange,sqrt
       ,zeros,array,matrix,asmatrix,real,imag
11 from matplotlib.pyplot import plot,show,xlabel,ylabel,title,
       legend,grid,axis,tight_layout
12 from scipy.special import factorial
13
14 def kjo(k):
15     return math.factorial(k)
16
17 def perpen(k,j,r0,ep1,ep2,ep3):
18     return ((ep2-ep1)*(ep1-ep3)*k*kjo(k+j))/((ep2+ep1)*((k+1)*ep1
           +k*ep3)*kjo(k)*kjo(j)*(2*r0)**(k+j+1))
19
20 def parallel(k,j,r0,ep1,ep2,ep3):
21     return ((ep2-ep1)*(ep1-ep3)*k*kjo(k+j))/((ep2+ep1)*((k+1)*ep1
           +k*ep3)*kjo(k+1)*kjo(j-1)*(2*r0)**(k+j+1))
```

```

def uhen(ep1,ep2,ep3):
    return (ep1-ep3)/(2*ep1+ep3)

k0 = 2 * pi / WLx  # vacuum wavenumber

qq=15  # order of multipoles
r=50  # radius of sphere
gap=1  # gap
d=gap+r  # d parameter
r0=d/r # r0 parameter

alpha_A=zeros(NumWLx, dtype=complex) # initialization of A
alpha_B=zeros(NumWLx, dtype=complex) # initialization of B

ep1=zeros(NumWLx, dtype=complex)
ep2=zeros(NumWLx, dtype=complex)
ep3=zeros(NumWLx, dtype=complex)
al=zeros([NumWLx,qq,qq], dtype=complex)
bl=zeros([NumWLx,qq,qq], dtype=complex)
fl=zeros([NumWLx,qq], dtype=complex)
Xal=zeros([NumWLx,qq], dtype=complex)
Xbl=zeros([NumWLx,qq], dtype=complex)
a11l=zeros([NumWLx,qq], dtype=complex)
b11l=zeros([NumWLx,qq], dtype=complex)

for i in range(NumWLx):
    ep1[i] = 1                  # dielectric constant of ambient
    ep2[i] = epAu[i]    # dielectric constant of sphere
    ep3[i] = epAu[i]    # dielectric constant of substrate
    for k in range(qq):   # A coefficient
        for j in range(qq):
             if k==j:
                 al[i,k,j]=1+perpen(k+1,j+1,r0,ep1[i],ep2[i],ep3[
                     i])
             else:
                 al[i,k,j]=perpen(k+1,j+1,r0,ep1[i],ep2[i],ep3[i
                     ])

    for k in range(qq):
        for j in range(qq):
            if k==j:
                bl[i,k,j]=1+parallel(k+1,j+1,r0,ep1[i],ep2[i],ep3
                    [i])
            else:
                bl[i,k,j]=parallel(k+1,j+1,r0,ep1[i],ep2[i],ep3[i
                    ])

    for k in range(qq):
            if k==0:
                fl[i,k]=uhen(ep1[i],ep2[i],ep3[i])
            else:
                fl[i,k]=0

    Xal[i]=np.linalg.solve(al[i],fl[i])
    Xbl[i]=np.linalg.solve(bl[i],fl[i])

```

```
    alpha_A[i]=-4*pi*r**3*ep1[i]*Xal[i,0]  # polarizability (normal)
    alpha_B[i]=-4*pi*r**3*ep1[i]*Xbl[i,0]  # polarizability (in-plane)

Csca_A = k0**4/(6*pi)*abs(alpha_A)**2  # scattering cross-section (normal)
Csca_B = k0**4/(6*pi)*abs(alpha_B)**2   # scattering cross-section (in-plane)
Cabs_A = k0*imag(alpha_A) # absorption cross-section (normal)
Cabs_B = k0*imag(alpha_B) # absorption cross-section (in-plane)

plt.figure(figsize=(8,6))
plot(WLx,Cabs_A, label=r"$C_{\rm abs}$")
axis([400,700,0,100000])
xlabel(r"wavelength (nm)",fontsize=12)
ylabel(r"$C_{\rm ext}$",fontsize=12)
legend(fontsize=20,loc='lower right')
tight_layout()
show()

the same applies hereinafter
```

The results of the calculations for the optical configuration of gold spheres on a gold substrate are shown in Figure 4.4. The sphere's radius is 50 nm, and the gap is 1 nm. The scattering efficiency $Q_{\mathrm{sca},z}$ and absorption efficiency

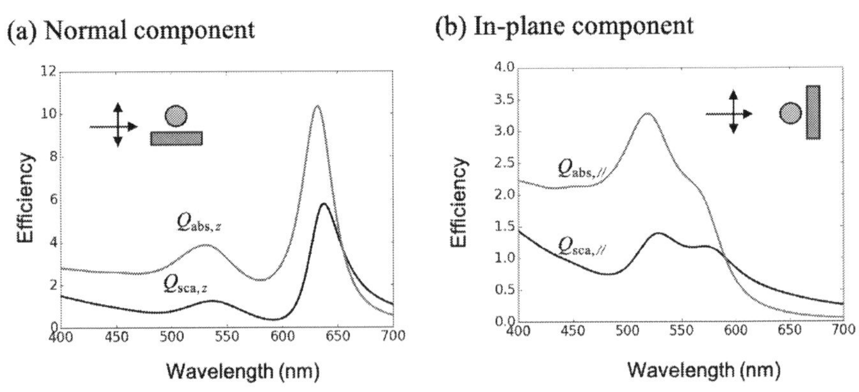

FIGURE 4.4
Calculated optical response of a sphere on a substrate. (a) Photoelectric field component perpendicular to the surface, and (b) scattering efficiency $Q_{\mathrm{sca},z}$ and absorption efficiency $Q_{\mathrm{abs},z}$ for photoelectric field component in the surface plane.

$Q_{\mathrm{abs},z}$ for the electric field component perpendicular to the surface are shown in Figure 4.4(a), as well as the scattering efficiency $Q_{\mathrm{sca},\|}$ and absorption efficiency $Q_{\mathrm{abs},\|}$. It can be seen that the component perpendicular to the surface produces a peak that is shifted to the long wavelength side, which is larger than the absorption peak of the isolated gold particle. This is a result of the interaction between the sphere and the substrate. Such a significant shift does not occur when the substrate is a dielectric such as glass. As shown in Figure 4.4(b), the in-plane component slightly shifts, but the displacement is not large. This indicates that the interaction with the substrate is not very large.

4.3 Bisphere

A method of calculating the optical response of a bisphere shown in Figure 4.5 has been proposed [16], and it can also be obtained using Eqs. (4.10) and (4.16) [14]. This is because if the substrate is an ideal metal, we can regard another sphere as a mirror image of the substrate. Note that the gap between the spheres is $2g$ in this case. In actual calculations, in Eqs. (4.10) and (4.16),

$$\frac{\epsilon_2 - \epsilon_1}{\epsilon_2 + \epsilon_1} = 1. \tag{4.17}$$

For the in-plane component, set

$$\frac{\epsilon_2 - \epsilon_1}{\epsilon_2 + \epsilon_1} = -1 \tag{4.18}$$

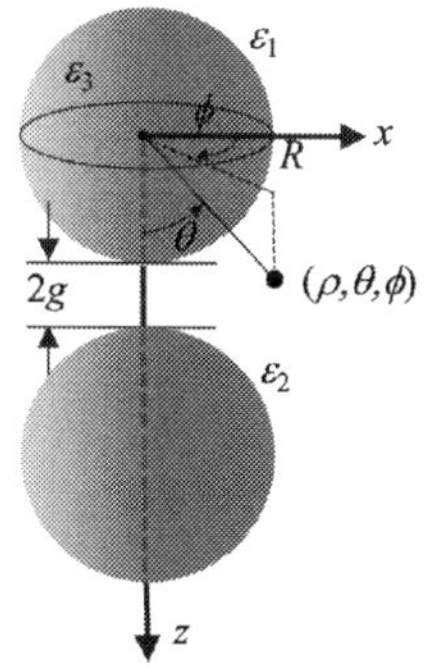

FIGURE 4.5
Optical geometry of a bisphere.

and make calculations. Namely, solve the following simultaneous equations for the appropriate number of undetermined coefficients to obtain the polarization ratio and calculate the scattering and absorption cross-sections.

$$\sum_{j=1}^{\infty}\left(\delta_{ij} + \frac{k(\epsilon_1 - \epsilon_3)}{((k+1)\epsilon_1 + k\epsilon_3)} \frac{(k+j)!}{k!j!(2r_0)^{k+j+1}}\right) A_{1j} = \frac{\epsilon_1 - \epsilon_3}{2\epsilon_1 + \epsilon_3}\delta_{k1} \quad (4.19)$$

$$\sum_{j=1}^{\infty}\left(\delta_{ij} - \frac{k(\epsilon_1 - \epsilon_3)}{((k+1)\epsilon_1 + k\epsilon_3)} \frac{(k+j)!}{(k+1)!(j-1)!(2r_0)^{k+j+1}}\right) B_{1j} = \frac{\epsilon_1 - \epsilon_3}{2\epsilon_1 + \epsilon_3}\delta_{k1} \quad (4.20)$$

The program in Appendix Program A.3 (Bisphere.py) shows the calculation results in Figure 4.6. The gold spheres have a 50 nm radius and a 1 nm gap. In this program, half of the gap is set as 0.5 nm. This condition can be compared with the calculated optical response of the sphere on the substrate shown in Figure 4.4. In the case of the sphere on the substrate described above, the absorption peak of the vertical component is around 635 nm, and the absorption efficiency is also significant. On the other hand, in the case of the bisphere, the absorption peak of the vertical component is around 590 nm, and the absorption efficiency is slightly lower than that of the sphere on the substrate. It seems that the dielectric constant of the substrate has an imaginary component, which causes a stronger interaction than that of the ideal metal. For example, when silver is used as the substrate, the absorption peak appears around 605 nm between the ideal metal and gold. For the horizontal component, the results are not much different from those calculated for isolated particles. This is likely due to the weak interaction between particles.

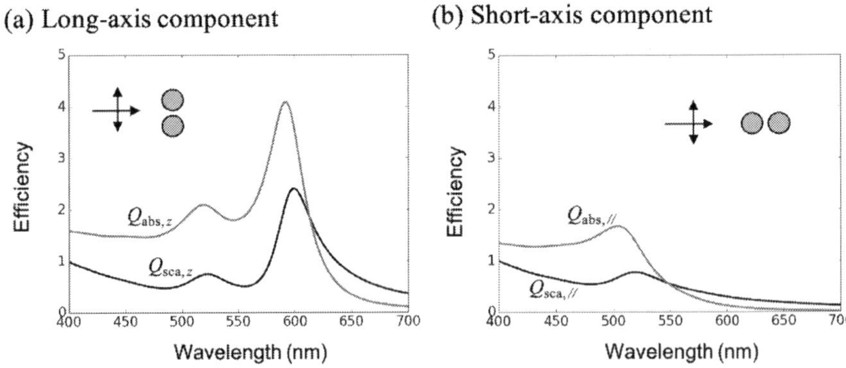

FIGURE 4.6
Absorption and scattering cross-sections calculated for a gold bisphere. The gap spacing was set to 1 nm.

4.4 Truncated sphere on a substrate

The optical response of a truncated sphere, as shown in Figure 4.7, can also be obtained by analytical calculations derived by Wind et al. [14]. Let Medium 1 be the surrounding medium, Medium 2 be the substrate, Medium 3 be the truncated sphere, and Medium 4 be the substrate. Let $\epsilon_i (i = 1 \sim 4)$ be their dielectric constants. The spherical coordinate system (ρ, θ, ϕ) is used. Medium 2 and Medium 4 are the same, i.e., $\epsilon_2 = \epsilon_4$. The cutting angle of the truncated sphere is then defined as θ_{sh}, as shown in Figure 4.7. In the case of a sphere, $\theta_{\mathrm{sh}} = 180°$, and in a hemisphere, $\theta_{\mathrm{sh}} = 90°$. The program defines $\theta_{\mathrm{a}} = 180° - \theta_{\mathrm{sh}}$ for ease of calculation. The distance D and ρ of the centre of the sphere from the substrate surface are normalized by the sphere radius R, and $r_0 = \cos\theta_{\mathrm{a}} = D/R$ and $r = \rho/R$, respectively. Also, let V_i be the potential at medium i that arises when an electric field E is applied.

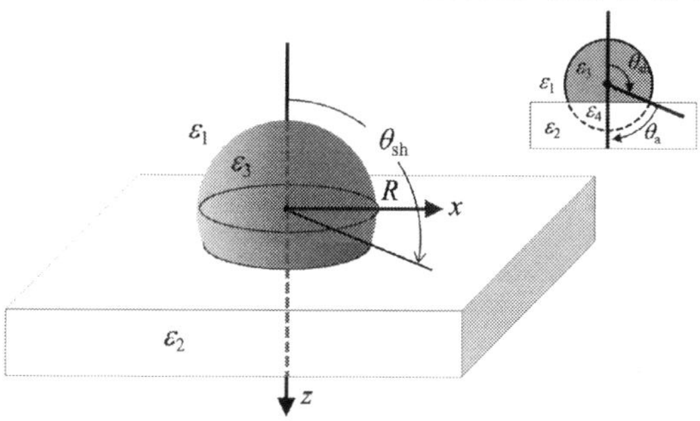

FIGURE 4.7
Optical geometry of a truncated sphere.

Suppose that the normalized potential $\psi_i (i = x, y, z)$ is $\psi_i = -V_i/(ER)$, then it is possible to use the multipole coefficients A_j^q and $\bar{A}_j^q$ and the potential is written as

$$\psi_x = r\sin\theta\cos\phi + \sum_{j=1}^{\infty} \frac{A_j^{\parallel} P_j^1(\cos\theta)\cos\phi}{r^{j+1}} + \bar{A}_j^{\parallel} V_j^1(r, \cos\theta)\cos\phi \tag{4.21}$$

$$\psi_y = r\sin\theta\sin\phi + \sum_{j=1}^{\infty} \frac{A_j^{\parallel} P_j^1(\cos\theta)\sin\phi}{r^{j+1}} + \bar{A}_j^{\parallel} V_j^1(r, \cos\theta)\sin\phi \tag{4.22}$$

$$\psi_z = r\cos\theta + \sum_{j=1}^{\infty} \frac{A_j^z P_j^0(\cos\theta)}{r^{j+1}} + \bar{A}_j^z V_j^0(r, \cos\theta). \tag{4.23}$$

Here, $P_j^m(\cos\theta)$ is a Legendre function, and $V_j^m(r, \cos\theta)$. $W_j^m(r, \cos\theta)$ on the mirror image are

$$V_j^m(r, \cos\theta) = \frac{P_j^m\left(\frac{r\cos\theta - 2r_0}{\sqrt{r^2 - 4rr_0\cos\theta + 4r_0^2}}\right)}{\left(\sqrt{r^2 - 4rr_0\cos\theta + 4r_0^2}\right)^{j+1}} \tag{4.24}$$

and

$$W_j^m(r,\cos\theta) = \left(\sqrt{r^2 - 4rr_0\cos\theta + 4r_0^2}\right)^j P_j^m\left(\frac{r\cos\theta - 2r_0}{\sqrt{r^2 - 4rr_0\cos\theta + 4r_0^2}}\right) \tag{4.25}$$

$\bar{A}_j^q$ has the following relationship with A_j^q:

$$\bar{A}_j^z = \frac{\epsilon_1 - \epsilon_2}{\epsilon_1 + \epsilon_2}(-1)^j A_j^z \tag{4.26}$$

$$\bar{A}_j^{\|} = \frac{\epsilon_1 - \epsilon_2}{\epsilon_1 + \epsilon_2}(-1)^{j+1} A_j^{\|}. \tag{4.27}$$

$q = \|$ is used to find the components of the multipole coefficients.

The multipole coefficients A_j^q and B_j^q are to be obtained by solving the following simultaneous equations:

$$\sum_{j=1}^{\infty}(C_{kj}^q A_j^q + D_{kj}^q B_j^q) = E_k^q \qquad (k = 1, 2, 3...) \tag{4.28}$$

$$\sum_{j=1}^{\infty}(F_{kj}^q A_j^q + G_{kj}^q B_j^q) = H_k^q \qquad (k = 1, 2, 3...). \tag{4.29}$$

The following equations give each coefficient in the simultaneous equations for perpendicular polarization:

$$\begin{aligned}
C_{kj}^z &= \frac{4\epsilon_1\delta_{kj}}{(\epsilon_1+\epsilon_2)(2k+1)} - \frac{\epsilon_1-\epsilon_2}{\epsilon_1+\epsilon_2}\int_{-1}^{r_0} dt P_k^0(t)(P_j^0(t) - (-1)^j V_j^0(1,t)) \\
D_{kj}^z &= -\frac{4\epsilon_3\delta_{kj}}{(\epsilon_2+\epsilon_3)(2k+1)} - \frac{\epsilon_2-\epsilon_3}{\epsilon_2+\epsilon_3}\int_{-1}^{r_0} dt P_k^0(t)(P_j^0(t) - (-1)^j W_j^0(1,t)) \\
E_k^z &= -\frac{2\epsilon_1\delta_{k1}}{3\epsilon_2} - (1 - \frac{\epsilon_1}{\epsilon_2})\int_{-1}^{r_0} dt P_k^0(t)(t - r_0) \\
F_{kj}^z &= -\frac{4\epsilon_1\epsilon_2(k+1)\delta_{kj}}{(\epsilon_1+\epsilon_2)(2k+1)} \\
&\quad -\frac{\epsilon_1(\epsilon_1-\epsilon_2)}{\epsilon_1+\epsilon_2}\int_{-1}^{r_0} dt P_k^0(t)\left((j+1)P_j^0(t) - (-1)^j \frac{\partial V_j^0(r,t)}{\partial r}\bigg|_{r=1}\right) \\
G_{kj}^z &= -\frac{4\epsilon_2\epsilon_3 k\delta_{kj}}{(\epsilon_2+\epsilon_3)(2k+1)} \\
&\quad +\frac{\epsilon_3(\epsilon_2-\epsilon_3)}{\epsilon_2+\epsilon_3}\int_{-1}^{r_0} dt P_k^0(t)\left(jP_j^0(t) + (-1)^j \frac{\partial W_j^0(r,t)}{\partial r}\bigg|_{r=1}\right) \\
H_k^z &= -\frac{2\epsilon_1\delta_{k1}}{3}.
\end{aligned} \tag{4.30}$$

For the in-plane component,

$$\begin{aligned}
C^{\parallel}_{kj} &= \frac{4\epsilon_1 k(k+1)\delta_{kj}}{(\epsilon_1+\epsilon_2)(2k+1)} - \frac{\epsilon_1-\epsilon_2}{\epsilon_1+\epsilon_2}\int_{-1}^{r_0} dt P_k^1(t)(P_j^1(t)+(-1)^j V_j^1(1,t)) \\
D^{\parallel}_{kj} &= -\frac{4\epsilon_3 k(k+1)\delta_{kj}}{(\epsilon_2+\epsilon_3)(2k+1)} \\
&\quad -\frac{\epsilon_2-\epsilon_3}{\epsilon_2+\epsilon_3}\int_{-1}^{r_0} dt P_k^1(t)(P_j^1(t)+(-1)^j W_j^1(1,t)) \\
E^{\parallel}_{k} &= -\frac{4\delta_{k1}}{3} \\
F^{\parallel}_{kj} &= -\frac{4\epsilon_1\epsilon_2 k(k+1)^2\delta_{kj}}{(\epsilon_1+\epsilon_2)(2k+1)} \\
&\quad -\frac{\epsilon_1(\epsilon_1-\epsilon_2)}{\epsilon_1+\epsilon_2}\int_{-1}^{r_0} dt P_k^1(t)\left((j+1)P_j^1(t)+(-1)^j \left.\frac{\partial V_j^1(r,t)}{\partial r}\right|_{r=1}\right) \\
G^{\parallel}_{kj} &= -\frac{4\epsilon_2\epsilon_3 k^2(k+1)\delta_{kj}}{(\epsilon_2+\epsilon_3)(2k+1)} \\
&\quad +\frac{\epsilon_3(\epsilon_2-\epsilon_3)}{\epsilon_2+\epsilon_3}\int_{-1}^{r_0} dt P_k^1(t)\left(jP_j^1(t)-(-1)^j \left.\frac{\partial W_j^1(r,t)}{\partial r}\right|_{r=1}\right) \\
H_{\parallel} &= -\frac{4\epsilon_2\delta_{k1}}{3} - (\epsilon_1-\epsilon_2)\int_{-1}^{r_0} dt P_k^1(t)P_1^1(t).
\end{aligned} \tag{4.31}$$

Program 8.4 in the Appendix is meant to calculate them. The results of the absorption efficiency spectra of the truncated spheres of various shapes are shown in Figure 4.8 [17]. In the case of $\theta_{\mathrm{a}} = 0°$ shown in Figure 4.8(a), the absorption efficiency for the light electric field normal to the surface $Q_{\mathrm{abs},z}$ and the absorption efficiency for the electric field in the in-plane direction of the surface $Q_{\mathrm{abs},\parallel}$ are almost the same as those of the sphere. The slight difference is due to the influence of the substrate. Since the in-plane component is sensitive to shape, the approximate shape of the truncated sphere can be determined by measuring the absorption spectrum at perpendicular incidence. Compared with cross-sectional transmission electron microscopy images, good agreement has been reported [18].

Next, the absorption efficiency spectrum of the truncated sphere on a substrate with different refractive indices is shown in Figure 4.9. Here, a hemisphere ($\theta_{sh} = 90°$) was considered. As the substrate's refractive index increases, $Q_{\mathrm{abs},\parallel}$ shifts significantly to the long wavelength side, thus increasing efficiency. On the other hand, the peak position of the absorption efficiency $Q_{\mathrm{abs},z}$ for the light electric field in the plane normal direction does not change much, and the absorption efficiency conversely decreases as the refractive index increases. This can be considered due to the substrate's mirror image effect. The island-like evaporated thin films of hemispherical structures may be used as a sensor for highly sensitive refractive index and bio-molecules.

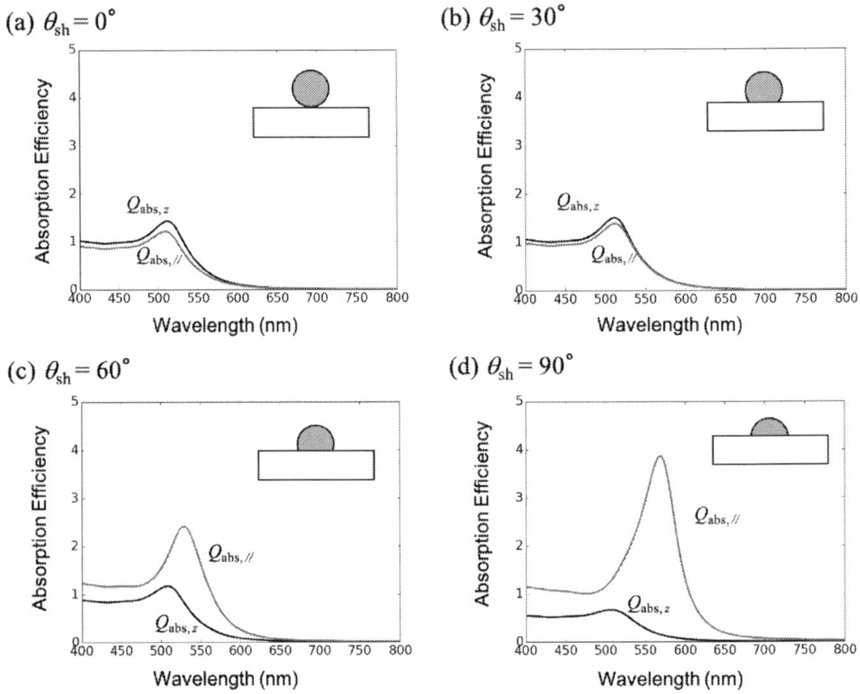

FIGURE 4.8
Comparison of calculated absorption efficiency spectra of cut spheres with various geometries. (a) $\theta_a = 0°$ (sphere), (b) $\theta_a = 30°$, (c) $\theta_a = 60°$, and (d) $\theta_a = 180°$ (hemisphere).

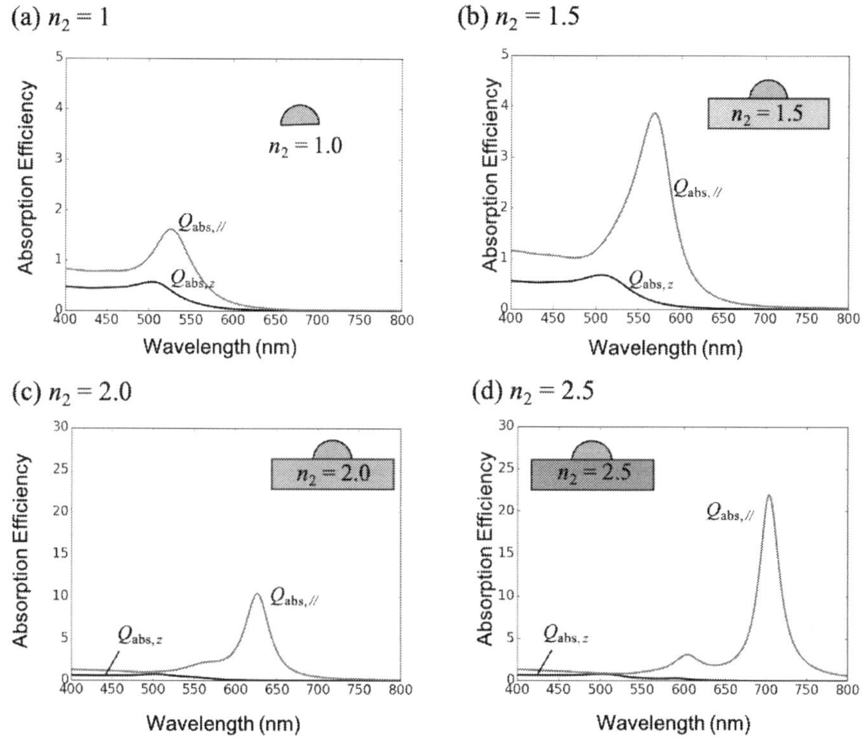

FIGURE 4.9
Calculated absorption efficiency spectra of cut spheres, compared by substrate refractive index n_2. (a) $n_2 = 1.0$, (b) $n_2 = 1.5$, (c) $n_2 = 2.0$, and (d) $n_2 = 2.5$.

5

Rigorous Coupled-Wave Analysis: RCWA

The rigorous coupled-wave analysis (RCWA) method was proposed to analyse the optical properties of diffraction gratings. In the early stage, there were problems of instability in calculation and slow convergence for TM polarization in metallic gratings, but both problems have been solved and the RCWA is now the most widely used analysis method for diffraction gratings.

5.1 Introduction

The rigorous coupled-wave analysis (RCWA) method was proposed by Moharam and Gaylord [19, 20, 21, 22]. The basic idea of the RCWA method is the same as the transfer matrix method for multilayers. First, the grating is divided into multiple layers as shown in Figure 5.1 and is approximated by a staircase shape. In this example, the layers are divided so that the thickness of each layer is equal, but the thicknesses of the layers do not have to be the same thickness. Within each layer, the dielectric constant is assumed to be uniform in the z-direction and modulated only in the x-direction. The distribution of the dielectric constant of layer l is presented as $\varepsilon^{(l)}(x)$. If the period of the grating is Λ, $\varepsilon^{(l)}(x+\Lambda) = \varepsilon^{(l)}(x)$. Layer 0 corresponds to the transmission side and layer L to the incident side. The dielectric constant of the layer L must be homogeneous. The RCWA method describes the light wave in each of these layers as a superposition of the eigenmodes and determines the amplitude of the light wave so that the boundary conditions are satisfied between the adjacent layers. Vectors $\boldsymbol{u}^{(l)}$ and $\boldsymbol{d}^{(l)}$ in Figure 5.1 are coefficient ones that give the amplitude of each eigenmode propagating in the $+z$- and $-z$-directions in layer l, respectively. The thickness of layer l is $h^{(l)}$. The difference from the multilayer case is that, due to the presence of diffracted light, the tangential component of the wave number is not a scalar value, but a vector whose elements are integers multiple of the lattice vector added to that of the incident light. In other words, if the in-plane wave number of the incident light is k_{x0}, that of the diffracted light is given by

$$k_{xm} = k_{x0} + mK, \tag{5.1}$$

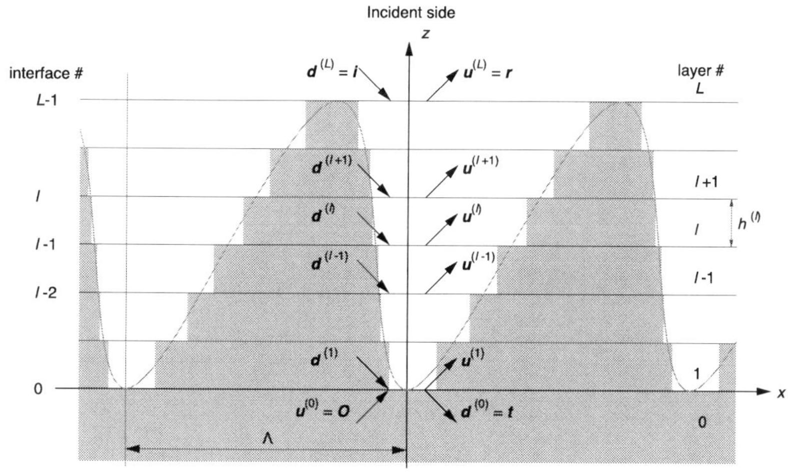

FIGURE 5.1
Modelling of geometry in the RCWA method. A periodic structure of arbitrary shape is divided into layers. When dividing the structure into layers, each layer should be homogeneous in the z-direction. In the figure, the thickness of each layer is equal, but it is not necessary to make them equal.

$$K = \frac{2\pi}{\Lambda}, \tag{5.2}$$

where K is the grating constant and m is the diffraction order.

5.1.1 TE polarization

First, let's consider the case of TE polarization. In this case, only the y component of the electric field, E_y should be considered. The other components of the electric field are zero, that is $E_x = E_z = 0$. Consider the electric field in a grating region, where the dielectric constant changes periodically. The electric field in layer l is also periodic and can be written as follows

$$E_y^{(l)} = \sum_m S_{ym}^{(l)}(z) \exp(ik_{xm}x), \tag{5.3}$$

where the origin of the z-coordinate is the lower interface of each layer. Only for layer 0, $z = 0$ is taken at the upper interface. Here, we obtain the wave equation for TE polarization with angular frequency ω. From Faraday's equation,

$$\boldsymbol{H} = \left(\frac{i}{\omega\mu_0}\right) \nabla \times \boldsymbol{E}, \tag{5.4}$$

$$\frac{\partial E_y^{(l)}}{\partial z} = i\omega\mu_0 H_x^{(l)}, \tag{5.5}$$

and

$$\frac{\partial E_y^{(l)}}{\partial x} = -i\omega\mu_0 H_z^{(l)} \tag{5.6}$$

are obtained. Also, from Ampere's equation,

$$\boldsymbol{E} = \left[\frac{-i}{\omega\varepsilon_0\varepsilon(x)}\right] \nabla \times \boldsymbol{H}, \tag{5.7}$$

$$\frac{\partial H_x^{(l)}}{\partial z} - \frac{\partial H_z^{(l)}}{\partial x} = i\omega\varepsilon_0\varepsilon^{(l)}(x)E_y^{(l)} \tag{5.8}$$

is obtained. Differentiating Eqs. (5.5) and (5.6) with respect to z and x, respectively, and substituting into Eq. (5.8),

$$\frac{1}{i\omega\mu_0}\frac{\partial^2 E_y^{(l)}}{\partial z^2} + \frac{1}{i\omega\mu_0}\frac{\partial^2 E_y^{(l)}}{\partial x^2} = i\omega\varepsilon_0\varepsilon^{(l)}(x)E_y^{(l)} \tag{5.9}$$

is obtained. Here, using the wave number $k_0 = \sqrt{\varepsilon_0\mu_0}\omega$ of light propagating in vacuum, the wave equation for TE polarization is rewritten as

$$\frac{\partial^2 E_y^{(l)}}{\partial z^2} + \frac{\partial^2 E_y^{(l)}}{\partial x^2} = -k_0^2\varepsilon^{(l)}(x)E_y^{(l)}. \tag{5.10}$$

Next, we represent the dielectric constant $\varepsilon^{(l)}(x)$ by its Fourier series,

$$\varepsilon^{(l)}(x) = \sum_p \varepsilon_p^{(l)} \exp(ipKx). \tag{5.11}$$

Substituting Eqs. (5.3) and (5.11) into Eq. (5.10), we obtain

$$\sum_m \frac{\partial^2 S_{ym}^{(l)}(z)}{\partial z^2} \exp(ik_{xm}) = \tag{5.12}$$
$$\sum_m k_{xm}^2 S_{ym}^{(l)}(z)\exp(ik_{xm}) - k_0^2 \sum_p \sum_m \varepsilon_{m-p}^{(l)} S_{yp}^{(l)}(z)\exp(ik_{xm}).$$

Therefore,

$$\frac{\partial^2 S_{ym}^{(l)}(z)}{\partial z^2} = k_{xm}^2 S_{ym}^{(l)}(z) - k_0^2 \sum_p \varepsilon_{m-p}^{(l)} S_{yp}^{(l)}(z). \tag{5.13}$$

If written in matrix form, we obtain

$$\frac{\partial^2 \boldsymbol{S}_y^{(l)}}{\partial z^2} = k_0^2\left(\mathbf{K}_x^2 - \mathbf{E}^{(l)}\right)\boldsymbol{S}_y^{(l)}, \tag{5.14}$$

where $\mathbf{K}_x$ is a diagonal matrix whose elements are k_{xm}/k_0. Matrix $\mathbf{E}^{(l)}$ is the

Toeplitz one whose elements are $\varepsilon^{(l)}_{(m-p)}$, i.e.

$$\mathbf{E}^{(l)} = \begin{bmatrix} \varepsilon^{(l)}_0 & \varepsilon^{(l)}_{-1} & \varepsilon^{(l)}_{-2} & \cdots \\ \varepsilon^{(l)}_1 & \varepsilon^{(l)}_0 & \varepsilon^{(l)}_{-1} & \cdots \\ \varepsilon^{(l)}_2 & \varepsilon^{(l)}_1 & \varepsilon^{(l)}_0 & \cdots \\ \vdots & \vdots & \vdots & \ddots \end{bmatrix}. \tag{5.15}$$

The solution of Eq. 5.14 is expressed as

$$S^{(l)}_{ym}(z) = \sum_{j=1}^{\infty} w^{(l)}_{mj} \left[u^{(l)}_j \exp(ik^{(l)}_{zj} z) + d^{(l)}_j \exp(-ik^{(l)}_{zj} z) \right], \tag{5.16}$$

where $-\left[k^{(l)}_{zj}\right]^2$ and $w^{(l)}_{mj}$ are the eigenvalue and the element of the eigenvector of matrix $k_0^2(\mathbf{K}_x^2 - \mathbf{E}^{(l)})$. It is important to note that j does not correspond to the diffraction order, but simply to the order of the eigenvalues. The quantity $k^{(l)}_{zj}$ corresponds to the z component of the wave number and obtained by reversing the sign of the eigenvalue of matrix $k_0^2(\mathbf{K}_x^2 - \mathbf{E}^{(l)})$ and taking its square root. If $k^{(l)}_{zj}$ is complex, we have to take the square root for which $\mathrm{Im}\left[k^{(l)}_{zj}\right] \geq 0$. This means that we employ evanescent waves that decay exponentially. However, care must be taken in the actual calculation even when there is no imaginary part in the dielectric constant of the medium. In this case, all eigenvalues are real numbers. Therefore, their square roots are real or pure imaginary. However, even when the eigenvalues are real numbers (corresponding to propagating light), a slight imaginary part that is not zero may be included due to less calculation accuracy. As a result, the square root also contains non-zero imaginary parts. When determining the sign of the square root by using the sign of the imaginary part, which should be zero, a problem arises. The solution to this problem is to use the real part of the square root as well. In other words, the sign should be taken to be $\mathrm{Re}\left[k^{(l)}_{zj}\right] + \mathrm{Im}\left[k^{(l)}_{zj}\right] \geq 0$. However, if the dielectric constant has an imaginary part, the above $\mathrm{Im}\left[k^{(l)}_{zj}\right] \geq 0$ condition must be used.

The coefficients $u^{(l)}_j$ and $d^{(l)}_j$ are determined by the boundary conditions. The tangential component H_x of the magnetic field is also necessary for the boundary conditions. Suppose that the magnetic field H_x can be written as follows,

$$H^{(l)}_x = \left(\frac{\varepsilon_0}{\mu_0}\right)^{1/2} \sum_m U^{(l)}_{xm}(z) \exp(ik_{xm}x). \tag{5.17}$$

Substituting Eqs. (5.16) and (5.17) into Eq. (5.5),

$$U^{(l)}_{xm}(z) = \sum_{j=1}^{\infty} v^{(l)}_{mj} \left[u^{(l)}_j \exp(ik^{(l)}_{zj} z) - d^{(l)}_j \exp(-ik^{(l)}_{zj} z) \right] \tag{5.18}$$

is obtained, where

$$v_{mj}^{(l)} = -\frac{1}{k_0} k_{zj}^{(l)} w_{mj}^{(l)}. \tag{5.19}$$

In matrix form,

$$\mathbf{V}^{(l)} = -\frac{1}{k_0} \mathbf{W}^{(l)} \mathbf{Q}^{(l)}, \tag{5.20}$$

where $\mathbf{V}^{(l)}$ and $\mathbf{W}^{(l)}$ are matrices whose elements are $v_{mj}^{(l)}$ and $w_{mj}^{(l)}$, respectively, and $\mathbf{Q}^{(l)}$ is a diagonal matrix whose elements are $k_{zj}^{(l)}$.

In matrix form, Eqs. (5.16) and (5.18) are

$$\boldsymbol{S}_y^{(l)}(z) = \mathbf{W}^{(l)} \begin{bmatrix} \phi_+^{(l)}(z) & \phi_-^{(l)}(z) \end{bmatrix} \begin{bmatrix} \boldsymbol{u}^{(l)} \\ \boldsymbol{d}^{(l)} \end{bmatrix}, \tag{5.21}$$

and

$$\boldsymbol{U}_x^{(l)}(z) = \mathbf{V}^{(l)} \begin{bmatrix} \phi_+^{(l)}(z) & -\phi_-^{(l)}(z) \end{bmatrix} \begin{bmatrix} \boldsymbol{u}^{(l)} \\ \boldsymbol{d}^{(l)} \end{bmatrix}, \tag{5.22}$$

respectively, where $\phi_\pm^{(l)}(z)$ is a diagonal matrix whose elements are $\exp(\pm i k_{zj}^{(l)} z)$. Equations (5.21) and (5.22) can be combined into one:

$$\begin{bmatrix} \boldsymbol{S}_y^{(l)}(z) \\ \boldsymbol{U}_x^{(l)}(z) \end{bmatrix} = \begin{bmatrix} \mathbf{W}^{(l)} & \mathbf{W}^{(l)} \\ \mathbf{V}^{(l)} & -\mathbf{V}^{(l)} \end{bmatrix} \begin{bmatrix} \phi_+^{(l)}(z) & \mathbf{O} \\ \mathbf{O} & \phi_-^{(l)}(z) \end{bmatrix} \begin{bmatrix} \boldsymbol{u}^{(l)} \\ \boldsymbol{d}^{(l)} \end{bmatrix}, \tag{5.23}$$

where $\mathbf{O}$ is zero matrix with all zero elements. Since the boundary condition between the layers l and $l+1$ is given by

$$\begin{bmatrix} \boldsymbol{S}_y^{(l+1)}(0) \\ \boldsymbol{U}_x^{(l+1)}(0) \end{bmatrix} = \begin{bmatrix} \boldsymbol{S}_y^{(l)}(h^{(l)}) \\ \boldsymbol{U}_x^{(l)}(h^{(l)}) \end{bmatrix}, \tag{5.24}$$

therefore,

$$\begin{bmatrix} \mathbf{W}^{(l+1)} & \mathbf{W}^{(l+1)} \\ \mathbf{V}^{(l+1)} & -\mathbf{V}^{(l+1)} \end{bmatrix} \begin{bmatrix} \boldsymbol{u}^{(l+1)} \\ \boldsymbol{d}^{(l+1)} \end{bmatrix} = \begin{bmatrix} \mathbf{W}^{(l)} & \mathbf{W}^{(l)} \\ \mathbf{V}^{(l)} & -\mathbf{V}^{(l)} \end{bmatrix} \begin{bmatrix} \mathbf{\Phi}_+^{(l)} & \mathbf{O} \\ \mathbf{O} & \mathbf{\Phi}_-^{(l)} \end{bmatrix} \begin{bmatrix} \boldsymbol{u}^{(l)} \\ \boldsymbol{d}^{(l)} \end{bmatrix}, \tag{5.25}$$

where $\mathbf{\Phi}_\pm^{(l)} = \phi_\pm^{(l)}(h^{(l)})$. This is the final form of the boundary condition.

5.1.2 TM polarization

Let us consider the case of TM polarization. In this case, we only need to consider H_y, the y component of the magnetic field. The magnetic field in the grating region can be written as follows:

$$H_y^{(l)} = \sum_m U_{ym}^{(l)}(z) \exp(i k_{xm} x). \tag{5.26}$$

Next, we consider the wave equation for TM polarization. From Eq. (5.7),

$$\frac{\partial H_y^{(l)}}{\partial z} = -i\omega\varepsilon_0\varepsilon(x)E_x^{(l)}, \tag{5.27}$$

$$\frac{\partial H_y^{(l)}}{\partial x} = i\omega\varepsilon_0\varepsilon(x)E_z^{(l)} \tag{5.28}$$

are obtained. Also, from Eq. (5.4),

$$i\omega\mu_0 H_y^{(l)} = \frac{\partial E_x^{(l)}}{\partial z} - \frac{\partial E_z^{(l)}}{\partial x} \tag{5.29}$$

is obtained. Substituting the derivative with respect of z in Eq. (5.27) and the derivative for x in Eq. (5.28) into Eq. (5.29),

$$\frac{\partial^2 H_y^{(l)}}{\partial z^2} = -\varepsilon^{(l)}(x)\left\{k_0^2 H_y^{(l)} + \frac{\partial}{\partial x}\left[\frac{1}{\varepsilon^{(l)}(x)}\frac{\partial H_y^{(l)}}{\partial x}\right]\right\}, \tag{5.30}$$

then,

$$\frac{1}{\varepsilon^{(l)}(x)}\frac{\partial^2 H_y^{(l)}}{\partial z^2} = -k_0^2 H_y^{(l)} - \frac{\partial}{\partial x}\left[\frac{1}{\varepsilon^{(l)}(x)}\frac{\partial H_y^{(l)}}{\partial x}\right] \tag{5.31}$$

is obtained. This is the wave equation for TM polarization.

Next, we express the inverse of the dielectric constant with the Fourier series,

$$\frac{1}{\varepsilon^{(l)}(x)} = \sum_p \tilde{\varepsilon}_p^{(l)} \exp(ipKx). \tag{5.32}$$

Substituting Eqs. (5.26) and (5.32) into Eq. (5.31),

$$\begin{aligned}&\sum_p\sum_m \tilde{\varepsilon}_{m-p}^{(l)} \frac{\partial^2 U_{yp}^{(l)}(z)}{\partial z^2}\exp(ik_{xm}x)\\ &= -k_0^2\sum_m U_{ym}^{(l)}(z)\exp(ik_{xm}x) - \frac{\partial}{\partial x}\left[\sum_p\sum_m \tilde{\varepsilon}_{m-p}^{(l)} ik_{xp}U_{yp}^{(l)}(z)\exp(ik_{xm}x)\right]\end{aligned} \tag{5.33}$$

is obtained. Performing the differentiation of the second term of the right-hand side of Eq. (5.33) with respect to x yields

$$\begin{aligned}&\sum_p\sum_m \tilde{\varepsilon}_{m-p}^{(l)} \frac{\partial^2 U_{yp}^{(l)}(z)}{\partial z^2}\exp(ik_{xm}x)\\ &= -k_0^2\sum_m U_{ym}^{(l)}(z)\exp(ik_{xm}x) + \sum_p\sum_m \tilde{\varepsilon}_{m-p}^{(l)} k_{xp}k_{xm}U_{yp}^{(l)}(z)\exp(ik_{xm}x).\end{aligned} \tag{5.34}$$

Therefore,

$$\sum_p \tilde{\varepsilon}_{m-p}^{(l)} \frac{\partial^2 U_{yp}^{(l)}(z)}{\partial z^2} = \sum_p \tilde{\varepsilon}_{m-p}^{(l)} k_{xp} k_{xm} U_{yp}^{(l)}(z) - k_0^2 U_{ym}^{(l)}(z) \tag{5.35}$$

is obtained. In matrix form, this yields

$$\frac{\partial^2 \boldsymbol{U}_y^{(l)}}{\partial z^2} = k_0^2 \mathbf{A}^{(l)^{-1}} \left(\mathbf{K}_x \mathbf{A}^{(l)} \mathbf{K}_x - \mathbf{I} \right) \boldsymbol{U}_y^{(l)}, \tag{5.36}$$

where $\mathbf{A}^{(l)}$ is the Toeplitz matrix of $\tilde{\varepsilon}_p^{(l)}$ and $\mathbf{I}$ is the identity matrix. On the other hand, if we start the Fourier series representation from Eq. (5.30), we obtain

$$\frac{\partial^2 \boldsymbol{U}_y^{(l)}}{\partial z^2} = k_0^2 \mathbf{E}^{(l)} \left(\mathbf{K}_x \mathbf{A}^{(l)} \mathbf{K}_x - \mathbf{I} \right) \boldsymbol{U}_y^{(l)}, \tag{5.37}$$

which is different form from Eq. (5.36).

Moharam stated that $\mathbf{A}^{(l)}$ in parentheses on the right-hand side of Eq. (5.36) is better replaced by $\mathbf{E}^{(l)^{-1}}$ in a private communication with Li [23]. Indeed, Moharam et al. [24] employed the following equation:

$$\frac{\partial^2 \boldsymbol{U}_y^{(l)}}{\partial z^2} = k_0^2 \mathbf{E}^{(l)} \left(\mathbf{K}_x \mathbf{E}^{(l)^{-1}} \mathbf{K}_x - \mathbf{I} \right) \boldsymbol{U}_y^{(l)}. \tag{5.38}$$

However, the reason is not stated. Even if this formula was used, the problem remained. The convergence was slower in the case of TM polarization in a metallic grating than in the case of TE polarization [22, 23].

Later, Granet and Guizal [25] and Lalanne [26] found that a better result could be obtained by replacing $\mathbf{A}^{(l)}$ in the parentheses in Eq. (5.36) by $\mathbf{E}^{(l)^{-1}}$. In other words,

$$\frac{\partial^2 \boldsymbol{U}_y^{(l)}}{\partial z^2} = k_0^2 \mathbf{A}^{(l)^{-1}} \left(\mathbf{K}_x \mathbf{E}^{(l)^{-1}} \mathbf{K}_x - \mathbf{I} \right) \boldsymbol{U}_y^{(l)}. \tag{5.39}$$

By using this equation, the convergence of the solution can be obtained for TM polarization in the same order as for TE polarization. However, this equation was obtained empirically, and no mathematical evidence for it was given. Subsequently, Li [27] showed the basis of Eq. (5.39). The details are discussed in the next section. On the other hand, if the layer thickness is very thin compared to the wavelength, the convergence of Eq. (5.37) is faster [28]. The reason for this is discussed in detail by Popov et al. [29].

The solution to Eq. (5.39) is expressed as

$$U_{ym}^{(l)}(z) = \sum_{j=1}^{\infty} w_{mj}^{(l)} \left[u_j^{(l)} \exp(ik_{zj}^{(l)} z) + d_j^{(l)} \exp(-ik_{zj}^{(l)} z) \right], \tag{5.40}$$

where $-\left[k_{zj}^{(l)}\right]^2$ and $w_{mj}^{(l)}$ are the eigenvalue and the element of the eigenvector of matrix $k_0^2\mathbf{A}^{(l)^{-1}}\left(\mathbf{K}_x\mathbf{E}^{(l)^{-1}}\mathbf{K}_x - \mathbf{I}\right)$. Here, $\mathrm{Im}\left[k_{zj}^{(l)}\right] \geq 0$ must be satisfied.

Next, let us consider the boundary conditions between the layers. Suppose that the tangent component E_x of the electric field in the grating region can be written as the following form.

$$E_x^{(l)} = \left(\frac{\mu_0}{\varepsilon_0}\right)^{1/2} \sum_m S_{xj}^{(l)}(z)\exp(ik_{xm}x). \tag{5.41}$$

Substituting Eqs. (5.26), (5.32), and (5.41) into Eq. (5.27) and rearranging, we obtain

$$\sum_m S_{xm}^{(l)}(z)\exp(ik_{xm}x) = \frac{1}{ik_0}\sum_p\sum_m \tilde{\varepsilon}_{m-p}^{(l)} \frac{\partial U_{yp}^{(l)}(z)}{\partial z}\exp(ik_{xm}x), \tag{5.42}$$

and then

$$S_{xm}^{(l)}(z) = \frac{1}{ik_0}\sum_p \tilde{\varepsilon}_{m-p}^{(l)} \frac{\partial U_{yp}^{(l)}(z)}{\partial z}. \tag{5.43}$$

Substituting Eq. (5.40) into this equation yields

$$S_{xm}^{(l)}(z) = \frac{1}{k_0}\sum_p \tilde{\varepsilon}_{m-p}^{(l)} \sum_{j-1}^{\infty} k_{zj}^{(l)} w_{pj}^{(l)} \left[u_j^{(l)}\exp(ik_{zj}^{(l)}z) - d_j^{(l)}\exp(-ik_{zj}^{(l)}z)\right] \tag{5.44}$$

$$S_{xm}^{(l)}(z) = \sum_{j=1}^{\infty} v_{mj}^{(l)}\left[u_j^{(l)}\exp(ik_{zj}^{(l)}z) - d_j^{(l)}\exp(-ik_{zj}^{(l)}z)\right], \tag{5.45}$$

where

$$v_{mj}^{(l)} = \frac{1}{k_0}\sum_p \tilde{\varepsilon}_{m-p}^{(l)} k_{zj}^{(l)} w_{pj}^{(l)}. \tag{5.46}$$

In a matrix form,

$$\mathbf{V}^{(l)} = \frac{1}{k_0}\mathbf{A}^{(l)}\mathbf{Q}^{(l)}\mathbf{W}^{(l)}. \tag{5.47}$$

Combining Eqs. (5.40) and (5.45) into a matrix form, we obtain

$$\begin{bmatrix} \boldsymbol{U}_y^{(l)} \\ \boldsymbol{S}_x^{(l)} \end{bmatrix} = \begin{bmatrix} \mathbf{W}^{(l)} & \mathbf{W}^{(l)} \\ \mathbf{V}^{(l)} & -\mathbf{V}^{(l)} \end{bmatrix} \begin{bmatrix} \phi_+^{(l)}(z) & \mathbf{O} \\ \mathbf{O} & \phi_-^{(l)}(z) \end{bmatrix} \begin{bmatrix} \boldsymbol{u}^{(l)} \\ \boldsymbol{d}^{(l)} \end{bmatrix}. \tag{5.48}$$

Using this equation, the boundary condition at the interface between layer l and layer $l+1$ is given by

$$\begin{bmatrix} \mathbf{W}^{(l+1)} & \mathbf{W}^{(l+1)} \\ \mathbf{V}^{(l+1)} & -\mathbf{V}^{(l+1)} \end{bmatrix} \begin{bmatrix} \boldsymbol{u}^{(l+1)} \\ \boldsymbol{d}^{(l+1)} \end{bmatrix} = \begin{bmatrix} \mathbf{W}^{(l)} & \mathbf{W}^{(l)} \\ \mathbf{V}^{(l)} & -\mathbf{V}^{(l)} \end{bmatrix} \begin{bmatrix} \Phi_+^{(l)} & \mathbf{O} \\ \mathbf{O} & \Phi_-^{(l)} \end{bmatrix} \begin{bmatrix} \boldsymbol{u}^{(l)} \\ \boldsymbol{d}^{(l)} \end{bmatrix}. \tag{5.49}$$

The form of this equation is the same as in Eq. (5.25). Note, however, that the order of $\boldsymbol{U}$ and $\boldsymbol{S}$ is switched compared to the case of TE polarization.

In addition, the component of the electric field in the z-direction is also obtained. This component can be written in the same way as follows

$$E_z^{(l)} = \left(\frac{\mu_0}{\varepsilon_0}\right)^{1/2} \sum_m S_{zm}^{(l)}(z) \exp(ik_{xm}x). \tag{5.50}$$

Substituting Eq. (5.50) into Eq. (5.28), we obtain

$$\sum_m S_{zm}^{(l)}(z) \exp(ik_{xm}x) = -\frac{1}{k_0} \sum_p \sum_m \tilde{\varepsilon}_{m-p}^{(l)} k_{xp} U_{yp}^{(l)}(z) \exp(ik_{xm}x), \tag{5.51}$$

and then

$$S_{zm}^{(l)}(z) = \sum_p \tilde{\varepsilon}_{m-p}^{(l)} k_{xp} U_{yp}^{(l)}(z). \tag{5.52}$$

5.1.3 Correct Fourier series

The basis for Eq. (5.39) is given by Li [27]. The problem is to find the correct Fourier series of a periodic function $h(x)$ given by the product of two periodic functions $f(s)$ and $g(x)$,

$$h(x) = f(x)g(x). \tag{5.53}$$

Let f_m and g_m be the Fourier coefficients of $f(x)$ and $g(x)$. The Fourier coefficient h_m of $h(x)$ is generally expressed as follows using Laurent's rule, the Fourier series version of the convolution theorem. The Fourier coefficient, h_m is given by

$$h_n = \sum_{m=-\infty}^{\infty} f_{m-n} g_m = \sum_{m=-\infty}^{\infty} g_{m-n} f_m. \tag{5.54}$$

This formula always gives the correct result when the series continues infinitely long. However, in actual calculations, m must be truncated at a finite value. If $f(x)$ and $g(x)$ are both piecewise continuous periodic functions and the positions of the discontinuity points do not coincide,

$$h_n = \sum_{m=-M}^{M} f_{m-n} g_m \tag{5.55}$$

is correct.

The problem arises when $f(x)$ and $g(x)$ are discontinuous at the same location. In this case, there is generally no correct way to express the coefficients [27]. However, there is a special case in the RCWA method where $f(x)$ and $g(x)$ are discontinuous at the same location, but the product $h(x) = f(x)g(x)$ is continuous. In this case, Eq. (5.55) is not the correct

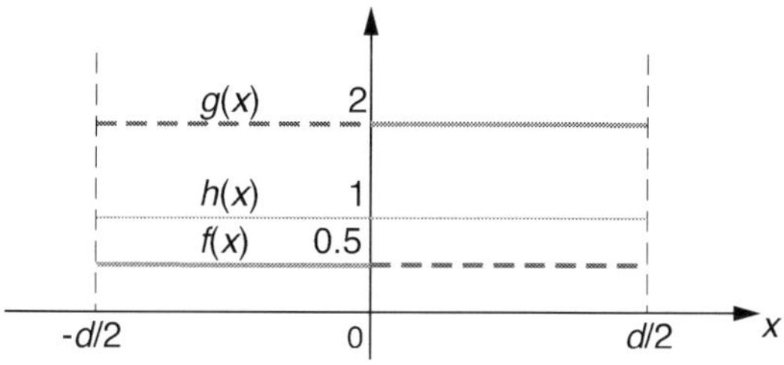

FIGURE 5.2
Pairs of functions that are discontinuous at the same location but whose products are continuous.

answer. An easy-to-understand example is given by Nevière and Popov [30]. As shown in Figure 5.2, we consider two functions

$$f(x) = \begin{cases} 0.5 & (-d/2 \leq x < 0) \\ 2 & (0 \leq x < d/2) \end{cases} \tag{5.56}$$

and

$$g(x) = \begin{cases} 2 & (-d/2 \leq x < 0) \\ 0.5 & (0 \leq x < d/2) \end{cases}. \tag{5.57}$$

In this case, the product of the two functions, $h(x) = f(x)g(x) = 1$ are continuous in the whole region. The zeroth order terms of the Fourier coefficients are both

$$f_0 = g_0 = 1.25. \tag{5.58}$$

Therefore, the product of the two functions is

$$f_0 g_0 = 1.5625. \tag{5.59}$$

On the other hand,

$$h_0 = [fg]_0 = 1. \tag{5.60}$$

So, the series truncated at the zeroth order would result in a large error. On the other hand, taking the inverse of $f(x)$, which is equal to $g(x)$, and computing the Fourier series, we obtain

$$\left[\frac{1}{f}\right]_0 = 1.25. \tag{5.61}$$

Furthermore, using the inverse of this, we obtain

$$\left[\frac{1}{f}\right]_0^{-1} g_0 = 1, \tag{5.62}$$

which agrees with h_0.

As can be seen by analogy from the above, if $f(x)$ and $g(x)$ are discontinuous at the same location but their product is continuous, the correct Fourier series representation is given by [27]

$$h_n = \sum_{m=-M}^{M} \left[\frac{1}{f}\right]_{n-m}^{-1} g_m. \tag{5.63}$$

Now, let us look at where the product of two such complementary functions appears in the eigenequations in the case of TM polarization. One is εE_x in Eq. (5.27). This corresponds to D_x and is continuous in the x-direction. The other is $(1/\varepsilon)(\partial H_y/\partial x)$ in Eq. (5.30). This corresponds to E_z, which is also continuous in the x-direction, as shown in Eq. (5.28). The result of the Fourier series representation, taking these considerations into account, is Eq. (5.39).

5.2 S (scattering) matrix method

Using the boundary conditions in Eqs. (5.25) and (5.49), the T (transmission) matrix method, as in the case of multilayers, provides the relationship between incident light, reflected diffracted light, and transmitted diffracted light. However, although this is mathematically correct, when actual calculations are performed, problems of instability may occur when the grating grooves (the thickness of layers) are deep. This is due to the presence of an exponentially increasing evanescent field in the calculation. The S (scattering) matrix method [27, 31] handles only exponentially decaying evanescent waves, and thus does not cause such instability. Another stable method using Enhanced Transmittance Matrix was proposed by Moharam et al. [32]. Here, the S matrix method is described.

5.2.1 T matrix, S matrix, and R matrix

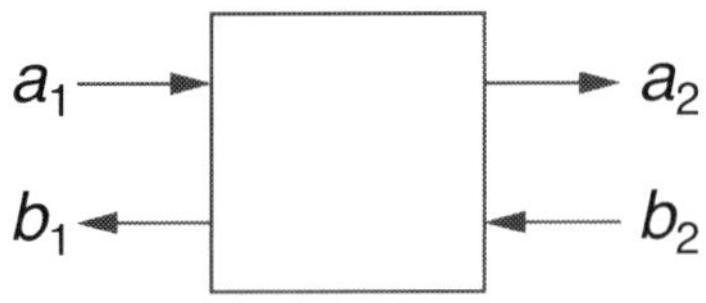

FIGURE 5.3
Response of the system (four-terminal circuits).

The typical matrices representing the response of the system (4-terminal circuit) shown in Figure 5.3 are the T matrix and the S matrix. The T matrix,

$\mathbf{T}$ is defined by

$$\begin{bmatrix} a_2 \\ b_2 \end{bmatrix} = \mathbf{T} \begin{bmatrix} a_1 \\ b_1 \end{bmatrix}. \tag{5.64}$$

This equation is easy to understand if we consider the left side of Figure 5.3 as input and the right side as output. However, it is difficult to understand if we consider the direction of the arrows. On the other hand, the S matrix, $\mathbf{S}$ is defined as

$$\begin{bmatrix} a_2 \\ b_1 \end{bmatrix} = \mathbf{S} \begin{bmatrix} a_1 \\ b_2 \end{bmatrix}. \tag{5.65}$$

The arrow pointing toward the system is the input and the arrow pointing away from it is the output. This is a physical image that is easy to understand. In fact, the elements of the S matrix are written as

$$\begin{bmatrix} a_2 \\ b_1 \end{bmatrix} = \begin{bmatrix} t_{21} & r_{22} \\ r_{11} & t_{12} \end{bmatrix} \begin{bmatrix} a_1 \\ b_2 \end{bmatrix}, \tag{5.66}$$

where t corresponds to the transmission coefficient and r corresponds to the reflection coefficient. In actual systems, inputs and outputs are often vectors rather than scalars. In such cases, t is the transmission matrix and r is the reflection matrix.

Incidentally, in Li's early paper on the RCWA method [33], he mistakenly wrote "R matrix" when he should have written "S matrix". The R matrix is the reactance matrix, which gives the relation between E and H as in the following equation [27]

$$\begin{bmatrix} E_1 \\ E_2 \end{bmatrix} = \mathbf{R} \begin{bmatrix} H_1 \\ H_2 \end{bmatrix}. \tag{5.67}$$

5.2.2 S matrix method

As shown in the early section, the T matrix of the entire system can be easily obtained by multiplying the T matrices of adjacent layers. In the S matrix method, however, the S matrix of the entire system is not so easily obtained. The S matrix of the entire system must be obtained using the following recurrence relation.

The S matrix from layer 0 to layer l, which is denoted by $\mathbf{S}^{0\rightleftharpoons l}$, is defined as

$$\begin{bmatrix} \boldsymbol{u}^{(l)} \\ \boldsymbol{d}^{(0)} \end{bmatrix} = \mathbf{S}^{0\rightleftharpoons l} \begin{bmatrix} \boldsymbol{u}^{(0)} \\ \boldsymbol{d}^{(l)} \end{bmatrix} = \begin{bmatrix} \mathbf{T}_{uu}^{0\rightleftharpoons l} & \mathbf{R}_{ud}^{0\rightleftharpoons l} \\ \mathbf{R}_{du}^{0\rightleftharpoons l} & \mathbf{T}_{dd}^{0\rightleftharpoons l} \end{bmatrix} \begin{bmatrix} \boldsymbol{u}^{(0)} \\ \boldsymbol{d}^{(l)} \end{bmatrix}. \tag{5.68}$$

Similarly, the S matrix from layer 0 to layer $l+1$, $\mathbf{S}^{0\rightleftharpoons l+1}$, is defined as

$$\begin{bmatrix} \boldsymbol{u}^{(l+1)} \\ \boldsymbol{d}^{(0)} \end{bmatrix} = \mathbf{S}^{0\rightleftharpoons l+1} \begin{bmatrix} \boldsymbol{u}^{(0)} \\ \boldsymbol{d}^{(l+1)} \end{bmatrix} = \begin{bmatrix} \mathbf{T}_{uu}^{0\rightleftharpoons l+1} & \mathbf{R}_{ud}^{0\rightleftharpoons l+1} \\ \mathbf{R}_{du}^{0\rightleftharpoons l+1} & \mathbf{T}_{dd}^{0\rightleftharpoons l+1} \end{bmatrix} \begin{bmatrix} \boldsymbol{u}^{(0)} \\ \boldsymbol{d}^{(l+1)} \end{bmatrix}. \tag{5.69}$$

The problem is to find the recursive equation that leads from $\mathbf{S}^{0\rightleftharpoons l}$ to $\mathbf{S}^{0\rightleftharpoons l+1}$, in other words, $\mathbf{T}_{uu}^{0\rightleftharpoons l+1}$, $\mathbf{R}_{ud}^{0\rightleftharpoons l+1}$, $\mathbf{R}_{du}^{0\rightleftharpoons l+1}$, and $\mathbf{T}_{dd}^{0\rightleftharpoons l+1}$ by $\mathbf{T}_{uu}^{0\rightleftharpoons l}$, $\mathbf{R}_{ud}^{0\rightleftharpoons l}$, $\mathbf{R}_{du}^{0\rightleftharpoons l}$,

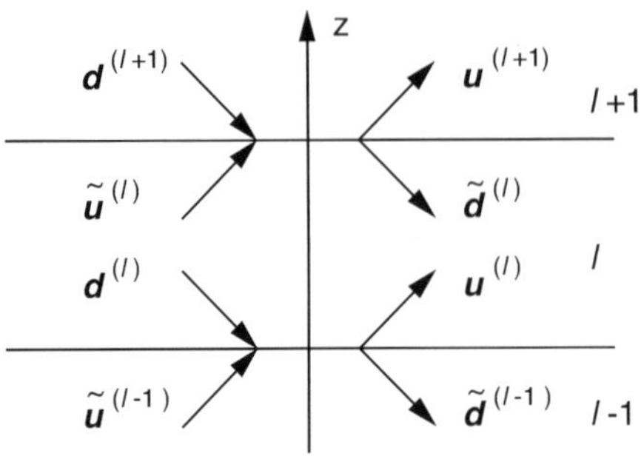

FIGURE 5.4
Definition of each coefficient vector used in the interface S matrix and the interface T matrix.

and $\mathbf{T}_{dd}^{0\rightleftharpoons l}$. To solve this problem, we introduce two new S matrices for two adjacent layers. One is the interface S matrix, $\mathrm{s}^{(l)}$, defined by (see Figure 5.4).

$$\begin{bmatrix} \boldsymbol{u}^{(l+1)} \\ \tilde{\boldsymbol{d}}^{(l)} \end{bmatrix} = \mathrm{s}^{(l)} \begin{bmatrix} \tilde{\boldsymbol{u}}^{(l)} \\ \boldsymbol{d}^{(l+1)} \end{bmatrix}, \tag{5.70}$$

where $\tilde{\boldsymbol{u}}^{(l)}$ and $\tilde{\boldsymbol{d}}^{(l)}$ are coefficient vectors defined at the location of the upper interface of layer l. The relationship between $\tilde{\boldsymbol{u}}$ and $\tilde{\boldsymbol{d}}$ is as follows, using the matrix representing propagation in the layer

$$\begin{bmatrix} \tilde{\boldsymbol{u}}^{(l)} \\ \tilde{\boldsymbol{d}}^{(l)} \end{bmatrix} = \begin{bmatrix} \boldsymbol{\Phi}_+^{(l)} & \mathbf{O} \\ \mathbf{O} & \boldsymbol{\Phi}_-^{(l)} \end{bmatrix} \begin{bmatrix} \boldsymbol{u}^{(l)} \\ \boldsymbol{d}^{(l)} \end{bmatrix}. \tag{5.71}$$

The other is the layer S matrix, $\tilde{\mathrm{s}}^{(l)}$, defined by

$$\begin{bmatrix} \boldsymbol{u}^{(l+1)} \\ \boldsymbol{d}^{(l)} \end{bmatrix} = \tilde{\mathrm{s}}^{(l)} \begin{bmatrix} \boldsymbol{u}^{(l)} \\ \boldsymbol{d}^{(l+1)} \end{bmatrix} = \begin{bmatrix} \tilde{\mathbf{t}}_{uu}^{(l)} & \tilde{\mathbf{r}}_{ud}^{(l)} \\ \tilde{\mathbf{r}}_{du}^{(l)} & \tilde{\mathbf{t}}_{dd}^{(l)} \end{bmatrix} \begin{bmatrix} \boldsymbol{u}^{(l)} \\ \boldsymbol{d}^{(l+1)} \end{bmatrix}. \tag{5.72}$$

Substituting Eq. (5.71) into Eq. (5.70), we obtain

$$\begin{bmatrix} \boldsymbol{u}^{(l+1)} \\ \boldsymbol{\Phi}_-^{(l)} \boldsymbol{d}^{(l)} \end{bmatrix} = \mathrm{s}^{(l)} \begin{bmatrix} \boldsymbol{\Phi}_+^{(l)} \boldsymbol{u}^{(l)} \\ \boldsymbol{d}^{(l+1)} \end{bmatrix}, \tag{5.73}$$

$$\begin{bmatrix} \mathbf{I} & \mathbf{O} \\ \mathbf{O} & \boldsymbol{\Phi}_-^{(l)} \end{bmatrix} \begin{bmatrix} \boldsymbol{u}^{(l+1)} \\ \boldsymbol{d}^{(l)} \end{bmatrix} = \mathrm{s}^{(l)} \begin{bmatrix} \boldsymbol{\Phi}_+^{(l)} & \mathbf{O} \\ \mathbf{O} & \mathbf{I} \end{bmatrix} \begin{bmatrix} \boldsymbol{u}^{(l)} \\ \boldsymbol{d}^{(l+1)} \end{bmatrix}, \tag{5.74}$$

$$\begin{bmatrix} \boldsymbol{u}^{(l+1)} \\ \boldsymbol{d}^{(l)} \end{bmatrix} = \begin{bmatrix} \mathbf{I} & \mathbf{O} \\ \mathbf{O} & {\boldsymbol{\Phi}_-^{(l)}}^{-1} \end{bmatrix} \mathrm{s}^{(l)} \begin{bmatrix} \boldsymbol{\Phi}_+^{(l)} & \mathbf{O} \\ \mathbf{O} & \mathbf{I} \end{bmatrix} \begin{bmatrix} \boldsymbol{u}^{(l)} \\ \boldsymbol{d}^{(l+1)} \end{bmatrix}. \tag{5.75}$$

Then, the following relation

$$\tilde{\mathrm{s}}^{(l)} = \begin{bmatrix} \mathbf{I} & \mathbf{O} \\ \mathbf{O} & {\boldsymbol{\Phi}_-^{(l)}}^{-1} \end{bmatrix} \mathrm{s}^{(l)} \begin{bmatrix} \boldsymbol{\Phi}_+^{(l)} & \mathbf{O} \\ \mathbf{O} & \mathbf{I} \end{bmatrix} \tag{5.76}$$

is obtained.

Next, we express the interface S matrix using the interface T matrix, which is defined as (see Figure 5.4),

$$\begin{bmatrix} \boldsymbol{u}^{(l+1)} \\ \boldsymbol{d}^{(l+1)} \end{bmatrix} = \begin{bmatrix} \mathbf{t}_{uu}^{(l)} & \mathbf{t}_{ud}^{(l)} \\ \mathbf{t}_{du}^{(l)} & \mathbf{t}_{dd}^{(l)} \end{bmatrix} \begin{bmatrix} \tilde{\boldsymbol{u}}^{(l)} \\ \tilde{\boldsymbol{d}}^{(l)} \end{bmatrix}. \tag{5.77}$$

From this equation,

$$\boldsymbol{u}^{(l+1)} = \mathbf{t}_{uu}^{(l)} \tilde{\boldsymbol{u}}^{(l)} + \mathbf{t}_{ud}^{(l)} \tilde{\boldsymbol{d}}^{(l)} \tag{5.78}$$

and

$$\boldsymbol{d}^{(l+1)} = \mathbf{t}_{du}^{(l)} \tilde{\boldsymbol{u}}^{(l)} + \mathbf{t}_{dd}^{(l)} \tilde{\boldsymbol{d}}^{(l)} \tag{5.79}$$

are obtained. Then, we obtain

$$\boldsymbol{u}^{(l+1)} - \mathbf{t}_{uu}^{(l)} \tilde{\boldsymbol{u}}^{(l)} = \mathbf{t}_{ud}^{(l)} \tilde{\boldsymbol{d}}^{(l)}, \tag{5.80}$$

$$-\mathbf{t}_{dd}^{(l)} \tilde{\boldsymbol{d}}^{(l)} = \mathbf{t}_{du}^{(l)} \tilde{\boldsymbol{u}}^{(l)} - \boldsymbol{d}^{(l+1)}, \tag{5.81}$$

$$\begin{bmatrix} \mathbf{I} & -\mathbf{t}_{ud}^{(l)} \\ \mathbf{O} & -\mathbf{t}_{dd}^{(l)} \end{bmatrix} \begin{bmatrix} \boldsymbol{u}^{(l+1)} \\ \tilde{\boldsymbol{d}}^{(l)} \end{bmatrix} = \begin{bmatrix} \mathbf{t}_{uu}^{(l)} & \mathbf{O} \\ \mathbf{t}_{du}^{(l)} & -\mathbf{I} \end{bmatrix} \begin{bmatrix} \tilde{\boldsymbol{u}}^{(l)} \\ \boldsymbol{d}^{(l+1)} \end{bmatrix}, \tag{5.82}$$

$$\begin{bmatrix} \boldsymbol{u}^{(l+1)} \\ \tilde{\boldsymbol{d}}^{(l)} \end{bmatrix} = \begin{bmatrix} \mathbf{t}_{uu}^{(l)} - \mathbf{t}_{ud}^{(l)} \mathbf{t}^{(l)}{}_{dd}^{-1} \mathbf{t}_{du}^{(l)} & \mathbf{t}_{ud}^{(l)} \mathbf{t} t^{(l)}{}_{dd}^{-1} \\ -\mathbf{t}^{(l)}{}_{dd}^{-1} \mathbf{t}_{du}^{(l)} & \mathbf{t}^{(l)}{}_{dd}^{-1} \end{bmatrix} \begin{bmatrix} \tilde{\boldsymbol{u}}^{(l)} \\ \boldsymbol{d}^{(l+1)} \end{bmatrix}. \tag{5.83}$$

As a result,

$$\mathbf{s}^{(l)} = \begin{bmatrix} \mathbf{t}_{uu}^{(l)} - \mathbf{t}_{ud}^{(l)} \mathbf{t}_{dd}^{(l)}{}^{-1} \mathbf{t}_{du}^{(l)} & \mathbf{t}_{ud}^{(l)} \mathbf{t}_{dd}^{(l)}{}^{-1} \\ -\mathbf{t}^{(l)}{}_{dd}^{-1} \mathbf{t}_{du}^{(l)} & \mathbf{t}^{(l)}{}_{dd}^{-1} \end{bmatrix} \tag{5.84}$$

is obtained.

We now return to the first problem. Transforming Eq. (5.72), we obtain

$$\begin{bmatrix} \mathbf{I} & -\tilde{\mathbf{r}}_{ud}^{(l)} \\ \mathbf{O} & -\tilde{\mathbf{t}}_{dd}^{(l)} \end{bmatrix} \begin{bmatrix} \boldsymbol{u}^{(l+1)} \\ \boldsymbol{d}^{(l+1)} \end{bmatrix} = \begin{bmatrix} \tilde{\mathbf{t}}_{uu}^{(l)} & \mathbf{O} \\ \tilde{\mathbf{r}}_{du}^{(l)} & -\mathbf{I} \end{bmatrix} \begin{bmatrix} \boldsymbol{u}^{(l)} \\ \boldsymbol{d}^{(l)} \end{bmatrix}. \tag{5.85}$$

Similarly, from Eq. (5.68)

$$\begin{bmatrix} \mathbf{I} & -\mathbf{R}_{ud}^{0\rightleftharpoons l} \\ \mathbf{O} & -\mathbf{T}_{dd}^{0\rightleftharpoons l} \end{bmatrix} \begin{bmatrix} \boldsymbol{u}^{(l)} \\ \boldsymbol{d}^{(l)} \end{bmatrix} = \begin{bmatrix} \mathbf{T}_{uu}^{0\rightleftharpoons l} & \mathbf{O} \\ \mathbf{R}_{du}^{0\rightleftharpoons l} & -\mathbf{I} \end{bmatrix} \begin{bmatrix} \boldsymbol{u}^{(0)} \\ \boldsymbol{d}^{(0)} \end{bmatrix} \tag{5.86}$$

is obtained. From these two equations, we obtain

$$\begin{aligned} &\begin{bmatrix} \tilde{\mathbf{t}}_{uu}^{(l)} & \mathbf{O} \\ \tilde{\mathbf{r}}_{du}^{(l)} & -\mathbf{I} \end{bmatrix}^{-1} \begin{bmatrix} \mathbf{I} & -\tilde{\mathbf{r}}_{ud}^{(l)} \\ \mathbf{O} & -\tilde{\mathbf{t}}_{dd}^{(l)} \end{bmatrix} \begin{bmatrix} \boldsymbol{u}^{(l+1)} \\ \boldsymbol{d}^{(l+1)} \end{bmatrix} \\ &= \begin{bmatrix} \mathbf{I} & -\mathbf{R}_{ud}^{0\rightleftharpoons l} \\ \mathbf{O} & -\mathbf{T}_{dd}^{0\rightleftharpoons l} \end{bmatrix}^{-1} \begin{bmatrix} \mathbf{T}_{uu}^{0\rightleftharpoons l} & \mathbf{O} \\ \mathbf{R}_{du}^{0\rightleftharpoons l} & -\mathbf{I} \end{bmatrix} \begin{bmatrix} \boldsymbol{u}^{(0)} \\ \boldsymbol{d}^{(0)} \end{bmatrix}. \end{aligned} \tag{5.87}$$

Using the following relationship,

$$\begin{bmatrix} \mathbf{A} & \mathbf{O} \\ \mathbf{B} & -\mathbf{I} \end{bmatrix}^{-1} = \begin{bmatrix} \mathbf{A}^{-1} & \mathbf{O} \\ \mathbf{B}\mathbf{A}^{-1} & -\mathbf{I} \end{bmatrix}, \tag{5.88}$$

the left-hand side of Eq. (5.87) is

$$\begin{aligned} &\begin{bmatrix} \tilde{\mathbf{t}}^{(l)}_{uu} & \mathbf{O} \\ \tilde{\mathbf{r}}^{(l)}_{du} & -\mathbf{I} \end{bmatrix}^{-1} \begin{bmatrix} \mathbf{I} & -\tilde{\mathbf{r}}^{(l)}_{ud} \\ \mathbf{O} & -\tilde{\mathbf{t}}^{(l)}_{dd} \end{bmatrix} \begin{bmatrix} \boldsymbol{u}^{(l+1)} \\ \boldsymbol{d}^{(l+1)} \end{bmatrix} \\ &= \begin{bmatrix} [\tilde{\mathbf{t}}^{(l)}_{uu}]^{-1} & 0 \\ \tilde{\mathbf{r}}^{(l)}_{du}[\tilde{\mathbf{t}}^{(l)}_{uu}]^{-1} & -1 \end{bmatrix}^{-1} \begin{bmatrix} \mathbf{I} & -\tilde{\mathbf{r}}^{(l)}_{ud} \\ \mathbf{O} & -\tilde{\mathbf{t}}^{(l)}_{dd} \end{bmatrix} \begin{bmatrix} \boldsymbol{u}^{(l+1)} \\ \boldsymbol{d}^{(l+1)} \end{bmatrix} \\ &= \begin{bmatrix} [\tilde{\mathbf{t}}^{(l)}_{uu}]^{-1} & -[\tilde{\mathbf{t}}^{(l)}_{uu}]^{-1}\tilde{\mathbf{r}}^{(l)}_{ud} \\ \tilde{\mathbf{r}}^{(l)}_{du}[\tilde{\mathbf{t}}^{(l)}_{uu}]^{-1} & -\tilde{\mathbf{r}}^{(l)}_{du}[\tilde{\mathbf{t}}^{(l)}_{uu}]^{-1}\tilde{\mathbf{r}}^{(l)}_{ud} + \tilde{\mathbf{t}}^{(l)}_{dd} \end{bmatrix} \begin{bmatrix} \boldsymbol{u}^{(l+1)} \\ \boldsymbol{d}^{(l+1)} \end{bmatrix}. \end{aligned} \tag{5.89}$$

Similarly using the following relation,

$$\begin{bmatrix} \mathbf{I} & -\mathbf{A} \\ \mathbf{O} & -\mathbf{B} \end{bmatrix}^{-1} = \begin{bmatrix} \mathbf{I} & \mathbf{O} \\ \mathbf{A}\mathbf{B}^{-1} & -\mathbf{B}^{-1} \end{bmatrix}, \tag{5.90}$$

the right-hand side of Eq. (5.87) becomes

$$\begin{aligned} &\begin{bmatrix} \mathbf{I} & -\mathbf{R}^{0\rightleftharpoons l}_{ud} \\ \mathbf{O} & -\mathbf{T}^{0\rightleftharpoons l}_{dd} \end{bmatrix}^{-1} \begin{bmatrix} \mathbf{T}^{0\rightleftharpoons l}_{uu} & \mathbf{O} \\ \mathbf{R}^{0\rightleftharpoons l}_{du} & -\mathbf{I} \end{bmatrix} \begin{bmatrix} \boldsymbol{u}^{(0)} \\ \boldsymbol{d}^{(0)} \end{bmatrix} \\ &= \begin{bmatrix} \mathbf{I} & -\mathbf{R}^{0\rightleftharpoons l}_{ud}(\mathbf{T}^{0\rightleftharpoons l}_{dd})^{-1} \\ \mathbf{O} & -(\mathbf{T}^{0\rightleftharpoons l}_{dd})^{-1} \end{bmatrix}^{-1} \begin{bmatrix} \mathbf{T}^{0\rightleftharpoons l}_{uu} & \mathbf{O} \\ \mathbf{R}^{0\rightleftharpoons l}_{du} & -\mathbf{I} \end{bmatrix} \begin{bmatrix} \boldsymbol{u}^{(0)} \\ \boldsymbol{d}^{(0)} \end{bmatrix} \\ &= \begin{bmatrix} \mathbf{T}^{0\rightleftharpoons l}_{uu} - \mathbf{R}^{0\rightleftharpoons l}_{ud}(\mathbf{T}^{0\rightleftharpoons l}_{dd})^{-1}\mathbf{R}^{0\rightleftharpoons l}_{du} & \mathbf{R}^{0\rightleftharpoons l}_{ud}(\mathbf{T}^{0\rightleftharpoons l}_{dd})^{-1} \\ -(\mathbf{T}^{0\rightleftharpoons l}_{dd})^{-1}\mathbf{R}^{0\rightleftharpoons l}_{du} & (\mathbf{T}^{0\rightleftharpoons l}_{dd})^{-1} \end{bmatrix} \begin{bmatrix} \boldsymbol{u}^{(0)} \\ \boldsymbol{d}^{(0)} \end{bmatrix}. \end{aligned} \tag{5.91}$$

Therefore, the following equation,

$$\begin{aligned} &\begin{bmatrix} [\tilde{\mathbf{t}}^{(l)}_{uu}]^{-1} & -[\tilde{\mathbf{t}}^{(l)}_{uu}]^{-1}\tilde{\mathbf{r}}^{(l)}_{ud} \\ \tilde{\mathbf{r}}^{(l)}_{du}[\tilde{\mathbf{t}}^{(l)}_{uu}]^{-1} & -\tilde{\mathbf{r}}^{(l)}_{du}[\tilde{\mathbf{t}}^{(l)}_{uu}]^{-1}\tilde{\mathbf{r}}^{(l)}_{ud} + \tilde{\mathbf{t}}^{(l)}_{dd} \end{bmatrix} \begin{bmatrix} \boldsymbol{u}^{(l+1)} \\ \boldsymbol{d}^{(l+1)} \end{bmatrix} \\ &= \begin{bmatrix} \mathbf{T}^{0\rightleftharpoons l}_{uu} - \mathbf{R}^{0\rightleftharpoons l}_{ud}(\mathbf{T}^{0\rightleftharpoons l}_{dd})^{-1}\mathbf{R}^{0\rightleftharpoons l}_{du} & \mathbf{R}^{0\rightleftharpoons l}_{ud}(\mathbf{T}^{0\rightleftharpoons l}_{dd})^{-1} \\ -(\mathbf{T}^{0\rightleftharpoons l}_{dd})^{-1}\mathbf{R}^{0\rightleftharpoons l}_{du} & (\mathbf{T}^{0\rightleftharpoons l}_{dd})^{-1} \end{bmatrix} \begin{bmatrix} \boldsymbol{u}^{(0)} \\ \boldsymbol{d}^{(0)} \end{bmatrix} \end{aligned} \tag{5.92}$$

is obtained. Rearranging the elements yields

$$\begin{aligned} &\begin{bmatrix} [\tilde{\mathbf{t}}^{(l)}_{uu}]^{-1} & -\mathbf{R}^{0\rightleftharpoons l}_{ud}(\mathbf{T}^{0\rightleftharpoons l}_{dd})^{-1} \\ \tilde{\mathbf{r}}^{(l)}_{du}\tilde{\mathbf{t}}^{(l)-1}_{uu} & -(\mathbf{T}^{0\rightleftharpoons l}_{dd})^{-1} \end{bmatrix} \begin{bmatrix} \boldsymbol{u}^{(l+1)} \\ \boldsymbol{d}^{(0)} \end{bmatrix} \\ &= \begin{bmatrix} \mathbf{T}^{0\rightleftharpoons l}_{uu} - \mathbf{R}^{0\rightleftharpoons l}_{ud}(\mathbf{T}^{0\rightleftharpoons l}_{dd})^{-1}\mathbf{R}^{0\rightleftharpoons l}_{du} & [\tilde{\mathbf{t}}^{(l)}_{uu}]^{-1}\tilde{\mathbf{r}}^{(l)}_{ud} \\ -(\mathbf{T}^{0\rightleftharpoons l}_{dd})^{-1}\mathbf{R}^{0\rightleftharpoons l}_{du} & \tilde{\mathbf{r}}^{(l)}_{du}[\tilde{\mathbf{t}}^{(l)}_{uu}]^{-1}\tilde{\mathbf{r}}^{(l)}_{ud} - \tilde{\mathbf{t}}^{(l)}_{dd} \end{bmatrix} \begin{bmatrix} \boldsymbol{u}^{(0)} \\ \boldsymbol{d}^{(l+1)} \end{bmatrix} \end{aligned} \tag{5.93}$$

and finally

$$
\begin{bmatrix} \boldsymbol{u}^{(l+1)} \\ \boldsymbol{d}^{(0)} \end{bmatrix} = \begin{bmatrix} [\tilde{\mathbf{t}}_{uu}^{(l)}]^{-1} & -\mathbf{R}_{ud}^{0\rightleftharpoons l}(\mathbf{T}_{dd}^{0\rightleftharpoons l})^{-1} \\ \tilde{\mathbf{r}}_{du}^{(l)}[\tilde{\mathbf{t}}_{uu}^{(l)}]^{-1} & -(\mathbf{T}_{dd}^{0\rightleftharpoons l})^{-1} \end{bmatrix}^{-1}
$$
$$
\times \begin{bmatrix} \mathbf{T}_{uu}^{0\rightleftharpoons l} - \mathbf{R}_{ud}^{0\rightleftharpoons l}(\mathbf{T}_{dd}^{0\rightleftharpoons l})^{-1}\mathbf{R}_{du}^{0\rightleftharpoons l} & [\tilde{\mathbf{t}}_{uu}^{(l)}]^{-1}\tilde{\mathbf{r}}_{ud}^{(l)} \\ -(\mathbf{T}_{dd}^{0\rightleftharpoons l})^{-1}\mathbf{R}_{du}^{0\rightleftharpoons l} & \tilde{\mathbf{r}}_{du}^{(l)}[\tilde{\mathbf{t}}_{uu}^{(l)}]^{-1}\tilde{\mathbf{r}}_{ud}^{(l)} - \tilde{\mathbf{t}}_{dd}^{(l)} \end{bmatrix} \begin{bmatrix} \boldsymbol{u}^{(0)} \\ \boldsymbol{d}^{(l+1)} \end{bmatrix} \quad (5.94)
$$

is obtained. The inverse of the block matrix is given by

$$
\begin{bmatrix} \mathbf{A} & \mathbf{B} \\ \mathbf{C} & \mathbf{D} \end{bmatrix}^{-1} = \begin{bmatrix} \mathbf{A}^{-1} + \mathbf{A}^{-1}\mathbf{B}\mathbf{S}^{-1}\mathbf{C}\mathbf{A}^{-1} & -\mathbf{A}^{-1}\mathbf{B}\mathbf{S}^{-1} \\ -\mathbf{S}^{-1}\mathbf{C}\mathbf{A}^{-1} & \mathbf{S}^{-1} \end{bmatrix}, \quad (5.95)
$$

where $\mathbf{S} = \mathbf{D} - \mathbf{C}\mathbf{A}^{-1}\mathbf{B}$, and the whole matrix and submatrices $\mathbf{A}$ and $\mathbf{D}$ must be square matrices. Using this relationship we obtain

$$
\begin{bmatrix} \tilde{\mathbf{t}}_{uu}^{(l)-1} & -\mathbf{R}_{ud}^{0\rightleftharpoons l}(\mathbf{T}_{dd}^{0\rightleftharpoons l})^{-1} \\ \tilde{\mathbf{r}}_{du}^{(l)}\tilde{\mathbf{t}}_{uu}^{(l)-1} & -(\mathbf{T}_{dd}^{0\rightleftharpoons l})^{-1} \end{bmatrix}^{-1}
$$
$$
= \begin{bmatrix} \tilde{\mathbf{t}}_{uu}^{(l)} + \tilde{\mathbf{t}}_{uu}^{(l)}\mathbf{R}_{ud}^{0\rightleftharpoons l}[\mathbf{I} - \tilde{\mathbf{r}}_{du}^{(l)}\mathbf{R}_{ud}^{0\rightleftharpoons l}]^{-1}\tilde{\mathbf{r}}_{du}^{(l)} & -\tilde{\mathbf{t}}_{uu}^{(l)}\mathbf{R}_{ud}^{0\rightleftharpoons l}[\mathbf{I} - \tilde{\mathbf{r}}_{du}^{(l)}\mathbf{R}_{ud}^{0\rightleftharpoons l}]^{-1} \\ \mathbf{T}_{dd}^{0\rightleftharpoons l}[\mathbf{I} - \tilde{\mathbf{r}}_{du}^{(l)}\mathbf{R}_{ud}^{0\rightleftharpoons l}]^{-1}\tilde{\mathbf{r}}_{du}^{(l)} & -\mathbf{T}_{dd}^{0\rightleftharpoons l}[\mathbf{I} - \tilde{\mathbf{r}}_{du}^{(l)}\mathbf{R}_{ud}^{0\rightleftharpoons l}]^{-1} \end{bmatrix}. \quad (5.96)
$$

The element of the first row and first column of the product of the matrices on the right-hand side of Eq. (5.94) should be equal to $\mathbf{T}_{uu}^{0\rightleftharpoons l+1}$ and

$$
\begin{aligned}
&\mathbf{T}_{uu}^{0\rightleftharpoons l+1} \\
&\quad=\{\tilde{\mathbf{t}}_{uu}^{(l)} + \tilde{\mathbf{t}}_{uu}^{(l)}\mathbf{R}_{ud}^{0\rightleftharpoons l}[\mathbf{I} - \tilde{\mathbf{r}}_{du}^{(l)}\mathbf{R}_{ud}^{0\rightleftharpoons l}]^{-1}\tilde{\mathbf{r}}_{du}^{(l)}\}\{\mathbf{T}_{uu}^{(l-1)} - \mathbf{R}_{ud}^{0\rightleftharpoons l}(\mathbf{T}_{dd}^{0\rightleftharpoons l})^{-1}\mathbf{R}_{du}^{0\rightleftharpoons l}\} \\
&\qquad + \tilde{\mathbf{t}}_{uu}^{(l)}\mathbf{R}_{ud}^{0\rightleftharpoons l}[\mathbf{I} - \tilde{\mathbf{r}}_{du}^{(l)}\mathbf{R}_{ud}^{0\rightleftharpoons l}]^{-1}(\mathbf{T}_{dd}^{0\rightleftharpoons l})^{-1}\mathbf{R}_{du}^{(l-1)} \\
&\quad=\tilde{\mathbf{t}}_{uu}^{(l)}\{\mathbf{I} + \mathbf{R}_{ud}^{0\rightleftharpoons l}[\mathbf{I} - \tilde{\mathbf{r}}_{du}^{(l)}\mathbf{R}_{ud}^{0\rightleftharpoons l}]^{-1}\tilde{\mathbf{r}}_{du}^{(l)}\}\mathbf{T}_{uu}^{0\rightleftharpoons l} \\
&\quad=\tilde{\mathbf{t}}_{uu}^{(l)}[\mathbf{I} - \mathbf{R}_{ud}^{0\rightleftharpoons l}\tilde{\mathbf{r}}_{du}^{(l)}]^{-1}\mathbf{T}_{uu}^{0\rightleftharpoons l}. \qquad (5.97)
\end{aligned}
$$

Here, we used the following relation

$$
\mathbf{I} + \mathbf{A}(\mathbf{I} - \mathbf{B}\mathbf{A})^{-1}\mathbf{B} = (\mathbf{I} - \mathbf{A}\mathbf{B})^{-1} \quad (5.98)
$$

to transform the last line of the above equation. By performing similar calculations, we finally obtain the following four sets of recursive formulas:

$$
\mathbf{T}_{uu}^{0\rightleftharpoons l+1} = \tilde{\mathbf{t}}_{uu}^{(l)}[\mathbf{I} - \mathbf{R}_{ud}^{0\rightleftharpoons l}\tilde{\mathbf{r}}_{du}^{(l)}]^{-1}\mathbf{T}_{uu}^{0\rightleftharpoons l}, \quad (5.99)
$$

$$
\mathbf{R}_{ud}^{0\rightleftharpoons l+1} = \tilde{\mathbf{r}}_{ud}^{(l)} + \tilde{\mathbf{t}}_{uu}^{(l)}\mathbf{R}_{ud}^{0\rightleftharpoons l}[\mathbf{I} - \tilde{\mathbf{r}}_{du}^{(l)}\mathbf{R}_{ud}^{0\rightleftharpoons l}]^{-1}\tilde{\mathbf{t}}_{dd}^{(l)}, \quad (5.100)
$$

$$
\mathbf{R}_{du}^{0\rightleftharpoons l+1} = \mathbf{R}_{du}^{0\rightleftharpoons l} + \mathbf{T}_{dd}^{0\rightleftharpoons l}\tilde{\mathbf{r}}_{du}^{(l)}[\mathbf{I} - \mathbf{R}_{ud}^{0\rightleftharpoons l}\tilde{\mathbf{r}}_{du}^{(l)}]^{-1}\mathbf{T}_{uu}^{0\rightleftharpoons l}, \quad (5.101)
$$

$$
\mathbf{T}_{dd}^{0\rightleftharpoons l+1} = \mathbf{T}_{dd}^{0\rightleftharpoons l}[\mathbf{I} - \tilde{\mathbf{r}}_{du}^{(l)}\mathbf{R}_{ud}^{0\rightleftharpoons l}]^{-1}\tilde{\mathbf{t}}_{dd}^{(l)}. \quad (5.102)
$$

5.2.3 Method without T matrix

Li [31] has given a recursive formula for the S-matrix that does not use the interface T-matrix. The equation under consideration is

$$\begin{bmatrix} \mathbf{W}_{11}^{(l+1)} & \mathbf{W}_{12}^{(l+1)} \\ \mathbf{W}_{21}^{(l+1)} & \mathbf{W}_{22}^{(l+1)} \end{bmatrix} \begin{bmatrix} \boldsymbol{u}^{(l+1)} \\ \boldsymbol{d}^{(l+1)} \end{bmatrix} = \begin{bmatrix} \mathbf{W}_{11}^{(l)} & \mathbf{W}_{12}^{(l)} \\ \mathbf{W}_{21}^{(l)} & \mathbf{W}_{22}^{(l)} \end{bmatrix} \begin{bmatrix} \boldsymbol{\Phi}_{+}^{(l)} & \mathbf{O} \\ \mathbf{O} & \boldsymbol{\Phi}_{-}^{(l)} \end{bmatrix} \begin{bmatrix} \boldsymbol{u}^{(l)} \\ \boldsymbol{d}^{(l)} \end{bmatrix}. \tag{5.103}$$

The recursive formulas of the S-matrix for this equation are

$$\mathbf{R}_{ud}^{0\rightleftharpoons l+1} = (\mathbf{Z}^{-1}\mathbf{X}_2)_1, \tag{5.104}$$

$$\mathbf{T}_{dd}^{0\rightleftharpoons l+1} = \tilde{\mathbf{T}}_{dd}^{0\rightleftharpoons l}(\mathbf{Z}^{-1}\mathbf{X}_2)_2, \tag{5.105}$$

$$\mathbf{T}_{uu}^{0\rightleftharpoons l+1} = (\mathbf{Z}^{-1}\mathbf{X}_1)_1, \tag{5.106}$$

$$\mathbf{R}_{du}^{0\rightleftharpoons l+1} = \mathbf{R}_{du}^{0\rightleftharpoons l} + \tilde{\mathbf{T}}_{dd}^{0\rightleftharpoons l}(Z^{-1}\mathbf{X}_1)_2, \tag{5.107}$$

where

$$\mathbf{Z} = \begin{bmatrix} \mathbf{W}_{11}^{(l+1)} & -\mathbf{W}_{11}^{(l)}\tilde{\mathbf{R}}_{ud}^{0\rightleftharpoons l} - \mathbf{W}_{12}^{(l)} \\ \mathbf{W}_{21}^{(l+1)} & -\mathbf{W}_{21}^{(l)}\tilde{\mathbf{R}}_{ud}^{0\rightleftharpoons l} - \mathbf{W}_{22}^{(l)} \end{bmatrix}, \tag{5.108}$$

$$\mathbf{X} = \begin{bmatrix} \mathbf{W}_{11}^{(l)}\tilde{\mathbf{T}}_{uu}^{0\rightleftharpoons l} & -\mathbf{W}_{12}^{(l+1)} \\ \mathbf{W}_{21}^{(l)}\tilde{\mathbf{T}}_{uu}^{0\rightleftharpoons l} & -\mathbf{W}_{22}^{(l+1)} \end{bmatrix} = [\mathbf{X}_1, \mathbf{X}_2], \tag{5.109}$$

$$\tilde{\mathbf{R}}_{ud}^{0\rightleftharpoons l} = \boldsymbol{\Phi}_{+}^{(l)}\mathbf{R}_{ud}^{0\rightleftharpoons l}[\boldsymbol{\Phi}_{-}^{(l)}]^{-1}, \tag{5.110}$$

$$\tilde{\mathbf{T}}_{dd}^{0\rightleftharpoons l} = \mathbf{T}_{dd}^{0\rightleftharpoons l}[\boldsymbol{\Phi}_{-}^{(l)}]^{-1}, \tag{5.111}$$

$$\tilde{\mathbf{T}}_{uu}^{0\rightleftharpoons l} = \boldsymbol{\Phi}_{+}^{(l)}\mathbf{T}_{uu}^{0\rightleftharpoons l}. \tag{5.112}$$

The subscripts 1 and 2 in Eqs. (5.104)–(5.107) refer to the upper and lower blocks of the matrix. The subscripts 1 and 2 in Eq. (5.109) refer to the left and right blocks of the matrix $\mathbf{X}$. In actual calculations, $\boldsymbol{\Phi}_{+}^{(l)}$ should be used instead of $[\boldsymbol{\Phi}_{-}^{(l)}]^{-1}$ using the relation $[\boldsymbol{\Phi}_{-}^{(l)}]^{-1} = \boldsymbol{\Phi}_{+}^{(l)}$, since the value of $\Phi_{-}^{(l)}$ itself may become very large and cause overflows.

In the case of the RCWA method, the matrices appearing in Eq. (5.103) have the following symmetry, that is,

$$\mathbf{W}_{11}^{(l)} = \mathbf{W}_{12}^{(l)} = \mathbf{W}_{1}^{(l)}, \tag{5.113}$$

$$\mathbf{W}_{21}^{(l)} = -\mathbf{W}_{22}^{(l)} = \mathbf{W}_{2}^{(l)}. \tag{5.114}$$

In this case, the recursive formulas become simpler [31]. Equation (5.108) is expressed as

$$\mathbf{Z} = \begin{bmatrix} \mathbf{W}_{1}^{(l+1)} & \mathbf{O} \\ \mathbf{O} & \mathbf{W}_{2}^{(l+1)} \end{bmatrix} \begin{bmatrix} \mathbf{I} & -\mathbf{F}^{(l)} \\ \mathbf{I} & \mathbf{G}^{(l)} \end{bmatrix}, \tag{5.115}$$

where

$$\mathbf{F}^{(l)} = \mathbf{Q}_1^{(l)}(\mathbf{I} + \tilde{\mathbf{R}}_{ud}^{0\rightleftharpoons l}), \tag{5.116}$$

$$\mathbf{G}^{(l)} = \mathbf{Q}_2^{(l)}(\mathbf{I} - \tilde{\mathbf{R}}_{ud}^{0\rightleftharpoons l}), \tag{5.117}$$

$$\mathbf{Q}_p^{(l)} = \mathbf{W}_p^{(l+1)-1}\mathbf{W}_p^{(l)}, (p = 1, 2). \tag{5.118}$$

The inverse of the second matrix on the right-hand side of Eq. (5.115) is easily obtained using the following relation:

$$\begin{bmatrix} \mathbf{I} & \mathbf{A} \\ \mathbf{I} & \mathbf{B} \end{bmatrix}^{-1} = \begin{bmatrix} -\mathbf{B} & \mathbf{A} \\ \mathbf{I} & -\mathbf{I} \end{bmatrix} (\mathbf{A} - \mathbf{B})^{-1}. \tag{5.119}$$

Then, we obtain

$$\mathbf{R}_{ud}^{0\rightleftharpoons l+1} = \mathbf{I} - 2\mathbf{G}^{(l)}\tau^{(l)}, \tag{5.120}$$

$$\mathbf{T}_{dd}^{0\rightleftharpoons l+1} = 2\tilde{\mathbf{T}}_{dd}^{0\rightleftharpoons l}\tau^{(l)}, \tag{5.121}$$

$$\mathbf{T}_{uu}^{0\rightleftharpoons l+1} = (\mathbf{F}^{(l)}\tau^{(l)}\mathbf{Q}_2^{(l)} + \mathbf{G}^{(l)}\tau^{(l)}\mathbf{Q}_1^{(l)})\tilde{\mathbf{T}}_{uu}^{0\rightleftharpoons l}, \tag{5.122}$$

$$\mathbf{R}_{du}^{0\rightleftharpoons l+1} = \mathbf{R}_{du}^{0\rightleftharpoons l} + \tilde{\mathbf{T}}_{dd}^{0\rightleftharpoons l}\tau^{(l)}(\mathbf{Q}_2^{(l)} - \mathbf{Q}_1^{(l)})\tilde{\mathbf{T}}_{uu}^{0\rightleftharpoons l}, \tag{5.123}$$

where

$$\tau^{(l)} = (\mathbf{F}^{(l)} + \mathbf{G}^{(l)})^{-1}. \tag{5.124}$$

Incidentally, using $\mathbf{Q}_q^{(l)} (q = 1, 2)$, the interface T matrix $\mathbf{t}^{(l)}$ is given by

$$\mathbf{t}^{(l)} = \frac{1}{2}\begin{bmatrix} \mathbf{Q}_1^{(l)} + \mathbf{Q}_2^{(l)} & \mathbf{Q}_1^{(l)} - \mathbf{Q}_2^{(l)} \\ \mathbf{Q}_1^{(l)} - \mathbf{Q}_2^{(l)} & \mathbf{Q}_1^{(l)} + \mathbf{Q}_2^{(l)} \end{bmatrix}. \tag{5.125}$$

5.2.4 Relationship between incident, reflected, and transmitted fields

First we consider the case of TE polarization. Consider the relationship between the incident electric field and $\boldsymbol{u}^{(L)}$ and $\boldsymbol{d}^{(L)}$ in layer L. The incident field is given by,

$$E_y^i = \exp[i(k_{x0}x - k_{z0}^L z)]. \tag{5.126}$$

The amplitude is taken as unity for convenience, since the coefficient vectors of the reflection and transmission diffraction fields obtained below directly indicate the diffraction coefficients. Comparing Eq. (5.126) with Eqs. (5.3) and (5.16), relationship

$$\boldsymbol{i}^e = \mathbf{W}_1^{(L)}\boldsymbol{d}^{(L)} \tag{5.127}$$

is obtained. Vector $\boldsymbol{i}^e$ is the coefficient vector of the incident electric field, where only the element corresponding to the wavenumber k_{x0} is unity and the remaining elements are all zero. Similarly, the relation between the coefficient vector $\boldsymbol{r}^e$ of the reflected electric field and $\boldsymbol{u}^{(L)}$ and $\boldsymbol{d}^{(L)}$ is

$$\boldsymbol{r}^e = \mathbf{W}_1^{(L)}\boldsymbol{u}^{(L)}. \tag{5.128}$$

Next, consider the relationship between the coefficient vectors of the transmitted electric field $\boldsymbol{t}^e$, $\boldsymbol{u}^{(0)}$, and $\boldsymbol{d}^{(0)}$ in layer 0. This relationship is

$$\boldsymbol{t}^e = \mathbf{W}_1^{(0)}\boldsymbol{d}^{(0)}. \tag{5.129}$$

Substituting these relations into Eq. (5.68),

$$\begin{bmatrix} [\mathbf{W}^{(L)}]^{-1}\boldsymbol{r}^e \\ [\mathbf{W}^{(0)}]^{-1}\boldsymbol{t}^e \end{bmatrix} = \begin{bmatrix} \mathbf{T}_{uu}^{0\rightleftharpoons L} & \mathbf{R}_{ud}^{0\rightleftharpoons L} \\ \mathbf{R}_{du}^{0\rightleftharpoons L} & \mathbf{T}_{dd}^{0\rightleftharpoons L} \end{bmatrix} \begin{bmatrix} \boldsymbol{o} \\ [\mathbf{W}^{(L)}]^{-1}\boldsymbol{i}^e \end{bmatrix}. \tag{5.130}$$

is obtained, since $\boldsymbol{u}^{(0)} = \boldsymbol{o}$ (where $\boldsymbol{o}$ is a zero vector with all zero elements). From this equation we obtain

$$\boldsymbol{r}^e = \mathbf{W}^{(L)}\mathbf{R}_{ud}^{0\rightleftharpoons L}[\mathbf{W}^{(L)}]^{-1}\boldsymbol{i}^e, \tag{5.131}$$

$$\boldsymbol{t}^e = \mathbf{W}^{(0)}\mathbf{T}_{dd}^{0\rightleftharpoons L}[\mathbf{W}^{(L)}]^{-1}\boldsymbol{i}^e. \tag{5.132}$$

Using these coefficient vectors, the (power) reflection and (power) transmission diffraction efficiencies for the m-th order are respectively,

$$R_m^{\text{TE}} = |r_m^e|^2, \tag{5.133}$$

$$T_m^{\text{TE}} = |t_m^e|^2 \frac{\text{Re}(k_{zm}^{(0)})}{\text{Re}(k_{z0}^{(L)})}. \tag{5.134}$$

Similarly, each coefficient vectors for the magnetic field in the case of TM polarization are

$$\boldsymbol{r}_h = \mathbf{W}^{(L)}\mathbf{R}_{ud}^{0\rightleftharpoons L}[\mathbf{W}^{(L)}]^{-1}\boldsymbol{i}^h, \tag{5.135}$$

$$\boldsymbol{t}_h = \mathbf{W}^{(0)}\mathbf{T}_{dd}^{0\rightleftharpoons L}[\mathbf{W}^{(L)}]^{-1}\boldsymbol{i}^h. \tag{5.136}$$

Thus, the m-th-order reflection and transmission diffraction efficiencies are respectively,

$$R_m^{\text{TM}} = |r_m^h|^2, \tag{5.137}$$

$$T_m^{\text{TM}} = |t_m^h|^2 \frac{\text{Re}(k_{zm}^{(0)}/\varepsilon^{(0)})}{\text{Re}(k_{z0}^{(L)}/\varepsilon^{(L)})}. \tag{5.138}$$

5.2.5 Fields in the grating region

In the T matrix method, the amplitude of each diffracted wave in the grating region can be calculated directly. However, the S matrix method requires some ingenuity to obtain these amplitudes. Consider the field in layer l. The partial S matrices

$$\begin{bmatrix} \boldsymbol{u}^{(l)} \\ \boldsymbol{d}^{(0)} \end{bmatrix} = \mathbf{S}^{0\rightleftharpoons l} \begin{bmatrix} \boldsymbol{o} \\ \boldsymbol{d}^{(l)} \end{bmatrix}, \tag{5.139}$$

$$\begin{bmatrix} \boldsymbol{u}^{(L)} \\ \boldsymbol{d}^{(l)} \end{bmatrix} = \mathbf{S}^{l\rightleftharpoons L} \begin{bmatrix} \boldsymbol{u}^{(l)} \\ \boldsymbol{d}^{(L)} \end{bmatrix}, \tag{5.140}$$

are used. From Eq. (5.139), we obtain

$$\boldsymbol{d}^{(0)} = \mathbf{T}_{ud}^{0\rightleftharpoons l}\boldsymbol{d}^{(l)}, \tag{5.141}$$

$$\boldsymbol{d}^{(l)} = [\mathbf{T}_{ud}^{0\rightleftharpoons l}]^{-1}\boldsymbol{d}^{(0)}. \tag{5.142}$$

Similarly, from Eq. (5.140), we obtain

$$\boldsymbol{u}^{(L)} = \mathbf{T}_{uu}^{l\rightleftharpoons L}\boldsymbol{u}^{(l)} + \mathbf{R}_{ud}^{l\rightleftharpoons L}\boldsymbol{d}^{(L)}, \tag{5.143}$$

$$\boldsymbol{u}^{(l)} = (\mathbf{T}_{uu}^{l\rightleftharpoons L})^{-1}(\boldsymbol{u}^{(L)} - \mathbf{R}_{ud}^{l\rightleftharpoons L}\boldsymbol{d}^{(L)}). \tag{5.144}$$

After obtaining the transmission coefficient vector $\boldsymbol{d}^{(0)}$ and reflection coefficient vector $\boldsymbol{u}^{(L)}$ using the S matrix of the entire system, $\boldsymbol{d}^{(l)}$ and $\boldsymbol{u}^{(l)}$ can be obtained by using the partial S matrix and the above equations. However, this method is unstable. This is because $\mathbf{T}_{dd}$ and $\mathbf{T}_{uu}$ may contain elements with very small absolute values, in which case their inverse matrices diverge. The following method can be used to avoid this instability [34].

From Eqs. (5.139) and (5.140), we obtain

$$\boldsymbol{u}^{(l)} = \mathbf{R}_{ud}^{0\rightleftharpoons l}\boldsymbol{d}^{(l)}, \tag{5.145}$$

$$\boldsymbol{d}^{(l)} = \mathbf{R}_{du}^{l\rightleftharpoons L}\boldsymbol{u}^{(l)} + \mathbf{T}_{dd}^{l\rightleftharpoons L}\boldsymbol{d}^{(L)}. \tag{5.146}$$

Substituting Eq. (5.146) into Eq. (5.145) yields

$$\boldsymbol{u}^{(l)} = \mathbf{R}_{ud}^{0\rightleftharpoons l}(\mathbf{R}_{du}^{l\rightleftharpoons L}\boldsymbol{u}^{(l)} + \mathbf{T}_{dd}^{l\rightleftharpoons L}\boldsymbol{d}^{(L)}), \tag{5.147}$$

$$\boldsymbol{u}^{(l)} = \mathbf{R}_{ud}^{0\rightleftharpoons l}\mathbf{R}_{du}^{l\rightleftharpoons L}\boldsymbol{u}^{(l)} + \mathbf{R}_{ud}^{0\rightleftharpoons l}\mathbf{T}_{dd}^{l\rightleftharpoons L}\boldsymbol{d}^{(L)}, \tag{5.148}$$

$$\boldsymbol{u}^{(l)}(\mathbf{I} - \mathbf{R}_{ud}^{0\rightleftharpoons l}\mathbf{R}_{du}^{l\rightleftharpoons L}) = \mathbf{R}_{ud}^{0\rightleftharpoons l}\mathbf{T}_{dd}^{l\rightleftharpoons L}\boldsymbol{d}^{(L)}, \tag{5.149}$$

$$\boldsymbol{u}^{(l)} = (\mathbf{I} - \mathbf{R}_{ud}^{0\rightleftharpoons l}\mathbf{R}_{du}^{l\rightleftharpoons L})^{-1}\mathbf{R}_{ud}^{0\rightleftharpoons l}\mathbf{T}_{dd}^{l\rightleftharpoons L}\boldsymbol{d}^{(L)}. \tag{5.150}$$

Note that $(\mathbf{I} - \mathbf{R}_{ud}^{0\rightleftharpoons l}\mathbf{R}_{du}^{l\rightleftharpoons L})^{-1}$ does not diverge even when $\mathbf{R}_{ud}^{0\rightleftharpoons l}$ and $\mathbf{R}_{du}^{l\rightleftharpoons L}$ contain elements with very small absolute values. Substituting Eq. (5.150) into Eq. (5.146) yields $\boldsymbol{d}^{(l)}$.

However, there is still a problem: calculating the field using $\boldsymbol{d}^{(l)}$ leads to instability, since we are calculating an exponentially increasing evanescent field. To avoid this instability, we can use the coefficient vector $\tilde{\boldsymbol{d}}^{(l)}$ at the upper interface of layer l shown in Figure 5.4. When the matrix is symmetric, The relationship of $\tilde{\boldsymbol{u}}^{(l)}$ and $\tilde{\boldsymbol{d}}^{(l)}$ to $\boldsymbol{u}^{(l+1)}$ and $\boldsymbol{d}^{(l+1)}$ is given by

$$\begin{bmatrix} \mathbf{W}^{(l+1)} & \mathbf{W}^{(l+1)} \\ \mathbf{V}^{(l+1)} & -\mathbf{V}^{(l+1)} \end{bmatrix} \begin{bmatrix} \boldsymbol{u}^{(l+1)} \\ \boldsymbol{d}^{(l+1)} \end{bmatrix} = \begin{bmatrix} \mathbf{W}^{(l)} & \mathbf{W}^{(l)} \\ \mathbf{V}^{(l)} & -\mathbf{V}^{(l)} \end{bmatrix} \begin{bmatrix} \tilde{\boldsymbol{u}}^{(l)} \\ \tilde{\boldsymbol{d}}^{(l)} \end{bmatrix}. \tag{5.151}$$

Using the following relation,

$$\begin{bmatrix} \mathbf{W}^{(l)} & \mathbf{W}^{(l)} \\ \mathbf{V}^{(l)} & -\mathbf{V}^{(l)} \end{bmatrix}^{-1} = \frac{1}{2}\begin{bmatrix} \mathbf{W}^{(l)^{-1}} & \mathbf{V}^{(l)^{-1}} \\ \mathbf{W}^{(l)^{-1}} & -\mathbf{V}^{(l)^{-1}} \end{bmatrix}, \tag{5.152}$$

we obtain

$$\begin{bmatrix} \tilde{\boldsymbol{u}}^{(l)} \\ \tilde{\boldsymbol{d}}^{(l)} \end{bmatrix} = \frac{1}{2} \begin{bmatrix} \mathbf{W}^{(l)^{-1}} & \mathbf{V}^{(l)^{-1}} \\ \mathbf{W}^{(l)^{-1}} & -\mathbf{V}^{(l)^{-1}} \end{bmatrix} \begin{bmatrix} \mathbf{W}^{(l+1)} & \mathbf{W}^{(l+1)} \\ \mathbf{V}^{(l+1)} & -\mathbf{V}^{(l+1)} \end{bmatrix} \begin{bmatrix} \boldsymbol{u}^{(l+1)} \\ \boldsymbol{d}^{(l+1)} \end{bmatrix}$$
$$= \frac{1}{2} \begin{bmatrix} \mathbf{W}^{(l)^{-1}}\mathbf{W}^{(l+1)} + \mathbf{V}^{(l)^{-1}}\mathbf{V}^{(l+1)} & \mathbf{W}^{(l)^{-1}}\mathbf{W}^{(l+1)} - \mathbf{V}^{(l)^{-1}}\mathbf{V}^{(l+1)} \\ \mathbf{W}^{(l)^{-1}}\mathbf{W}^{(l+1)} - \mathbf{V}^{(l)^{-1}}\mathbf{V}^{(l+1)} & \mathbf{W}^{(l)^{-1}}\mathbf{W}^{(l+1)} + \mathbf{V}^{(l)^{-1}}\mathbf{V}^{(l+1)} \end{bmatrix}$$
$$\times \begin{bmatrix} \boldsymbol{u}^{(l+1)} \\ \boldsymbol{d}^{(l+1)} \end{bmatrix}. \tag{5.153}$$

Using $\boldsymbol{u}^{(l)}$ and $\tilde{\boldsymbol{d}}^{(l)}$ obtained in this way, the amplitude in the layer l can be calculated stably. For example, in the case of TM polarization,

$$U_{ym}^{(l)}(z) = \sum_j w_{mj}^{(l)} \{u_j^{(l)} \exp(ik_{zj}^{(l)} z) + \tilde{d}_j^{(l)} \exp[ik_{zj}^{(l)}(h^{(l)} - z)]\}. \tag{5.154}$$

5.2.6 Recursive calculation from the incident side of the S matrix

In order to calculate the field in the grating region with the above method, the S matrix, $S^{l\rightleftharpoons L}$ must be obtained. This can be obtained recursively from the incident side.

The S matrix from layer $l+1$ to layer L is expressed as

$$\begin{bmatrix} \boldsymbol{u}^{(L)} \\ \boldsymbol{d}^{(l+1)} \end{bmatrix} = \begin{bmatrix} \mathbf{T}_{uu}^{l+1\rightleftharpoons L} & \mathbf{R}_{ud}^{l+1\rightleftharpoons L} \\ \mathbf{R}_{du}^{l+1\rightleftharpoons L} & \mathbf{T}_{dd}^{l+1\rightleftharpoons L} \end{bmatrix} \begin{bmatrix} \boldsymbol{u}^{(l+1)} \\ \boldsymbol{d}^{(L)} \end{bmatrix}. \tag{5.155}$$

The S matrix from layer l to layer L is expressed as

$$\begin{bmatrix} \boldsymbol{u}^{(L)} \\ \boldsymbol{d}^{(l)} \end{bmatrix} = \begin{bmatrix} \mathbf{T}_{uu}^{l\rightleftharpoons L} & \mathbf{R}_{ud}^{l\rightleftharpoons L} \\ \mathbf{R}_{du}^{l\rightleftharpoons L} & \mathbf{T}_{dd}^{l\rightleftharpoons L} \end{bmatrix} \begin{bmatrix} \boldsymbol{u}^{(l)} \\ \boldsymbol{d}^{(L)} \end{bmatrix}. \tag{5.156}$$

From Eq. (5.155),

$$\begin{bmatrix} \mathbf{I} & -\mathbf{R}_{ud}^{l+1\rightleftharpoons L} \\ \mathbf{O} & -\mathbf{T}_{dd}^{l+1\rightleftharpoons L} \end{bmatrix} \begin{bmatrix} \boldsymbol{u}^{(L)} \\ \boldsymbol{d}^{(L)} \end{bmatrix} = \begin{bmatrix} \mathbf{T}_{uu}^{l+1\rightleftharpoons L} & \mathbf{O} \\ \mathbf{R}_{du}^{l+1\rightleftharpoons L} & -\mathbf{I} \end{bmatrix} \begin{bmatrix} \boldsymbol{u}^{(l+1)} \\ \boldsymbol{d}^{(l+1)} \end{bmatrix} \tag{5.157}$$

is obtained. From Eqs. (5.85) and (5.157),

$$\begin{bmatrix} \mathbf{T}_{uu}^{l+1\rightleftharpoons L} & \mathbf{O} \\ \mathbf{R}_{du}^{l+1\rightleftharpoons L} & -\mathbf{I} \end{bmatrix}^{-1} \begin{bmatrix} \mathbf{I} & -\mathbf{R}_{ud}^{l+1\rightleftharpoons L} \\ \mathbf{O} & -\mathbf{T}_{dd}^{l+1\rightleftharpoons L} \end{bmatrix} \begin{bmatrix} \boldsymbol{u}^{(L)} \\ \boldsymbol{d}^{(L)} \end{bmatrix}$$
$$= \begin{bmatrix} \mathbf{I} & -\tilde{\mathbf{r}}_{ud}^{(l)} \\ \mathbf{O} & -\tilde{\mathbf{t}}_{dd}^{(l)} \end{bmatrix}^{-1} \begin{bmatrix} \tilde{\mathbf{t}}_{uu}^{(l)} & \mathbf{O} \\ \tilde{\mathbf{r}}_{du}^{(l)} & -\mathbf{I} \end{bmatrix} \begin{bmatrix} \boldsymbol{u}^{(l)} \\ \boldsymbol{d}^{(l)} \end{bmatrix} \tag{5.158}$$

is obtained. Comparing Eq. (5.158) with Eq. (5.87), we see that it is just the following replacements

$$\tilde{\mathbf{r}}_{ud}^{(l)} \to \mathbf{R}_{ud}^{l+1\rightleftharpoons L}, \tag{5.159}$$

$$\tilde{\mathbf{t}}_{dd}^{(l)} \to \mathbf{T}_{dd}^{l+1\rightleftharpoons L}, \tag{5.160}$$

$$\tilde{\mathbf{r}}_{du}^{(l)} \to \mathbf{R}_{du}^{l+1\rightleftharpoons L}, \tag{5.161}$$

$$\tilde{\mathbf{t}}_{uu}^{(l)} \to \mathbf{T}_{uu}^{l+1\rightleftharpoons L}, \tag{5.162}$$

$$\mathbf{R}_{ud}^{0\rightleftharpoons l-1} \to \tilde{\mathbf{r}}_{ud}^{(l)}, \tag{5.163}$$

$$\mathbf{T}_{dd}^{0\rightleftharpoons l-1} \to \tilde{\mathbf{t}}_{dd}^{(l)}, \tag{5.164}$$

$$\mathbf{R}_{du}^{0\rightleftharpoons l-1} \to \tilde{\mathbf{r}}_{du}^{(l)}, \tag{5.165}$$

$$\mathbf{T}_{uu}^{0\rightleftharpoons l-1} \to \tilde{\mathbf{t}}_{uu}^{(l)}. \tag{5.166}$$

Applying these replacements to Eq. (5.87) and using Eq. (5.156), we obtain

$$\mathbf{T}_{uu}^{l\rightleftharpoons L} = \mathbf{T}_{uu}^{l+1\rightleftharpoons L}[\mathbf{I} - \tilde{\mathbf{r}}_{ud}^{(l)}\mathbf{R}_{du}^{l+1\rightleftharpoons L}]^{-1}\tilde{\mathbf{t}}_{uu}^{(l)}, \tag{5.167}$$

$$\mathbf{R}_{ud}^{l\rightleftharpoons L} = \mathbf{R}_{ud}^{l+1\rightleftharpoons L} + \mathbf{T}_{uu}^{l+1\rightleftharpoons L}\tilde{\mathbf{r}}_{ud}^{(l)}[\mathbf{I} - \mathbf{R}_{du}^{l+1\rightleftharpoons L}\tilde{\mathbf{r}}_{ud}^{(l)}]^{-1}\mathbf{T}_{dd}^{l+1\rightleftharpoons L}, \tag{5.168}$$

$$\mathbf{R}_{du}^{l\rightleftharpoons L} = \tilde{\mathbf{r}}_{du}^{(l)} + \tilde{\mathbf{t}}_{dd}^{(l)}\mathbf{R}_{du}^{l+1\rightleftharpoons L}[\mathbf{I} - \tilde{\mathbf{r}}_{ud}^{(l)}\mathbf{R}_{du}^{l+1\rightleftharpoons L}]^{-1}\tilde{\mathbf{t}}_{uu}^{(l)}, \tag{5.169}$$

$$\mathbf{T}_{dd}^{l\rightleftharpoons L} = \tilde{\mathbf{t}}_{dd}^{(l)}[\mathbf{I} - \mathbf{R}_{du}^{l+1\rightleftharpoons L}\tilde{\mathbf{r}}_{ud}^{(l)}]^{-1}\mathbf{T}_{dd}^{l+1\rightleftharpoons L}. \tag{5.170}$$

5.3 Two-dimensional grating

The RCWA method for a two-dimensional (2D) grating can be calculated basically in the same way as for a one-dimensional (1D) grating. Since the plane of incidence on a one-dimensional grating is the plane containing the grating vector, the polarization components in the x- and y-directions are not coupled, and only the electric field in the y-direction for TE polarization and the magnetic field in the y-direction for TM polarization should be considered. Of course, when the plane of incidence is other than the plane containing the grating vector (conical diffraction), coupling occurs between these two. For more information on conical diffraction, see Refs. [24, 26]. However, in a two-dimensional grating, the two are always coupled, so both polarizations cannot be treated separately. Another difference is that diffracted light in two directions, x and y, must be considered; in a 1D grating, the various coefficients representing diffracted light can be expressed as 1D vectors, but in a 2D grating, these coefficients become a second-order tensor, which would be computationally difficult. Therefore, in a 2D grating, the elements of this second-order tensor are rearranged into a single column and treated as a vector. These two points are the difference from the case of a 1D grating. As a result, the amount of computation is much larger than for a 1D grating.

5.3.1 Two-dimensional grating in Cartesian coordinate system

Consider a 2D periodic structure with period Λ_x in the x-direction and period Λ_y in the y-direction. The electric and magnetic fields in the grating domain can be written as follows

$$\boldsymbol{E}^{(l)} = \sum_{m,n}[S^{(l)}_{xmn}(z)\hat{\boldsymbol{x}} + S^{(l)}_{ymn}(z)\hat{\boldsymbol{y}} + S^{(l)}_{zmn}(z)\hat{\boldsymbol{z}}]\exp[i(k_{xm}x + k_{yn}y)], \quad (5.171)$$

$$\boldsymbol{H}^{(l)} = i\left(\frac{\varepsilon_0}{\mu_0}\right)^{1/2}\sum_{m,n}[U^{(l)}_{xmn}(z)\hat{\boldsymbol{x}}+U^{(l)}_{ymn}(z)\hat{\boldsymbol{y}}+U^{(l)}_{zmn}(z)\hat{\boldsymbol{z}}]\exp[i(k_{xm}x+k_{yn}y)]. \quad (5.172)$$

Note that the imaginary unit i at the beginning of the right-hand side of Eq. (5.172) is only to prevent the imaginary unit from appearing in the derivation of the equation, and has no intrinsic meaning. Here, $\hat{\boldsymbol{x}}$, $\hat{\boldsymbol{y}}$ and $\hat{\boldsymbol{z}}$ are unit vectors. m and n are diffraction orders in the x- and y-directions, respectively. k_{xm} and k_{yn} are the x and y components of the wave vector of the diffracted wave. In the following, (l) at the right shoulder of the variable meaning layer l is omitted to avoid complication in the equation.

If the in-plane wave vector of the incident wave are $[k_{x0}, k_{y0}]$, the in-plane wave vector of the diffracted waves are given by

$$k_{xm} = k_{x0} + mK_x, \quad (5.173)$$

$$k_{yn} = k_{y0} + nK_y, \quad (5.174)$$

where $[K_x, K_y]$ is the lattice vector:

$$K_x = \frac{2\pi}{\Lambda_x}, \quad (5.175)$$

$$K_y = \frac{2\pi}{\Lambda_y}. \quad (5.176)$$

From Maxwell's equation (Faraday's equation, Eq. (5.4))

$$\frac{\partial E_z}{\partial y} - \frac{\partial E_y}{\partial z} = i\omega\mu_0 H_x, \quad (5.177)$$

$$\frac{\partial E_x}{\partial z} - \frac{\partial E_z}{\partial x} = i\omega\mu_0 H_y, \quad (5.178)$$

$$\frac{\partial E_y}{\partial x} - \frac{\partial E_x}{\partial y} = i\omega\mu_0 H_z \quad (5.179)$$

are obtained. Substituting Eqs. (5.171) and (5.172) into Eqs. (5.177), (5.178), and (5.179),

$$i\mathbf{K}_y\boldsymbol{S}_z - \boldsymbol{S}'_y = -\boldsymbol{U}_x \quad (5.180)$$

$$\boldsymbol{S}'_x - i\mathbf{K}_x\boldsymbol{S}_z = -\boldsymbol{U}_y \quad (5.181)$$

$$i\mathbf{K}_x \boldsymbol{S}_y - i\mathbf{K}_y \boldsymbol{S}_x = -\boldsymbol{U}_z \tag{5.182}$$

are obtained in matrix form, where prime $'$ expresses the derivative with respect to $k_0 z$, and $\mathbf{K}_x$ and $\mathbf{K}_y$ are diagonal matrices of k_{xm}/k_0 and k_{yn}/k_0, respectively. Specifically, $\mathbf{K}_x$ is a block diagonal matrix consisting of $(2N+1)$ matrices $\overline{\mathbf{K}}_x$, that is,

$$\mathbf{K}_x = \begin{bmatrix} \overline{\mathbf{K}}_x & & & \mathbf{O} \\ & \overline{\mathbf{K}}_x & & \\ & & \ddots & \\ \mathbf{O} & & & \overline{\mathbf{K}}_x \end{bmatrix} \tag{5.183}$$

and

$$\overline{\mathbf{K}}_x = \frac{1}{k_0} \begin{bmatrix} k_{x,-N} & & & & & & \mathbf{O} \\ & \ddots & & & & & \\ & & k_{x,-1} & & & & \\ & & & k_{x,0} & & & \\ & & & & k_{x,1} & & \\ & & & & & \ddots & \\ \mathbf{O} & & & & & & k_{x,N} \end{bmatrix}, \tag{5.184}$$

where N is the maximum number of diffraction orders to be used in the calculation. The matrix, $\mathbf{K}_y$ is

$$\mathbf{K}_y = \begin{bmatrix} \overline{\mathbf{K}}_{y,-N} & & & & & & \mathbf{O} \\ & \ddots & & & & & \\ & & \overline{\mathbf{K}}_{y,-1} & & & & \\ & & & \overline{\mathbf{K}}_{y,0} & & & \\ & & & & \overline{\mathbf{K}}_{y,1} & & \\ & & & & & \ddots & \\ \mathbf{O} & & & & & & \overline{\mathbf{K}}_{y,N} \end{bmatrix}, \tag{5.185}$$

and

$$\overline{\mathbf{K}}_{y,n} = \frac{k_{yn}}{k_0}\mathbf{I}, \tag{5.186}$$

where the number of diagonal elements of $\overline{\mathbf{K}}_{y,n}$ is $(2N+1)$. In addition, $\boldsymbol{S}_a (a = x, y)$ is expressed as

$$\boldsymbol{S}_a = \begin{bmatrix} \overline{\boldsymbol{S}}_{a,-N} \\ \vdots \\ \overline{\boldsymbol{S}}_{a,-1} \\ \overline{\boldsymbol{S}}_{a,0} \\ \overline{\boldsymbol{S}}_{a,1} \\ \vdots \\ \overline{\boldsymbol{S}}_{a,N} \end{bmatrix}, \tag{5.187}$$

where

$$\overline{\boldsymbol{S}}_{a,n} = \begin{bmatrix} S_{a,n,-N} \\ \vdots \\ S_{a,n,-1} \\ S_{a,n,0} \\ S_{a,n,1} \\ \vdots \\ S_{a,n,N} \end{bmatrix}. \tag{5.188}$$

From Maxwell's equations (Ampere's equation, Eq. (5.7)),

$$\frac{\partial H_z}{\partial y} - \frac{\partial H_y}{\partial z} = -i\omega\varepsilon_0\varepsilon(x,y)E_x, \tag{5.189}$$

$$\frac{\partial H_x}{\partial z} - \frac{\partial H_z}{\partial x} = -i\omega\varepsilon_0\varepsilon(x,y)E_y, \tag{5.190}$$

$$\frac{\partial H_y}{\partial x} - \frac{\partial H_x}{\partial y} = -i\omega\varepsilon_0\varepsilon(x,y)E_z \tag{5.191}$$

are obtained. Here the dielectric constant $\varepsilon(x, y)$ is expressed by a two-dimensional Fourier series,

$$\varepsilon(x,y) = \sum_{p,q} \varepsilon_{p,q} \exp[i(pK_x x + qK_y y)]). \tag{5.192}$$

Substituting Eqs. (5.171), (5.172), and (5.192) into Eqs. (5.189), (5.190) and (5.191),

$$i\mathbf{K}_y\boldsymbol{U}_z - \boldsymbol{U}'_y = -\mathbf{E}\boldsymbol{S}_x, \tag{5.193}$$

$$\boldsymbol{U}'_x - i\mathbf{K}_x\boldsymbol{U}_z = -\mathbf{E}\boldsymbol{S}_y, \tag{5.194}$$

$$i\mathbf{K}_x\boldsymbol{U}_y - i\mathbf{K}_y\boldsymbol{U}_x = -\mathbf{E}\boldsymbol{S}_z \tag{5.195}$$

are obtained in matrix form. Matrix $\mathbf{E}$ is a two-dimensional Toeplitz matrix of Fourier coefficients $\varepsilon_{p,q}^{(l)}$. Specifically this is given by

$$\mathbf{E} = \begin{bmatrix} \mathbf{E}_0 & \mathbf{E}_{-1} & \mathbf{E}_{-2} & \cdots & \mathbf{E}_{-2N} \\ \mathbf{E}_1 & \mathbf{E}_0 & \mathbf{E}_{-1} & \cdots & \mathbf{E}_{-2N+1} \\ \mathbf{E}_2 & \mathbf{E}_1 & \mathbf{E}_0 & \cdots & \mathbf{E}_{-2N+2} \\ \vdots & \vdots & \vdots & \ddots & \vdots \\ \mathbf{E}_{2N} & \mathbf{E}_{2N-1} & \mathbf{E}_{2N-2} & \cdots & \mathbf{E}_0 \end{bmatrix} \tag{5.196}$$

which is the Toeplitz matrix of the submatrix $\mathbf{E}_n$. Matrix $\mathbf{E}_n$ is also a Toeplitz matrix and is given by

$$\mathbf{E}_n = \begin{bmatrix} \varepsilon_{0,n} & \varepsilon_{-1,n} & \cdots & \varepsilon_{-2N,n} \\ \varepsilon_{1,n} & \varepsilon_{0,n} & \cdots & \varepsilon_{-2N+1,n} \\ \vdots & \vdots & \ddots & \vdots \\ \varepsilon_{2N,n} & \varepsilon_{2N-1,n} & \cdots & \varepsilon_{0,n} \end{bmatrix}. \tag{5.197}$$

Eliminating $\boldsymbol{S}_z$ and $\boldsymbol{U}_z$ from Eqs. (5.180), (5.181), (5.182), (5.193), (5.194), and (5.195), we obtain

$$\begin{bmatrix} \boldsymbol{S}'_y \\ \boldsymbol{S}'_x \\ \boldsymbol{U}'_y \\ \boldsymbol{U}'_x \end{bmatrix} = \begin{bmatrix} \mathbf{O} & \mathbf{O} & \mathbf{K}_y\mathbf{E}^{-1}\mathbf{K}_x & \mathbf{I}-\mathbf{K}_y\mathbf{E}^{-1}\mathbf{K}_y \\ \mathbf{O} & \mathbf{O} & \mathbf{K}_x\mathbf{E}^{-1}\mathbf{K}_x-\mathbf{I} & -\mathbf{K}_x\mathbf{E}^{-1}\mathbf{K}_y \\ \mathbf{K}_x\mathbf{K}_y & \mathbf{E}-\mathbf{K}_y^2 & \mathbf{O} & \mathbf{O} \\ \mathbf{K}_x^2-\mathbf{E} & -\mathbf{K}_x\mathbf{K}_y & \mathbf{O} & \mathbf{O} \end{bmatrix} \begin{bmatrix} \boldsymbol{S}_y \\ \boldsymbol{S}_x \\ \boldsymbol{U}_y \\ \boldsymbol{U}_x \end{bmatrix}. \tag{5.198}$$

This equation can be separated into two equations

$$\begin{bmatrix} \boldsymbol{S}'_y \\ \boldsymbol{S}'_x \end{bmatrix} = \mathbf{F} \begin{bmatrix} \boldsymbol{U}_y \\ \boldsymbol{U}_x \end{bmatrix}, \tag{5.199}$$

and

$$\begin{bmatrix} \boldsymbol{U}'_y \\ \boldsymbol{U}'_x \end{bmatrix} = \mathbf{G} \begin{bmatrix} \boldsymbol{S}_y \\ \boldsymbol{S}_x \end{bmatrix}, \tag{5.200}$$

where

$$\mathbf{F} = \begin{bmatrix} \mathbf{K}_y\mathbf{E}^{-1}\mathbf{K}_x & \mathbf{I}-\mathbf{K}_y\mathbf{E}^{-1}\mathbf{K}_y \\ \mathbf{K}_x\mathbf{E}^{-1}\mathbf{K}_x-\mathbf{I} & -\mathbf{K}_x\mathbf{E}^{-1}\mathbf{K}_y \end{bmatrix}, \tag{5.201}$$

$$\mathbf{G} = \begin{bmatrix} \mathbf{K}_x\mathbf{K}_y & \mathbf{E}-\mathbf{K}_y^2 \\ \mathbf{K}_x^2-\mathbf{E} & -\mathbf{K}_x\mathbf{K}_y \end{bmatrix}. \tag{5.202}$$

Substituting Eq. (5.200) into the derivative of both sides of Eq. (5.199) with respect to $k_0 z$, we obtain

$$\begin{bmatrix} \boldsymbol{S}''_y \\ \boldsymbol{S}''_x \end{bmatrix} = \mathbf{FG} \begin{bmatrix} \boldsymbol{S}_y \\ \boldsymbol{S}_x \end{bmatrix}, \tag{5.203}$$

$$\mathbf{FG} = \begin{bmatrix} \mathbf{K}_x^2 + (\mathbf{K}_y\mathbf{E}^{-1}\mathbf{K}_y - \mathbf{I})\mathbf{E} & \mathbf{K}_y(\mathbf{E}^{-1}\mathbf{K}_x\mathbf{E} - \mathbf{K}_x) \\ \mathbf{K}_x(\mathbf{E}^{-1}\mathbf{K}_y\mathbf{E} - \mathbf{K}_y) & \mathbf{K}_y^2 + (\mathbf{K}_x\mathbf{E}^{-1}\mathbf{K}_x - \mathbf{I})\mathbf{E} \end{bmatrix}. \tag{5.204}$$

Similarly, substituting Eq. (5.199) into the derivative of both sides of Eq. (5.200) with respect to $k_0 z$ yields

$$\begin{bmatrix} \boldsymbol{U}''_y \\ \boldsymbol{U}''_x \end{bmatrix} = \mathbf{GF} \begin{bmatrix} \boldsymbol{U}_y \\ \boldsymbol{U}_x \end{bmatrix}, \tag{5.205}$$

$$\mathbf{GF} = \begin{bmatrix} \mathbf{K}_y^2 + \mathbf{E}(\mathbf{K}_x\mathbf{E}^{-1}\mathbf{K}_x - \mathbf{I}) & (\mathbf{K}_x - \mathbf{E}\mathbf{K}_x\mathbf{E}^{-1})\mathbf{K}_y \\ (\mathbf{K}_y - \mathbf{E}\mathbf{K}_y\mathbf{E}^{-1})\mathbf{K}_x & \mathbf{K}_x^2 + \mathbf{E}(\mathbf{K}_y\mathbf{E}^{-1}\mathbf{K}_y - \mathbf{I}) \end{bmatrix}. \tag{5.206}$$

For a homogeneous layer without grating and with dielectric constant ε

$$\mathbf{F} = \begin{bmatrix} \frac{1}{\varepsilon}\mathbf{K}_y\mathbf{K}_x & \mathbf{I} - \frac{1}{\varepsilon}\mathbf{K}_y^2 \\ \frac{1}{\varepsilon}\mathbf{K}_x^2 - \mathbf{I} & -\frac{1}{\varepsilon}\mathbf{K}_x\mathbf{K}_y \end{bmatrix}, \tag{5.207}$$

$$\mathbf{G} = \begin{bmatrix} \mathbf{K}_x\mathbf{K}_y & \varepsilon\mathbf{I} - \mathbf{K}_y^2 \\ \mathbf{K}_x^2 - \varepsilon\mathbf{I} & -\mathbf{K}_x\mathbf{K}_y \end{bmatrix}, \tag{5.208}$$

then,

$$\mathbf{FG} = \begin{bmatrix} \mathbf{K}_x^2 + \mathbf{K}_y^2 - \varepsilon\mathbf{I} & \mathbf{O} \\ \mathbf{O} & \mathbf{K}_x^2 + \mathbf{K}_y^2 - \varepsilon\mathbf{I} \end{bmatrix}. \tag{5.209}$$

Since this matrix is a diagonal matrix, we can omit the eigenvalue calculations required below.

Let q_j^2 be the eigenvalue of the matrix $\mathbf{FG}$ in Eq. (5.203) and w_{jk} be the element of the eigenvector (where $j, k = 1, 2, \ldots, 2(2N+1)^2$), $\boldsymbol{S}_x$ and $\boldsymbol{S}_y$ are given by

$$S_{y,k}(z) = \sum_{j=1}^{2(2N+1)^2} w_{jk}[u_j \exp(k_0 q_j z) + d_j \exp(-k_0 q_j z)], \text{for } k = 1, \ldots, (2N+1)^2, \tag{5.210}$$

$$S_{x,k}(z) = \sum_{j=1}^{2(2N+1)^2} w_{jk}[u_j \exp(k_0 q_j z) + d_j \exp(-k_0 q_j z)], \quad \text{for } k = 1, \ldots, (2N+1)^2, \tag{5.211}$$

where $S_{x,k}$ and $S_{y,k}$ are the k-th element of vectors $\boldsymbol{S}_x$ and $\boldsymbol{S}_y$, respectively. Note that q_j is used here as a convention, but this quantity corresponds to ik_{zj}/k_0 in the case of a 1D grating.

Equations (5.210) and (5.211) can be written in matrix form as

$$\begin{bmatrix} \boldsymbol{S}_y \\ \boldsymbol{S}_x \end{bmatrix} = \begin{bmatrix} \mathbf{W} & \mathbf{W} \end{bmatrix} \begin{bmatrix} \boldsymbol{\Phi}_+ & \mathbf{O} \\ \mathbf{O} & \boldsymbol{\Phi}_- \end{bmatrix} \begin{bmatrix} \boldsymbol{u} \\ \boldsymbol{d} \end{bmatrix}. \tag{5.212}$$

Substituting Eq. (5.212) into Eq. (5.199) yields

$$\begin{bmatrix} \boldsymbol{U}_y \\ \boldsymbol{U}_x \end{bmatrix} = \mathbf{F}^{-1} \begin{bmatrix} \mathbf{QW} & -\mathbf{QW} \end{bmatrix} \begin{bmatrix} \boldsymbol{\Phi}_+ & \mathbf{O} \\ \mathbf{O} & \boldsymbol{\Phi}_- \end{bmatrix} \begin{bmatrix} \boldsymbol{u} \\ \boldsymbol{d} \end{bmatrix}, \tag{5.213}$$

and finally the following equation is obtained:

$$\begin{bmatrix} \boldsymbol{S}_y \\ \boldsymbol{S}_x \\ \boldsymbol{U}_y \\ \boldsymbol{U}_x \end{bmatrix} = \begin{bmatrix} \mathbf{W} & \mathbf{W} \\ \mathbf{F}^{-1}\mathbf{QW} & -\mathbf{F}^{-1}\mathbf{QW} \end{bmatrix} \begin{bmatrix} \boldsymbol{\Phi}_+ & \mathbf{O} \\ \mathbf{O} & \boldsymbol{\Phi}_- \end{bmatrix} \begin{bmatrix} \boldsymbol{u} \\ \boldsymbol{d} \end{bmatrix}. \tag{5.214}$$

In Eq. (5.214), with

$$\boldsymbol{S} = \begin{bmatrix} \boldsymbol{S}_y \\ \boldsymbol{S}_x \end{bmatrix}, \tag{5.215}$$

$$\boldsymbol{U} = \begin{bmatrix} \boldsymbol{U}_y \\ \boldsymbol{U}_x \end{bmatrix}, \tag{5.216}$$

we obtain the following equation:

$$\begin{bmatrix} \boldsymbol{S} \\ \boldsymbol{U} \end{bmatrix} = \begin{bmatrix} \mathbf{W} & \mathbf{W} \\ \mathbf{V} & -\mathbf{V} \end{bmatrix} \begin{bmatrix} \boldsymbol{\Phi}_+ & \mathbf{O} \\ \mathbf{O} & \boldsymbol{\Phi}_- \end{bmatrix} \begin{bmatrix} \boldsymbol{u} \\ \boldsymbol{d} \end{bmatrix}, \tag{5.217}$$

where $\mathbf{V} = \mathbf{F}^{-1}\mathbf{QW}$. This equation has the same form as that for a 1D grating. Therefore, the same manner can be used to obtain the boundary conditions as for the S matrix in the symmetric case.

5.3.2 Improvement of convergence

When considering convergence, Lalanne [28] suggested that

$$\mathbf{G}^{\dagger} = \begin{bmatrix} \mathbf{K}_x\mathbf{K}_y & \mathbf{A}^{-1} - \mathbf{K}_y^2 \\ \mathbf{K}_x^2 - \mathbf{E} & -\mathbf{K}_x\mathbf{K}_y \end{bmatrix} \tag{5.218}$$

should be used instead of $\mathbf{G}$, as in the case of TM polarization in a 1D grating. However, this has not been very successful. On the other hand, Li [35] proposed the correct way to obtain the Toeplitz matrix of the Fourier series of the distribution of the dielectric constant. Here we introduce new symbols $\lceil\cdot\rceil$ and $\lfloor\cdot\rfloor$ as follows:

$$\lceil\varepsilon\rceil_{mn} = \frac{1}{\Lambda_x}\int_0^{\Lambda_x} \varepsilon(x, y)\exp[-i(m-n)K_x x]dx, \tag{5.219}$$

$$\lfloor\varepsilon\rfloor_{mn} = \frac{1}{\Lambda_y}\int_0^{\Lambda_y} \varepsilon(x, y)\exp[-i(m-n)K_y y]dy. \tag{5.220}$$

The obtained $\lceil\cdot\rceil$ and $\lfloor\cdot\rfloor$ are functions of y and x, respectively. We further define $\lfloor\lceil\cdot\rceil\rfloor$ and $\lceil\lfloor\cdot\rfloor\rceil$ as follows,

$$\lfloor\lceil\varepsilon\rceil\rfloor_{mn,pq} = \lfloor\{\lceil 1/\varepsilon\rceil^{-1}\}_{mp}\rfloor_{nq} = \frac{1}{\Lambda_y}\int_0^{\Lambda_y} \{\lceil 1/\varepsilon\rceil^{-1}\}_{mp}(y)\exp[-i(n-q)K_y y]dy, \tag{5.221}$$

$$\lceil\lfloor\varepsilon\rfloor\rceil_{mn,pq} = \lceil\{\lfloor 1/\varepsilon\rfloor^{-1}\}_{nq}\rceil_{mp} = \frac{1}{\Lambda_x}\int_0^{\Lambda_x} \{\lfloor 1/\varepsilon\rfloor^{-1}\}_{nq}(x)\exp[-i(m-p)K_x x]dx. \tag{5.222}$$

Using these, we introduce an alternative $\mathbf{G}$ defined by

$$\mathbf{G}^{\ddagger} = \begin{bmatrix} \mathbf{K}_x\mathbf{K}_y & \lfloor\lceil\varepsilon\rceil\rfloor - \mathbf{K}_y^2 \\ \mathbf{K}_x^2 - \lceil\lfloor\varepsilon\rfloor\rceil & -\mathbf{K}_x\mathbf{K}_y \end{bmatrix}. \tag{5.223}$$

Next, we show the specific form of the matrix. We will discuss the case of $\lfloor\lceil\varepsilon\rceil\rfloor$. First, $1/\varepsilon(x, y)$ is sampled equally spaced in two dimensions on the unit cell. The number of samplings is $M \times M$. However, $M > 4N + 1$ must be satisfied in order to construct the Toeplitz matrix. Then, at each sampling point $y = y_p$ in the y-direction, calculate the Fourier coefficients from the $-2N$th order to the $2N$th order with respect to x, create the Toeplitz matrix, and obtain its inverse matrix $\boldsymbol{\alpha}(y_p)$. Then, the elements $\alpha_{mn}(y_p)$ of this inverse matrix are arranged in a single row. For all y_p, we obtain the following matrix:

$$\lceil 1/\varepsilon \rceil^{-1} =$$
$$\begin{bmatrix} \alpha_{11}(y_1) & \alpha_{12}(y_1) & \dots & \alpha_{1,2N+1}(y_1) & \alpha_{21}(y_1) & \dots & \alpha_{2N+1,2N+1}(y_1) \\ \alpha_{11}(y_2) & \alpha_{12}(y_2) & \dots & \alpha_{1,2N+1}(y_2) & \alpha_{21}(y_2) & \dots & \alpha_{2N+1,2N+1}(y_2) \\ \vdots & \vdots & & \vdots & \vdots & & \vdots \\ \vdots & \vdots & & \vdots & \vdots & & \vdots \\ \alpha_{11}(y_M) & \alpha_{12}(y_M) & \dots & \alpha_{1,2N+1}(y_M) & \alpha_{21}(y_M) & \dots & \alpha_{2N+1,2N+1}(y_M) \end{bmatrix}. \tag{5.224}$$

Next, the Fourier coefficients are computed with respect to y (in the vertical direction of the above matrix) and left to the $\pm 2N$th order, we obtain

$$\mathcal{F}_y\left(\lceil 1/\varepsilon \rceil^{-1}\right) = \begin{bmatrix} \beta_{11}^{(-2N)} & \beta_{12}^{(-2N)} & \dots & \beta_{1,2N+1}^{(-2N)} & \beta_{21}^{(-2N)} & \dots & \beta_{2N+1,2N+1}^{(-2N)} \\ \vdots & \vdots & \ddots & \vdots & \vdots & \ddots & \vdots \\ \beta_{11}^{(-1)} & \beta_{12}^{(-1)} & \dots & \beta_{1,2N+1}^{(-1)} & \beta_{21}^{(-1)} & \dots & \beta_{2N+1,2N+1}^{(-1)} \\ \beta_{11}^{(0)} & \beta_{12}^{(0)} & \dots & \beta_{1,2N+1}^{(0)} & \beta_{21}^{(0)} & \dots & \beta_{2N+1,2N+1}^{(0)} \\ \beta_{11}^{(1)} & \beta_{12}^{(1)} & \dots & \beta_{1,2N+1}^{(1)} & \beta_{21}^{(1)} & \dots & \beta_{2N+1,2N+1}^{(1)} \\ \vdots & \vdots & \ddots & \vdots & \vdots & \ddots & \vdots \\ \beta_{11}^{(2N)} & \beta_{12}^{(2N)} & \dots & \beta_{1,2N+1}^{(2N)} & \beta_{21}^{(2N)} & \dots & \beta_{2N+1,2N+1}^{(2N)} \end{bmatrix}, \tag{5.225}$$

where $\mathcal{F}_y$ denotes the discrete Fourier transform in the y-direction. Matrix $\lfloor\lceil\varepsilon\rceil\rfloor$ is obtained by rearranging these elements as follows:

$$\lfloor\lceil\varepsilon\rceil\rfloor = \begin{bmatrix} \boldsymbol{\beta}^{(0)} & \boldsymbol{\beta}^{(-1)} & \dots & \boldsymbol{\beta}^{(-2N)} \\ \boldsymbol{\beta}^{(1)} & \boldsymbol{\beta}^{(0)} & \dots & \boldsymbol{\beta}^{(-2N+1)} \\ \vdots & \vdots & \ddots & \vdots \\ \boldsymbol{\beta}^{(2N)} & \boldsymbol{\beta}^{(2N-1)} & \dots & \boldsymbol{\beta}^{(0)} \end{bmatrix}, \tag{5.226}$$

where

$$\boldsymbol{\beta}^{(n)} = \begin{bmatrix} \beta_{11}^{(n)} & \beta_{12}^{(n)} & \dots & \beta_{1,2N+1}^{(n)} \\ \beta_{21}^{(n)} & \beta_{22}^{(n)} & \dots & \beta_{2,2N+1}^{(n)} \\ \vdots & \vdots & \ddots & \vdots \\ \beta_{2N+1,1}^{(n)} & \beta_{2N+1,2}^{(n)} & \dots & \beta_{2N+1,2N+1}^{(n)} \end{bmatrix}. \tag{5.227}$$

Matrix $\lceil\lfloor\varepsilon\rfloor\rceil$ is obtained by rearranging the order of the operations on x and on y in a similar calculation. That is

$$\mathcal{F}_x\left(\lfloor 1/\varepsilon \rfloor^{-1}\right) =$$
$$\begin{bmatrix} \gamma_{11}^{(-2N)} & \cdots & \gamma_{11}^{(-1)} & \gamma_{11}^{(0)} & \gamma_{11}^{(1)} & \cdots & \gamma_{11}^{(2N)} \\ \gamma_{12}^{(-2N)} & \cdots & \gamma_{12}^{(-1)} & \gamma_{12}^{(0)} & \gamma_{12}^{(1)} & \cdots & \gamma_{12}^{(2N)} \\ \vdots & \ddots & \vdots & \vdots & \vdots & \ddots & \vdots \\ \gamma_{1,2N+1}^{(-2N)} & \cdots & \gamma_{1,2N+1}^{(-1)} & \gamma_{1,2N+1}^{(0)} & \gamma_{1,2N+1}^{(1)} & \cdots & \gamma_{1,2N+1}^{(2N)} \\ \gamma_{21}^{(-2N)} & \cdots & \gamma_{21}^{(-1)} & \gamma_{21}^{(0)} & \gamma_{21}^{(1)} & \cdots & \gamma_{21}^{(2N)} \\ \vdots & \ddots & \vdots & \vdots & \vdots & \ddots & \vdots \\ \gamma_{2N+1,2N+1}^{(-2N)} & \cdots & \gamma_{2N+1,2N+1}^{(-1)} & \gamma_{2N+1,2N+1}^{(0)} & \gamma_{2N+1,2N+1}^{(1)} & \cdots & \gamma_{2N+1,2N+1}^{(2N)} \end{bmatrix}. \tag{5.228}$$

Finally,

$$\lceil \lfloor \varepsilon \rfloor \rceil = \begin{bmatrix} \gamma_{11} & \gamma_{12} & \cdots & \gamma_{1,2N+1} \\ \gamma_{21} & \gamma_{22} & \cdots & \gamma_{2,2N+1} \\ \vdots & \vdots & \ddots & \vdots \\ \gamma_{2N+1,1} & \gamma_{2N+1,2} & \cdots & \gamma_{2N+1,2N+1} \end{bmatrix}, \tag{5.229}$$

$$\gamma_{mn} = \begin{bmatrix} \gamma_{mn}^{(0)} & \gamma_{mn}^{(-1)} & \cdots & \gamma_{mn}^{(-2N)} \\ \gamma_{mn}^{(1)} & \gamma_{mn}^{(0)} & \cdots & \gamma_{mn}^{(-2N+1)} \\ \vdots & \vdots & \ddots & \vdots \\ \gamma_{mn}^{(2N)} & \gamma_{mn}^{(2N-1)} & \cdots & \gamma_{mn}^{(0)} \end{bmatrix}. \tag{5.230}$$

5.3.3 Two-dimensional gratings in an oblique coordinate system

In the case of a triangular lattice, an oblique coordinate system as shown in Figure 5.5 is more convenient than a Cartesian coordinate system. When the same diffraction order is used in the calculation of triangular lattice, the results using the oblique coordinate system are $\sqrt{3}$ times higher accuracy than those using the Cartesian coordinate system.

The RCWA method in an oblique coordinate system has been proposed by Li [35]. Consider the coordinate system shown in Figure 5.5, where x_3 is perpendicular to the paper surface. In the case of a triangular lattice, $d_1 = d_2$ and $\zeta = 30°$, but this method can handle other than $\zeta = 30°$. In the Cartesian coordinate system, the basis vectors of the real space and those of the reciprocal lattice space are in the same direction. However, in an oblique coordinate system, they are not in the same direction. Here, the concepts of covariant vectors and contravariant vectors are used. Covariant basis vectors are represented by $\boldsymbol{b}_1$, $\boldsymbol{b}_2$, and $\boldsymbol{b}_3$. On the other hand, the contravariant basis

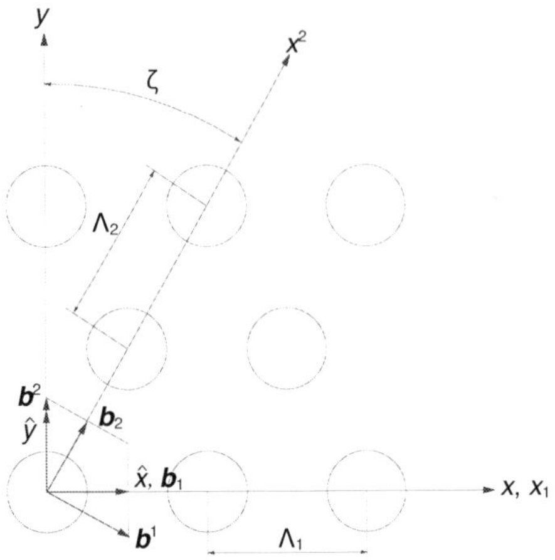

FIGURE 5.5
Oblique coordinate system.

vectors are represented by $\boldsymbol{b}^1$, $\boldsymbol{b}^2$, and $\boldsymbol{b}^3$ (see Figure 5.5). The relation between the two is as follows:

$$\boldsymbol{b}_i \cdot \boldsymbol{b}^j = \delta_i^j, \tag{5.231}$$

where δ_i^j is Kronecker's delta. Using these basis vectors, any vector $\boldsymbol{A}$ can be expressed as follows:

$$\boldsymbol{A} = x_1\boldsymbol{b}^1 + x_2\boldsymbol{b}^2 + x_3\boldsymbol{b}^3 = x^1\boldsymbol{b}_1 + x^2\boldsymbol{b}_2 + x^3\boldsymbol{b}_3. \tag{5.232}$$

As a rule, superscript vectors have subscript coefficients and subscript vectors have superscript coefficients.

The coordinate vectors are represented by covariant basis vectors, and electric field, magnetic field, and wave vectors are represented by contravariant basis vectors. That is, the wave vector $\boldsymbol{k}$ is represented by

$$\boldsymbol{k} = \alpha\boldsymbol{b}^1 + \beta\boldsymbol{b}^2 + \gamma\boldsymbol{b}^3. \tag{5.233}$$

The relationship between the wave vector $[k_1, k_2, k_3]$ in the oblique coordinate system and the wave vector $[k_x, k_y, k_z]$ in the Cartesian coordinate system is given by

$$k_x = \alpha, \tag{5.234}$$

$$k_y = (\beta - \alpha\sin\zeta)/\cos\zeta = \beta\sec\zeta - \alpha\tan\zeta, \tag{5.235}$$

$$k_z = \gamma. \tag{5.236}$$

From these relations,

$$|\boldsymbol{k}|^2 = \frac{\alpha^2 + \beta^2 - 2\alpha\beta\sin\zeta}{\cos^2\zeta} + \gamma^2 \tag{5.237}$$

is obtained.

The electric and magnetic fields are written in layer l as follows:

$$\boldsymbol{E}^{(l)} = \sum_{m,n}[S_{1mn}^{(l)}(z)\hat{\boldsymbol{x}}_1 + S_{2mn}^{(l)}(x_3)\hat{\boldsymbol{x}}_2 + S_{3mn}^{(l)}(x_3)\hat{\boldsymbol{x}}_3]\exp[i(k_{1m}x_1 + k_{2n}x_2)], \tag{5.238}$$

$$\boldsymbol{H}^{(l)} = i\left(\frac{\varepsilon_0}{\mu_0}\right)^{1/2} \times \sum_{m,n}[U_{1mn}^{(l)}(x_3)\hat{\boldsymbol{x}}_1 + U_{2mn}^{(l)}(x_3)\hat{\boldsymbol{x}}_2 + U_{3mn}^{(l)}(x_3)\hat{\boldsymbol{x}}_3]\exp[i(k_{1m}x_1 + k_{2n}x_2)], \tag{5.239}$$

where $k_{1m} = k_{10} + mK_1$, $k_{2n} = k_{20} + nK_2$, $K_1 = 2\pi/\Lambda_1$, and $K_2 = 2\pi/\Lambda_2$. k_{10} and k_{20} are the components of the in-plane wave vector of the incident light. As before, the superscript (l) on the right shoulder is omitted hereafter. From Maxwell's equations (Faraday's equation, Eq. (5.4)),

$$\frac{\partial E_3}{\partial x_2} - \frac{\partial E_2}{\partial x_3} = i\omega\mu_0 \sec\zeta(H_1 - \sin\zeta H_2), \tag{5.240}$$

$$\frac{\partial E_1}{\partial x_3} - \frac{\partial E_3}{\partial x_1} = i\omega\mu_0 \sec\zeta(H_2 - \sin\zeta H_1), \tag{5.241}$$

$$\frac{\partial E_2}{\partial x_1} - \frac{\partial E_1}{\partial x_2} = i\omega\mu_0 \cos\zeta H_3 \tag{5.242}$$

are obtained. Substituting Eqs. (5.238) and (5.239) into Eqs. (5.240), (5.41), and (5.242), we obtain the following matrix equations:

$$i\mathbf{K}_2\boldsymbol{S}_3 - \boldsymbol{S}_2' = -\sec\zeta(\boldsymbol{U}_1 - \sin\zeta\boldsymbol{U}_2), \tag{5.243}$$

$$\boldsymbol{S}_1' - i\mathbf{K}_1\boldsymbol{S}_3 = -\sec\zeta(\boldsymbol{U}_2 - \sin\zeta\boldsymbol{U}_1), \tag{5.244}$$

$$i\mathbf{K}_1\boldsymbol{S}_2 - i\mathbf{K}_2\boldsymbol{S}_1 = -\cos\zeta\boldsymbol{U}_3, \tag{5.245}$$

where prime $'$ denotes the derivative with respect to k_0x_3. Matrices $\mathbf{K}_1$ and $\mathbf{K}_2$ are diagonal ones of k_{1m}/k_0 and k_{2n}/k_0, respectively.

From Maxwell's equations (Ampere's equation, Eq. (5.7)),

$$\frac{\partial H_3}{\partial x_2} - \frac{\partial H_2}{\partial x_3} = -i\omega\varepsilon_0\varepsilon(x_1, x_2)\sec\zeta(E_1 - \sin\zeta E_2), \tag{5.246}$$

$$\frac{\partial H_1}{\partial x_3} - \frac{\partial H_3}{\partial x_1} = -i\omega\varepsilon_0\varepsilon(x_1, x_2)\sec\zeta(E_2 - \sin\zeta E_1), \tag{5.247}$$

$$\frac{\partial H_2}{\partial x_1} - \frac{\partial H_1}{\partial x_2}, = -i\omega\varepsilon_0\varepsilon(x_1, x_2)\cos\zeta E_3 \tag{5.248}$$

are obtained. Next, we express the dielectric constant $\varepsilon(x_1, x_2)$ as its Fourier series as

$$\varepsilon(x_1, x_2) = \sum_{p,q}\varepsilon_{p,q}\exp[i(pK_1x_1 + qK_2x_2)]. \tag{5.249}$$

Substituting Eqs. (5.238), (5.239), and (5.249) into Eqs. (5.246), (5.247), and (5.248), we obtain

$$i\mathbf{K}_2\boldsymbol{U}_3 - \boldsymbol{U}_2' = -\mathbf{E}\sec\zeta(\boldsymbol{S}_1 - \sin\zeta\boldsymbol{S}_2) \tag{5.250}$$

$$\boldsymbol{U}_1' - i\mathbf{K}_1\boldsymbol{U}_3 = -\mathbf{E}\sec\zeta(\boldsymbol{S}_2 - \sin\zeta\boldsymbol{S}_1) \tag{5.251}$$

$$i\mathbf{K}_1\boldsymbol{U}_2 - i\mathbf{K}_2\boldsymbol{U}_1 = -\mathbf{E}\cos\zeta\boldsymbol{S}_3 \tag{5.252}$$

in matrix form, where the matrix $\mathbf{E}$ is obtained by replacing $\varepsilon_{p,q}$ in Eqs. (5.196) and (5.197) by $\varepsilon_{p,q}$ in Eq. (5.249).

From Eqs. (5.243), (5.244), and (5.252), eliminating $\boldsymbol{S}_3$ yields

$$\cos\zeta\begin{bmatrix}\boldsymbol{S}_2'\\ \boldsymbol{S}_1'\end{bmatrix} = \mathbf{F}\begin{bmatrix}\boldsymbol{U}_2\\ \boldsymbol{U}_1\end{bmatrix} = \begin{bmatrix}\mathbf{K}_2\mathbf{E}^{-1}\mathbf{K}_1 - \sin\zeta\mathbf{I} & \mathbf{I} - \mathbf{K}_2\mathbf{E}^{-1}\mathbf{K}_2\\ \mathbf{K}_1\mathbf{E}^{-1}\mathbf{K}_1 - \mathbf{I} & \sin\zeta\mathbf{I} - \mathbf{K}_1\mathbf{E}^{-1}\mathbf{K}_2\end{bmatrix}\begin{bmatrix}\boldsymbol{U}_2\\ \boldsymbol{U}_1\end{bmatrix}. \tag{5.253}$$

Similarly, from Eqs. (5.245), (5.250), and (5.251), eliminating $\boldsymbol{U}_3$ yields

$$\cos\zeta\begin{bmatrix}\boldsymbol{U}_2'\\ \boldsymbol{U}_1'\end{bmatrix} = \mathbf{G}\begin{bmatrix}\boldsymbol{S}_2\\ \boldsymbol{S}_1\end{bmatrix} = \begin{bmatrix}\mathbf{K}_1\mathbf{K}_2 - \sin\zeta\mathbf{E} & \mathbf{E} - \mathbf{K}_2^2\\ \mathbf{K}_1^2 - \mathbf{E} & \sin\zeta\mathbf{E} - \mathbf{K}_1\mathbf{K}_2\end{bmatrix}\begin{bmatrix}\boldsymbol{S}_2\\ \boldsymbol{S}_1\end{bmatrix}. \tag{5.254}$$

Substituting Eq. (5.254) into the derivative of both sides of Eq. (5.253) with respect to k_0x_3, we obtain

$$\cos^2\zeta\begin{bmatrix}\boldsymbol{S}_2''\\ \boldsymbol{S}_1''\end{bmatrix} = \mathbf{FG}\begin{bmatrix}\boldsymbol{S}_2\\ \boldsymbol{S}_1\end{bmatrix}. \tag{5.255}$$

For a homogeneous dielectric constant layer with no modulation,

$$\mathbf{F} = \begin{bmatrix}\frac{1}{\varepsilon}\mathbf{K}_1\mathbf{K}_2 - \sin\zeta\mathbf{I} & \mathbf{I} - \frac{1}{\varepsilon}\mathbf{K}_2^2\\ \frac{1}{\varepsilon}\mathbf{K}_1^2 - \mathbf{I} & \sin\zeta\mathbf{I} - \frac{1}{\varepsilon}\mathbf{K}_1\mathbf{K}_2\end{bmatrix}, \tag{5.256}$$

$$\mathbf{G} = \begin{bmatrix}\mathbf{K}_1\mathbf{K}_2 - \varepsilon\sin\zeta\mathbf{I} & \varepsilon\mathbf{I} - \mathbf{K}_2^2\\ \mathbf{K}_1^2 - \varepsilon\mathbf{I} & \varepsilon\sin\zeta\mathbf{I} - \mathbf{K}_1\mathbf{K}_2\end{bmatrix}, \tag{5.257}$$

$$\mathbf{FG} = \begin{bmatrix}\mathbf{K}_1^2 + \mathbf{K}_2^2 - 2\sin\zeta\mathbf{K}_1\mathbf{K}_2 - \varepsilon\cos^2\zeta\mathbf{I} & \mathbf{O}\\ \mathbf{O} & \mathbf{K}_1^2 + \mathbf{K}_2^2 - 2\sin\zeta\mathbf{K}_1\mathbf{K}_2 - \varepsilon\cos^2\zeta\mathbf{I}\end{bmatrix}. \tag{5.258}$$

To improve convergence, instead of $\mathbf{G}$

$$\mathbf{G}^\ddagger = \begin{bmatrix}\mathbf{K}_1\mathbf{K}_2 - \sin\zeta\mathbf{A}^{-1} & (\cos^2\zeta\lfloor\lceil\varepsilon\rceil\rfloor + \sin^2\zeta\mathbf{A}^{-1}) - \mathbf{K}_2^2\\ \mathbf{K}_1^2 - (\cos^2\zeta\lceil\lfloor\varepsilon\rfloor\rceil + \sin^2\zeta\mathbf{A}^{-1}) & \sin\zeta\mathbf{A}^{-1} - \mathbf{K}_1\mathbf{K}_2\end{bmatrix} \tag{5.259}$$

can be used, where the matrix $\mathbf{A}$ is obtained by replacing $\varepsilon_{p,q}$ in Eqs. (5.196) and (5.197) by the Fourier coefficients $\tilde{\varepsilon}_{p,q}$ of the reciprocal of the dielectric constant given by

$$\frac{1}{\varepsilon(x_1, x_2)} = \sum_{p,q} \tilde{\varepsilon}_{p,q} \exp[i(pK_1x_1 + qK_2x_2)]. \tag{5.260}$$

Let $q_j^2 \cos^2 \zeta$ be the eigenvalues of the matrix $\mathbf{FG}$ in Eq. (5.255) and w_{jk} the eigenvectors $(j, k = 1, 2, \dots, 2(2N+1)^2)$, $\boldsymbol{S}_1$ and $\boldsymbol{S}_2$ are given by

$$S_{2,k}(x_3) = \sum_{j=1}^{2(2N+1)^2} w_{jk}[u_j \exp(k_0 q_j x_3) + d_j \exp(-k_0 q_j x_3)], \quad \text{for } k = 1, \dots, (2N+1)^2, \tag{5.261}$$

$$S_{1,k}(x_3) = \sum_{j=1}^{2(2N+1)^2} w_{jk}[u_j \exp(k_0 q_j x_3) + d_j \exp(-k_0 q_j x_3)], \quad \text{for } k = 1, \dots, (2N+1)^2, \tag{5.262}$$

respectively. The above two can be written in matrix form as

$$\begin{bmatrix} \boldsymbol{S}_2 \\ \boldsymbol{S}_1 \end{bmatrix} = \begin{bmatrix} \mathbf{W} & \mathbf{W} \end{bmatrix} \begin{bmatrix} \boldsymbol{\Phi}_+ & \mathbf{O} \\ \mathbf{O} & \boldsymbol{\Phi}_- \end{bmatrix} \begin{bmatrix} \boldsymbol{u} \\ \boldsymbol{d} \end{bmatrix}. \tag{5.263}$$

Substituting Eq. (5.263) into Eq. (5.253) yields

$$\begin{bmatrix} \boldsymbol{U}_2 \\ \boldsymbol{U}_1 \end{bmatrix} = \cos\zeta \mathbf{F}^{-1} \begin{bmatrix} \mathbf{QW} & -\mathbf{QW} \end{bmatrix} \begin{bmatrix} \boldsymbol{\Phi}_+ & \mathbf{O} \\ \mathbf{O} & \boldsymbol{\Phi}_- \end{bmatrix} \begin{bmatrix} \boldsymbol{u} \\ \boldsymbol{d} \end{bmatrix} \tag{5.264}$$

and finally the following equation is obtained:

$$\begin{bmatrix} \boldsymbol{S}_2 \\ \boldsymbol{S}_1 \\ \boldsymbol{U}_2 \\ \boldsymbol{U}_1 \end{bmatrix} = \begin{bmatrix} \mathbf{W} & \mathbf{W} \\ \cos\zeta \mathbf{F}^{-1}\mathbf{QW} & -\cos\zeta \mathbf{F}^{-1}\mathbf{QW} \end{bmatrix} \begin{bmatrix} \boldsymbol{\Phi}_+ & \mathbf{O} \\ \mathbf{O} & \boldsymbol{\Phi}_- \end{bmatrix} \begin{bmatrix} \boldsymbol{u} \\ \boldsymbol{d} \end{bmatrix}. \tag{5.265}$$

In Eq. (5.265), letting

$$\boldsymbol{S} = \begin{bmatrix} \boldsymbol{S}_2 \\ \boldsymbol{S}_1 \end{bmatrix} \tag{5.266}$$

$$\boldsymbol{U} = \begin{bmatrix} \boldsymbol{U}_2 \\ \boldsymbol{U}_1 \end{bmatrix}, \tag{5.267}$$

we obtain the following equation

$$\begin{bmatrix} \boldsymbol{S} \\ \boldsymbol{U} \end{bmatrix} = \begin{bmatrix} \mathbf{W} & \mathbf{W} \\ \cos\zeta \mathbf{F}^{-1}\mathbf{QW} & -\cos\zeta \mathbf{F}^{-1}\mathbf{QW} \end{bmatrix} \begin{bmatrix} \boldsymbol{\Phi}_+ & \mathbf{O} \\ \mathbf{O} & \boldsymbol{\Phi}_- \end{bmatrix} \begin{bmatrix} \boldsymbol{u} \\ \boldsymbol{d} \end{bmatrix}. \tag{5.268}$$

The following procedure is the same as before.

5.4 Limitations of RCWA

In the RCWA method, calculations are performed by approximating arbitrary grating shapes with staircase shapes. For example, can a sinusoidal or saw-tooth grating be calculated with a sufficiently small error if the thickness of one layer is sufficiently small and the number of layers is sufficiently large? This problem was discussed by Popov et al. [29]. This problem becomes more pronounced in the case of TM polarization in metallic gratings.

They approximated an aluminium diffraction grating with a sinusoidal surface shape by a staircase with different number of layers and examined the convergence of the solutions. They found that the order of the Fourier series (diffraction order) required for convergence increases as the number of layers increases, i.e. as the thickness of one layer decreases. The reason for this is that in TM polarization, the discontinuity of the electric field causes the charge to concentrate at the grating edges, generating a peak in the electric field (localized surface plasmon). As the layer becomes thinner, the width of this peak becomes smaller. Therefore, the order of the Fourier series required to represent it also increases. This problem arises because in staircase approximation, the interface always includes right-angle corners.

5.5 Example of program code

An example program (rcwa.py) for a 1D RCWA is shown in Program A.5 in the Appendix. The function `Rcwa1d(pol, lambda0, kx0, period, layer, norder)` is the body of the RCWA. Each argument is

`pol`: Polarization of the incident light, `'s'` or `'p'`.
`lambda0`: Wavelength of the incident light in vacuum, in μm.
`kx0`: In-plane wave number of incident light, in $1/\mu$m.
`period`: The period of the grating, in μm.
`layer`: The structure of the grating (see below).
`norder`: total diffraction order taken into account in the calculation ($2N+1$ for $\pm N$ orders).

The `layer` that gives the structure of the grating (1-D periodic structure) is given by a double list structure as follows:

$$\texttt{layer} = ((d_0, n_{00}, w_{00}), (d_1, n_{10}, w_{10}, n_{11}, w_{11}, \ldots), (d_2, n_{20}, w_{20}, n_{21}, w_{21}, \ldots), \ldots)$$

where d_l is the thickness of the l-th layer, n_{li} is the (complex) refractive index of the i-th medium composing one period of the l-th layer, w_{li} is the amount of

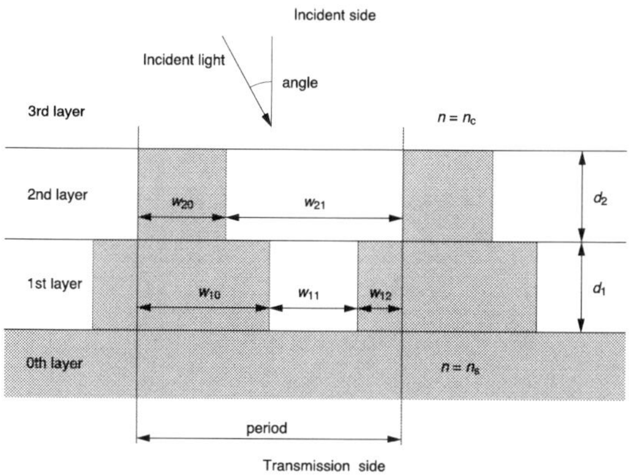

FIGURE 5.6
Model and parameters of a one-dimensional periodic structure (diffraction grating).

the width of the i-th medium composing one period of the l-th layer normalized by the period. The starting point of one period can be anywhere, but must be the same for all layers. In the example grating shown in Figure 5.6,

$$\texttt{layer} = ((0, n_s, 1), (d_1, n_s, w_{10}, n_c, w_{11}, n_s, w_{12}),$$
$$(d_2, n_s, w_{20}, n_c, w_{21}), (0, n_c, 1)).$$

The thickness and the second and the following media in the first and the last layers are ignored.

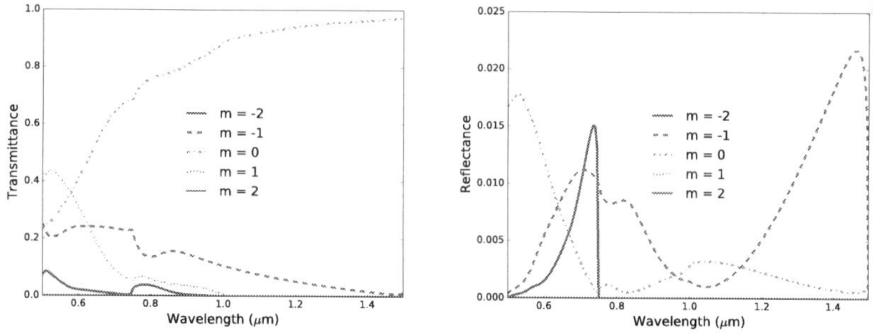

FIGURE 5.7
Calculated wavelength dependence of transmission and reflection diffraction efficiencies.

The calculated results for an example structure is shown in Figure 5.7. The assumed structure is a one-dimensional grating with $d_1 = d_2 = 0.25$ μm, $n_s = 1.5$, $n_c = 1.0$, $w_{10} = 1/2$, $w_{11} = 1/3$, $w_{12} = 1/6$, $w_{20} = 1/3$, $w_{21} = 2/3$, and

$\Lambda = 1$ μm. The diffraction order taken into calculation is set to $2N + 1 = 21$. The incident light is TM(p) polarized at an incident angle 30°. Note that the calculation cannot be performed by RCWA when the in-plane wave numbers of the incident and diffracted light coincide with the grating vector, because the matrix becomes singular. Thus, the wavelength of the incident light was slightly shifted from 0.5 μm to avoid this.

6

FDTD (Finite Difference Time Domain) Method

The FDTD (Finite Difference Time Domain) method is a numerical method for calculating electromagnetic fields invented by Yee. This method has the following advantages:

(1) Modelling is independent of object geometry.

(2) Time response can be obtained.

(3) It is easy to program and suitable for parallel computation.

In exchange for these advantages, the FDTD method also has the following weaknesses

(4) Large memory capacity and long computation time are required.

(5) Accuracy is not so high.

6.1 Introduction

In the FDTD method, the electric and magnetic fields, as well as the permittivity and permeability, are discretized and represented by values at points on a special lattice called the Yee grid, where the points defining the electric and magnetic fields are mutually exclusive [36]. Also, in time discretization, the time defining the electric field and the time defining the magnetic field are different from each other. The FDTD method calculates the time evolution for the initial electromagnetic field given on these grid points.

An example of the Yee grid in the one-dimensional (1D) case is shown in Figure 6.1. It is the case with only E_x electric field component in x-direction and H_y magnetic field component in y-direction on z-axis as

$$E_x|_k^n = E_x(k\Delta z, n\Delta t), \tag{6.1}$$

$$H_y|_{k+\frac{1}{2}}^{n+\frac{1}{2}} = H_y\left[\left(k+\frac{1}{2}\right)\Delta z, \left(n+\frac{1}{2}\right)\Delta t\right], \tag{6.2}$$

where Δz and Δt are the sampling intervals in z-direction and time. In FDTD, the electric field $E_x|_k^n$ is calculated from the electric field $E_x|_k^{n-1}$ and the magnetic fields $H_y|_{k-1/2}^{n-1/2}$ and $H_y|_{k+1/2}^{n-1/2}$. The magnetic field $H_y|_{k+1/2}^{n+1/2}$ is calculated

DOI: 10.1201/9781003357674-6

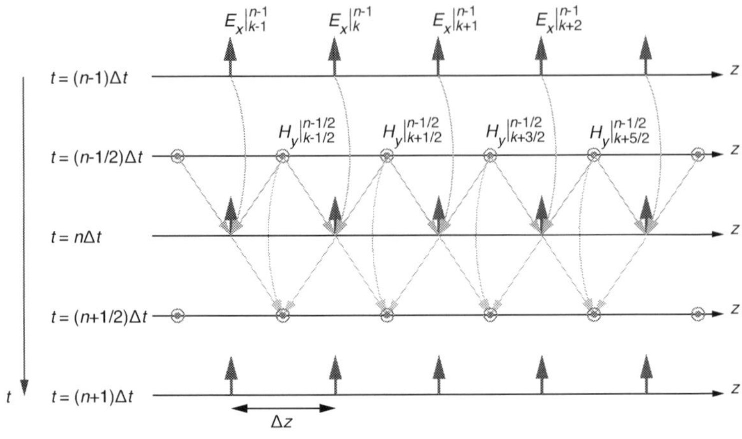

FIGURE 6.1
One-dimensional FDTD.

from the magnetic field $H_y|_{k+1/2}^{n-1/2}$ and the electric fields $E_x|_k^n$ and $E_x|_{k+1}^n$. The same calculation is repeated to obtain the time evolution of the fields.

6.2 Discretization and time evolution

The Yee grid for the general three-dimensional (3D) space is as shown in Figure 6.2. Ampere's law and Faraday's law of Maxwell's equations

$$\nabla \times \boldsymbol{H} = \varepsilon_0 \varepsilon \frac{\partial \boldsymbol{E}}{\partial t} + \sigma \boldsymbol{E} \tag{6.3}$$

$$\nabla \times \boldsymbol{E} = -\mu_0 \mu \frac{\partial \boldsymbol{H}}{\partial t} \tag{6.4}$$

are used for time evolution, where $\boldsymbol{E}$ and $\boldsymbol{H}$ are the electric and magnetic field vectors, respectively, and ε_0, μ_0, ε, μ, and σ are the vacuum permittivity, vacuum permeability, relative permittivity, relative permeability, and conductivity, respectively. In FDTD, the differential operations in these equations with respect to time and space are replaced by central differences. If we take out the x component from Eq. (6.3), we obtain

$$\varepsilon_0 \varepsilon \frac{\partial E_x}{\partial t} + \sigma E_x = \frac{\partial H_z}{\partial y} - \frac{\partial H_y}{\partial z}. \tag{6.5}$$

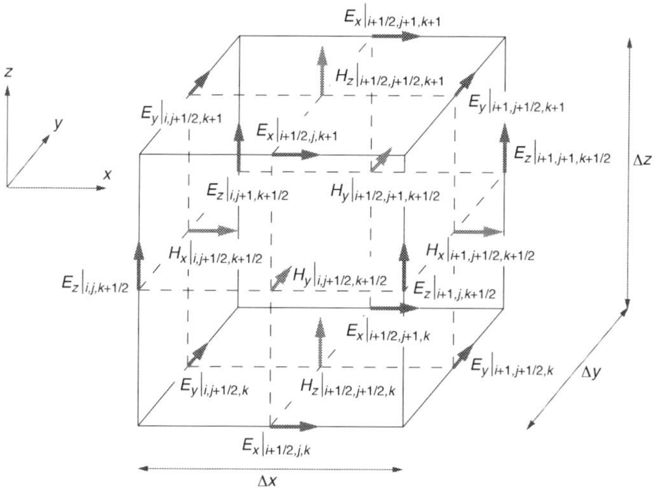

FIGURE 6.2
Yee grid for 3D space.

Replacing the derivative of Eq. (6.5) by the central difference, we obtain

$$\varepsilon_0\varepsilon\frac{E_x|^n_{i+\frac{1}{2},j,k}-E_x|^{n-1}_{i+\frac{1}{2},j,k}}{\Delta t}+\sigma\frac{E_x|^n_{i+\frac{1}{2},j,k}+E_x|^{n-1}_{i+\frac{1}{2},j,k}}{2}$$
$$=\frac{H_z|^{n-\frac{1}{2}}_{i+\frac{1}{2},j+\frac{1}{2},k}-H_z|^{n-\frac{1}{2}}_{i+\frac{1}{2},j-\frac{1}{2},k}}{\Delta y}-\frac{H_y|^{n-\frac{1}{2}}_{i+\frac{1}{2},j,k+\frac{1}{2}}-H_y|^{n-\frac{1}{2}}_{i+\frac{1}{2},j,k-\frac{1}{2}}}{\Delta z}, \tag{6.6}$$

$$\begin{aligned}E_x|^n_{i+\frac{1}{2},j,k}=&\left(\frac{2\varepsilon_0\varepsilon-\sigma\Delta t}{2\varepsilon_0\varepsilon+\sigma\Delta t}\right)E_x|^{n-1}_{i+\frac{1}{2},j,k}\\&+\left[\frac{2\Delta t}{(2\varepsilon_0\varepsilon+\sigma\Delta t)\Delta y}\right]\left(H_z|^{n-\frac{1}{2}}_{i+\frac{1}{2},j+\frac{1}{2},k}-H_z|^{n-\frac{1}{2}}_{i+\frac{1}{2},j-\frac{1}{2},k}\right)\\&-\left[\frac{2\Delta t}{(2\varepsilon_0\varepsilon+\sigma\Delta t)\Delta z}\right]\left(H_y|^{n-\frac{1}{2}}_{i+\frac{1}{2},j,k+\frac{1}{2}}-H_y|^{n-\frac{1}{2}}_{i+\frac{1}{2},j,k-\frac{1}{2}}\right).\end{aligned} \tag{6.7}$$

Similarly for y and z,

$$\begin{aligned}E_y|^n_{i,j+\frac{1}{2},k}=&\left(\frac{2\varepsilon_0\varepsilon-\sigma\Delta t}{2\varepsilon_0\varepsilon+\sigma\Delta t}\right)E_y|^{n-1}_{i,j+\frac{1}{2},k}\\&+\left[\frac{2\Delta t}{(2\varepsilon_0\varepsilon+\sigma\Delta t)\Delta z}\right]\left(H_x|^{n-\frac{1}{2}}_{i,j+\frac{1}{2},k+\frac{1}{2}}-H_x|^{n-\frac{1}{2}}_{i,j+\frac{1}{2},k-\frac{1}{2}}\right)\\&-\left[\frac{2\Delta t}{(2\varepsilon_0\varepsilon+\sigma\Delta t)\Delta x}\right]\left(H_z|^{n-\frac{1}{2}}_{i+\frac{1}{2},j+\frac{1}{2},k}-H_z|^{n-\frac{1}{2}}_{i-\frac{1}{2},j+\frac{1}{2},k}\right),\end{aligned} \tag{6.8}$$

$$E_z\Big|_{i,j,k+\frac{1}{2}}^{n} = \left(\frac{2\varepsilon_0\varepsilon - \sigma\Delta t}{2\varepsilon_0\varepsilon + \sigma\Delta t}\right) E_z\Big|_{i,j,k+\frac{1}{2}}^{n-1} + \left[\frac{2\Delta t}{(2\varepsilon_0\varepsilon + \sigma\Delta t)\Delta x}\right]\left(H_y\Big|_{i+\frac{1}{2},j,k+\frac{1}{2}}^{n-\frac{1}{2}} - H_y\Big|_{i-\frac{1}{2},j,k+\frac{1}{2}}^{n-\frac{1}{2}}\right) - \left[\frac{2\Delta t}{(2\varepsilon_0\varepsilon + \sigma\Delta t)\Delta y}\right]\left(H_x\Big|_{i,j+\frac{1}{2},k+\frac{1}{2}}^{n-\frac{1}{2}} - H_x\Big|_{i,j-\frac{1}{2},k+\frac{1}{2}}^{n-\frac{1}{2}}\right). \tag{6.9}$$

The same rule is used to discretize Eq. (6.4). If we take out the x component, we obtain

$$\mu_0\mu\frac{\partial H_x}{\partial t} = -\left(\frac{\partial E_z}{\partial y} - \frac{\partial E_y}{\partial z}\right), \tag{6.10}$$

$$\mu_0\mu\frac{H_x\Big|_{i+\frac{1}{2},j+\frac{1}{2},k}^{n+\frac{1}{2}} - H_x\Big|_{i+\frac{1}{2},j+\frac{1}{2},k}^{n-\frac{1}{2}}}{\Delta t} = -\left(\frac{E_z\Big|_{i+1,j+\frac{1}{2},k}^{n} - E_z\Big|_{i,j+\frac{1}{2},k}^{n}}{\Delta y} - \frac{E_y\Big|_{i+\frac{1}{2},j+1,k}^{n} - E_y\Big|_{i+\frac{1}{2},j,k}^{n}}{\Delta z}\right), \tag{6.11}$$

$$H_x\Big|_{i,j+\frac{1}{2},k+\frac{1}{2}}^{n+\frac{1}{2}} = H_x\Big|_{i,j+\frac{1}{2},k+\frac{1}{2}}^{n-\frac{1}{2}} - \left(\frac{\Delta t}{\mu_0\mu\Delta y}\right)\left(E_z\Big|_{i,j+1,k+\frac{1}{2}}^{n} - E_z\Big|_{i,j,k+\frac{1}{2}}^{n}\right) + \left(\frac{\Delta t}{\mu_0\mu\Delta z}\right)\left(E_y\Big|_{i,j+\frac{1}{2},k+1}^{n} - E_y\Big|_{i,j+\frac{1}{2},k}^{n}\right). \tag{6.12}$$

Similarly for the y and z components,

$$H_y\Big|_{i+\frac{1}{2},j,k+\frac{1}{2}}^{n+\frac{1}{2}} = H_y\Big|_{i+\frac{1}{2},j,k+\frac{1}{2}}^{n-\frac{1}{2}} - \left(\frac{\Delta t}{\mu_0\mu\Delta z}\right)\left(E_x\Big|_{i+\frac{1}{2},j,k+1}^{n} - E_x\Big|_{i+\frac{1}{2},j,k}^{n}\right) + \left(\frac{\Delta t}{\mu_0\mu\Delta x}\right)\left(E_z\Big|_{i+1,j,k+\frac{1}{2}}^{n} - E_z\Big|_{i,j,k+\frac{1}{2}}^{n}\right), \tag{6.13}$$

$$H_z\Big|_{i+\frac{1}{2},j+\frac{1}{2},k}^{n+\frac{1}{2}} = H_z\Big|_{i+\frac{1}{2},j+\frac{1}{2},k}^{n-\frac{1}{2}} - \left(\frac{\Delta t}{\mu_0\mu\Delta x}\right)\left(E_y\Big|_{i+1,j+\frac{1}{2},k}^{n} - E_y\Big|_{i,j+\frac{1}{2},k}^{n}\right) + \left(\frac{\Delta t}{\mu_0\mu\Delta y}\right)\left(E_x\Big|_{i+\frac{1}{2},j+1,k}^{n} - E_x\Big|_{i+\frac{1}{2},j,k}^{n}\right). \tag{6.14}$$

In FDTD, the time evolution of the electric field is calculated using the Eqs. (6.7)–(6.9), and the time evolution of the magnetic field is calculated using the Eqs. (6.12)–(6.14). These calculations are repeated to obtain the time evolution of the electromagnetic field distribution in 3D space.

6.2.1 On the computer

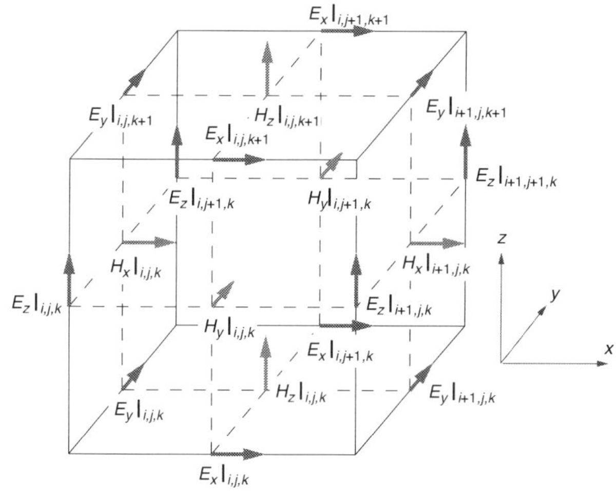

FIGURE 6.3
Yee lattice on computer array.

In actual calculations, the electromagnetic field, permittivity, and permeability are stored in arrays in the program code. However, arrays do not have half-integer subscripts, so they must be stored in arrays with integer subscripts. Usually, it is common to store them in arrays with subscripts except for 1/2. The Yee lattice in this case is shown in Figure 6.3. With this subscript, Eqs. (6.7)–(6.9) are:

$$\begin{aligned} E_x|_{i,j,k}^{n} = &\left(\frac{2\varepsilon_0\varepsilon - \sigma\Delta t}{2\varepsilon_0\varepsilon + \sigma\Delta t}\right) E_x|_{i,j,k}^{n-1} \\ &+ \left[\frac{2\Delta t}{(2\varepsilon_0\varepsilon + \sigma\Delta t)\Delta y}\right]\left(H_z|_{i,j,k}^{n-\frac{1}{2}} - H_z|_{i,j-1,k}^{n-\frac{1}{2}}\right) \\ &- \left[\frac{2\Delta t}{(2\varepsilon_0\varepsilon + \sigma\Delta t)\Delta}\right]\left(H_y|_{i,j,k}^{n-\frac{1}{2}} - H_y|_{i,j,k-1}^{n-\frac{1}{2}}\right), \end{aligned} \tag{6.15}$$

$$\begin{aligned} E_y|_{i,j,k}^{n} = &\left(\frac{2\varepsilon_0\varepsilon - \sigma\Delta t}{2\varepsilon_0\varepsilon + \sigma\Delta t}\right) E_y|_{i,j,k}^{n-1} \\ &+ \left[\frac{2\Delta t}{(2\varepsilon_0\varepsilon + \sigma\Delta t)\Delta z}\right]\left(H_x|_{i,j,k}^{n-\frac{1}{2}} - H_x|_{i,j,k-1}^{n-\frac{1}{2}}\right) \\ &- \left[\frac{2\Delta t}{(2\varepsilon_0\varepsilon + \sigma\Delta t)\Delta x}\right]\left(H_z|_{i,j,k}^{n-\frac{1}{2}} - H_z|_{i-1,j,k}^{n-\frac{1}{2}}\right), \end{aligned} \tag{6.16}$$

$$
\begin{aligned}
E_z|_{i,j,k}^{n} = &\left(\frac{2\varepsilon_0\varepsilon - \sigma\Delta t}{2\varepsilon_0\varepsilon + \sigma\Delta t}\right) E_z|_{i,j,k}^{n-1} \\
&+ \left[\frac{2\Delta t}{(2\varepsilon_0\varepsilon + \sigma\Delta t)\Delta x}\right]\left(H_y|_{i,j,k}^{n-\frac{1}{2}} - H_y|_{i-1,j,k}^{n-\frac{1}{2}}\right) \\
&- \left[\frac{2\Delta t}{(2\varepsilon_0\varepsilon + \sigma\Delta t)\Delta y}\right]\left(H_x|_{i,j,k}^{n-\frac{1}{2}} - H_x|_{i,j-1,k}^{n-\frac{1}{2}}\right),
\end{aligned} \tag{6.17}
$$

and Eqs. (6.12)–(6.14) are

$$
\begin{aligned}
H_x|_{i,j,k}^{n+\frac{1}{2}} = &H_x|_{i,j,k}^{n-\frac{1}{2}} \\
&- \left(\frac{\Delta t}{\mu_0\mu\Delta y}\right)\left(E_z|_{i,j+1,k}^{n} - E_z|_{i,j,k}^{n}\right) \\
&+ \left(\frac{\Delta t}{\mu_0\mu\Delta z}\right)\left(E_y|_{i,j,k+1}^{n} - E_y|_{i,j,k}^{n}\right),
\end{aligned} \tag{6.18}
$$

$$
\begin{aligned}
H_y|_{i,j,k}^{n+\frac{1}{2}} = &H_y|_{i,j,k}^{n-\frac{1}{2}} \\
&- \left(\frac{\Delta t}{\mu_0\mu\Delta z}\right)\left(E_x|_{i,j,k+1}^{n} - E_x|_{i,j,k}^{n}\right) \\
&+ \left(\frac{\Delta t}{\mu_0\mu\Delta x}\right)\left(E_z|_{i+1,j,k}^{n} - E_z|_{i,j,k}^{n}\right),
\end{aligned} \tag{6.19}
$$

$$
\begin{aligned}
H_z|_{i,j,k}^{n+\frac{1}{2}} = &H_z|_{i,j,k}^{n-\frac{1}{2}} \\
&- \left(\frac{\Delta t}{\mu_0\mu\Delta x}\right)\left(E_y|_{i+1,j,k}^{n} - E_y|_{i,j,k}^{n}\right) \\
&+ \left(\frac{\Delta t}{\mu_0\mu\Delta y}\right)\left(E_x|_{i,j+1,k}^{n} - E_x|_{i,j,k}^{n}\right).
\end{aligned} \tag{6.20}
$$

As time series data, only the last two temporal fields, $\boldsymbol{E}|^{n-1}$, $\boldsymbol{E}|^{n}$, $\boldsymbol{H}|^{n-\frac{1}{2}}$, and $\boldsymbol{H}|^{n+\frac{1}{2}}$ are required to be stored.

6.2.2 Cell size and time step

In actual calculations, an important question is how large the cell size and time step should be used. Naturally, the Nyquist sampling theorem must be satisfied, so the cell size must be finer than one-half the shortest wavelength in the computational domain. The smaller the cell size, the smaller the error that is called the grid dispersion, and the better the accuracy. In actual calculations, it is sufficient to make the cell size less than one-tenth of the shortest wavelength. However, when dealing with nano-region structures such as for plasmonics, this size is not sufficient to represent fine shapes, and cell sizes of 10 nm, 5 nm, or even smaller are often used.

Now, once the cell size is determined, a corresponding time step is required. In order for the solution to be stable with respect to time evolution, the relationship between time step and cell size must satisfy a condition for stability called the Courant condition [37]:

$$\Delta t \leq \frac{1}{v\sqrt{\frac{1}{\Delta x^2} + \frac{1}{\Delta y^2} + \frac{1}{\Delta z^2}}}, \tag{6.21}$$

where v is the maximum phase velocity of light in the medium. The Δt must satisfy this equation. When dealing with metals such as in plasmonics, special considerations must be made. The phase velocity in a medium is given by $v = c/\mathrm{Re}(n)$, where n is the refractive index of the medium and c is the speed of light in vacuum. In a metal-free system, $\mathrm{Re}(n) \geq 1$ is usual and the fastest phase velocity is the speed of light in vacuum. Therefore, Δt should be considered with respect to the speed of light in vacuum. However, when a metal is included, for example, silver in the visible region, $\mathrm{Re}(n) \sim 0.05$. In other words, the phase velocity of light in silver is 20 times faster than that in vacuum (but the attenuation is extremely fast). Therefore, Δt that satisfies the Courant condition must also be set to 1/20 or less compared to the case without metal. However, this condition can be relaxed to the Courant condition for vacuum by using the treatment for dispersive media described later.

6.2.3 Placement of an object on Yee grid

First, let us discuss the E- and H-cells. As shown in Figure 6.4, an E-cell is a unit cell where the electric field is defined at the midpoint of the edge, and an H-cell is a unit cell where the magnetic field is defined at the midpoint of the edge.

Now, to place an object, the permittivity and the permeability of the object must be set on the Yee grid according to the shape and location of the

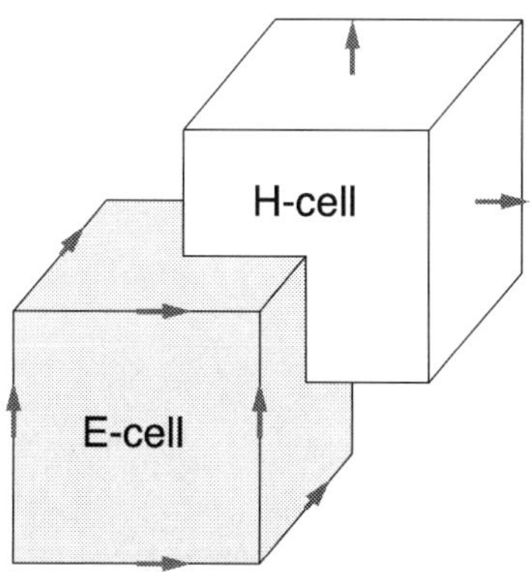

FIGURE 6.4
E-cell and H-cell; arrows indicate electric field and their defined positions.

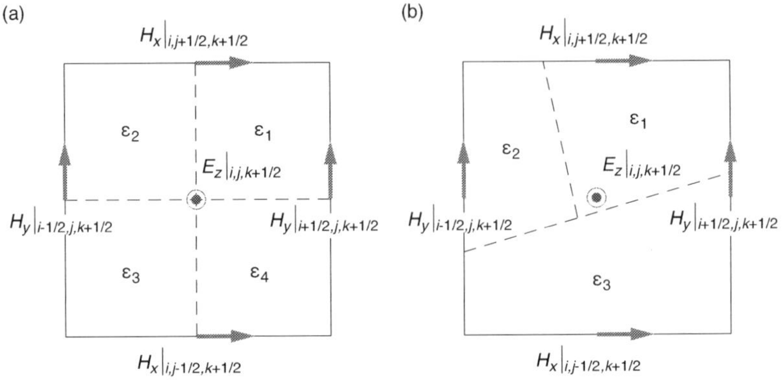

FIGURE 6.5
Distribution of permittivity: (a) the case where objects with four different permittivities come into contact at the defined position of E_z, (b) more general case.

object. Usually, the permittivity of the object at the location where the electric field is defined and the permeability of the object at the location where the magnetic field is defined are used. However, the following problem arises here. As an example, consider the permittivity at the location of E_z as shown in Figure 6.5. Figure 6.5(a) is the case where objects with four different relative permittivities come into contact at the defined position of $E_z|_{i,j,k+1/2}$. In this case, the permittivity,

$$\varepsilon_{i,j,k+1/2} = \frac{1}{4}(\varepsilon_1 + \varepsilon_2 + \varepsilon_3 + \varepsilon_4) \tag{6.22}$$

can be used. In a general case such as Figure 6.5(b), it is sufficient to use the average of the permittivity with the area occupied by the object with each permittivity as the weight. However, it is cumbersome to program to calculate the permittivity with such a method when the permittivity of the actual object is set to the Yee grid. Therefore, it is often taken to approximate the object shape as a collection of E-cells and to set the permittivity at all defined positions of the electric field belonging to the E-cells to the same value. In this case, the ideal permittivity at the boundary of objects with different permittivity should be the average value as described above, but in practice complicated problems remain, for example, how to program the permittivity at the boundary between a non-dispersive object and an object following Drude dispersion. Therefore, a simpler approach is to use the permittivity of the last object placed instead of the average permittivity at the boundary.

Since the FDTD method is not a very accurate calculation method, there is no noticeable decrease in accuracy by using such a rough approximation. To improve the accuracy, it is effective to reduce the cell size.

6.2.4 Perfect electric conductor and perfect magnetic conductor

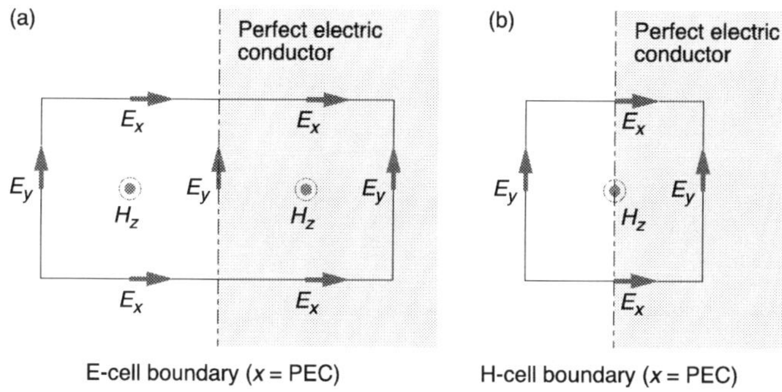

FIGURE 6.6
Perfect electric conductor with (a) E-cell boundary and (b) H-cell boundary. The perfect electric conductor occupies in $x \geq$ PEC.

Inside a perfect conductor, both the electric and magnetic fields are zero. Furthermore, at the surface of a perfect electrical conductor (PEC), the direction of the electric field is always perpendicular to the surface and its tangential component is zero. Similarly, on the surface of a perfect magnetic conductor, the direction of the magnetic field is always perpendicular to the surface and its tangential component is zero. We will show how the electromagnetic field at the surface of these perfect conductors is treated in the time evolution, using examples of a PEC.

If the surface of a perfect electrical conductor is at $x =$ PEC and coincides with the E-cell boundary, as shown in Figure 6.6(a), simply,

$$E_y|_{\mathrm{PEC}} = 0, \quad E_z|_{\mathrm{PEC}} = 0. \tag{6.23}$$

On the other hand, if the H-cell boundary coincides with the surface of the PEC, as shown in Figure 6.6(b), a little ingenuity is required. This is because a part of E_z and E_y, which are necessary for the time evolution of the tangential components of the magnetic field, H_y and H_z, on the surface of the PEC, are contained in the PEC and cannot be calculated. To solve this problem, the mirror image effect due to the surface of the PEC, that is:

$$E_y|_{\mathrm{PEC}+\frac{1}{2}} = -E_y|_{\mathrm{PEC}-\frac{1}{2}}, \tag{6.24}$$

$$E_z|_{\mathrm{PEC}+\frac{1}{2}} = -E_z|_{\mathrm{PEC}-\frac{1}{2}}, \tag{6.25}$$

are used. Substituting these relationships into Eqs. (6.13) and (6.14), on a

PEC surface perpendicular to the x-axis, we obtain

$$H_y\Big|_{i+\frac{1}{2},j,k+\frac{1}{2}}^{n+\frac{1}{2}} = H_y\Big|_{i+\frac{1}{2},j,k+\frac{1}{2}}^{n-\frac{1}{2}} - \left(\frac{\Delta t}{\mu_0\mu\Delta z}\right)\left(E_x\Big|_{i+\frac{1}{2},j,k+1}^{n} - E_x\Big|_{i+\frac{1}{2},j,k}^{n}\right) - 2\left(\frac{\Delta t}{\mu_0\mu\Delta x}\right)E_z\Big|_{i,j,k+\frac{1}{2}}^{n}, \tag{6.26}$$

$$H_z\Big|_{i+\frac{1}{2},j+\frac{1}{2},k}^{n+\frac{1}{2}} = H_z\Big|_{i+\frac{1}{2},j+\frac{1}{2},k}^{n-\frac{1}{2}} + 2\left(\frac{\Delta t}{\mu_0\mu\Delta x}\right)E_y\Big|_{i,j+\frac{1}{2},k}^{n} + \left(\frac{\Delta t}{\mu_0\mu\Delta y}\right)\left(E_x\Big|_{i+\frac{1}{2},j+1,k}^{n} - E_x\Big|_{i+\frac{1}{2},j,k}^{n}\right). \tag{6.27}$$

The same approach can be used for the case of perfect magnetic conductors.

6.2.5 Reduction of computational complexity using the symmetry of the system

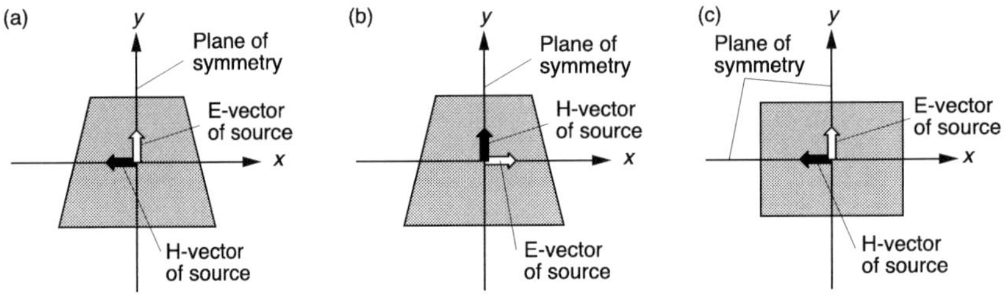

FIGURE 6.7
Symmetry of the system.

If there is symmetry throughout the system, the computational complexity can be reduced to 1/2, 1/4, or even 1/8. Assume that the permittivity and the permeability distributions of the object and medium are mirror symmetric about the plane containing the centre of the system. Both the electric and magnetic fields cannot have mirror symmetry at the same time. This is because they constitute a right-handed system, so if the electric field is symmetric, the magnetic field will be antisymmetric. And conversely, if the magnetic field is symmetric, then the electric field is antisymmetric. As an example, consider the case where the object and the incident electric field are mirror symmetric with respect to the $x = 0$ plane as shown in Figure 6.7(a). In this case, the magnetic field is antisymmetric about the $x = 0$ plane, and the relation between the electric and magnetic fields are as follows:

$$\begin{aligned} E_x(-x,y,z) &= -E_x(x,y,z), \\ E_y(-x,y,z) &= E_y(x,y,z), \\ E_z(-x,y,z) &= E_z(x,y,z), \end{aligned} \tag{6.28}$$

$$\begin{aligned} H_x(-x,y,z) &= H_x(x,y,z), \\ H_y(-x,y,z) &= -H_y(x,y,z), \\ H_z(-x,y,z) &= -H_z(x,y,z). \end{aligned} \tag{6.29}$$

Thus, in the $x = 0$ plane, the antisymmetric field is zero. Specifically,

$$\begin{aligned} E_x(0,y,z) &= 0, \\ H_y(0,y,z) &= 0, \\ H_z(0,y,z) &= 0. \end{aligned} \tag{6.30}$$

This condition is automatically satisfied by placing a perfect magnetic conductor (PMC) on one side of the $x = 0$ plane. On the other hand, if the electric field is antisymmetric as shown in Figure 6.7(b), the opposite is true,

$$\begin{aligned} H_x(0,y,z) &= 0, \\ E_y(0,y,z) &= 0, \\ E_z(0,y,z) &= 0. \end{aligned} \tag{6.31}$$

This condition is automatically satisfied by placing a PEC on one side of the $x = 0$ plane. In both cases, we can see that we only need to calculate the half $x \geq 0$ or $x \leq 0$ region of the system. Furthermore, if the $x = 0$ and $y = 0$ planes are both centres of mirror symmetry, as shown in Figure 6.7(c), we can reduce the amount of calculation to a quarter by placing perfect magnetic and perfect electric conductors in the respective symmetry plane.

To set the perfect magnetic and perfect electric conductors on one side of the $x = 0$ plane, the H-cell boundary must coincide with $x = 0$ in the former case and the E-cell boundary in the latter. E-cells and H-cells are defined as Figure 6.4 for the defined location of the electric field. However, if we want to calculate symmetric and antisymmetric cases for the same system, we will need to shift the position of the Yee grid to the object for each symmetry. If this is not desired, then the same ingenuity as described in the previous section is required.

Consider the case where the E-cell boundary is aligned with the $x = 0$ plane. If the electric field is antisymmetric, we only need to place a perfect electric conductor in the $x = 0$ plane. Next, consider the case where the electric field is symmetric. As mentioned above, the conditions that must be satisfied in the $x = 0$ plane are

$$E_x = H_y = H_z = 0. \tag{6.32}$$

However, in the Yee grid these values are not defined in the $x = 0$ plane. Therefore, we can only use the values at $x = \pm\Delta x/2$. From symmetry, we obtain

$$H_y\left(-\frac{\Delta x}{2}, y, z\right) = -H_y\left(\frac{\Delta x}{2}, y, z\right), \tag{6.33}$$

$$H_z\left(-\frac{\Delta x}{2}, y, z\right) = -H_z\left(\frac{\Delta x}{2}, y, z\right). \tag{6.34}$$

Consider the case where the object is defined in the region of $x \geq 0$ ($i \geq 0$). Using the relationship between Eqs. (6.33) and (6.34), and between Eqs. (6.8) and (6.9), we obtain

$$\begin{aligned} E_y\big|^{n}_{0,j+\frac{1}{2},k} &= \left(\frac{2\varepsilon_0\varepsilon - \sigma\Delta t}{2\varepsilon_0\varepsilon + \sigma\Delta t}\right) E_y\big|^{n-1}_{0,j+\frac{1}{2},k} \\ &+ \left[\frac{2\Delta t}{(2\varepsilon_0\varepsilon + \sigma\Delta t)\Delta z}\right]\left(H_x\big|^{n-\frac{1}{2}}_{0,j+\frac{1}{2},k+\frac{1}{2}} - H_x\big|^{n-\frac{1}{2}}_{0,j+\frac{1}{2},k-\frac{1}{2}}\right) \\ &- \left[\frac{2\Delta t}{(2\varepsilon_0\varepsilon + \sigma\Delta t)\Delta x}\right]\left(2H_z\big|^{n-\frac{1}{2}}_{\frac{1}{2},j+\frac{1}{2},k}\right), \end{aligned} \tag{6.35}$$

$$\begin{aligned} E_z\big|^{n}_{0,j,k+\frac{1}{2}} &= \left(\frac{2\varepsilon_0\varepsilon - \sigma\Delta t}{2\varepsilon_0\varepsilon + \sigma\Delta t}\right) E_z\big|^{n-1}_{0,j,k+\frac{1}{2}} \\ &+ \left[\frac{2\Delta t}{(2\varepsilon_0\varepsilon + \sigma\Delta t)\Delta x}\right]\left(2H_y\big|^{n-\frac{1}{2}}_{\frac{1}{2},j,k+\frac{1}{2}}\right) \\ &- \left[\frac{2\Delta t}{(2\varepsilon_0\varepsilon + \sigma\Delta t)\Delta y}\right]\left(H_x\big|^{n-\frac{1}{2}}_{0,j+\frac{1}{2},k+\frac{1}{2}} - H_x\big|^{n-\frac{1}{2}}_{0,j-\frac{1}{2},k+\frac{1}{2}}\right). \end{aligned} \tag{6.36}$$

We can calculate E_y and E_z at $x = 0$ according to these equations.

6.3 Dispersive medium

When dealing with dielectric materials, it is often not so problematic if the permittivity is constant in the frequency region of interest. However, in the case of metals, the permittivity (ideally) follows a Drude dispersion, so the dispersion (frequency dependence of the permittivity) cannot be ignored in most cases. Also, in single-frequency calculations, one might think that a medium with negative relative permittivity, such as a metal, can be treated in the form σ/ω, using the electrical conductivity σ and the angular frequency ω without considering dispersion, but the field diverges quickly in the actual calculation of time evolution.

The most commonly used methods for dealing with dispersive media are the Recursive Convolution (RC) method, the Piecewise Linear Recursive Convolution (PLRC) method, which is an extension of the RC method, and the Auxiliary Differential Equation (ADE) method. The ADE method, which has a wide range of applications, is described here. The Drude and Lorentz dispersions are discussed.

6.3.1 Drude dispersion

The relative permittivity ε is given by

$$\varepsilon(\omega) = \varepsilon_\infty + \chi(\omega) \tag{6.37}$$

where ε_∞ is the relative permittivity in the limit of frequency infinity, and $\chi(\omega)$ is the electric susceptibility. In the Drude dispersion medium the electric susceptibility $\chi(\omega)$ is given by,

$$\chi(\omega) = -\frac{\omega_p^2}{\omega^2 + i\Gamma\omega} \tag{6.38}$$

where ω_p is the plasma frequency, and Γ is the dumping constant. The relation between the polarization P and the electric susceptibility χ is

$$\boldsymbol{P} = \varepsilon_0 \chi \boldsymbol{E}, \tag{6.39}$$

and the relationship between the polarization current $\boldsymbol{J}$ and the polarization P is

$$\boldsymbol{J} = \frac{\partial \boldsymbol{P}}{\partial t}. \tag{6.40}$$

From these relationships,

$$\boldsymbol{J} = \varepsilon_0 \chi \frac{\partial \boldsymbol{E}}{\partial t} \tag{6.41}$$

is obtained.

Here we describe the method developed by Okoniewski et al. [38]. Substituting Eq. (6.38) into Eq. (6.41), we obtain

$$\omega^2 \boldsymbol{J}(\omega) + i\omega\Gamma \boldsymbol{J}(\omega) = -\varepsilon_0 \omega_p^2 \frac{\partial \boldsymbol{E}}{\partial t}. \tag{6.42}$$

Using the fact that the time-dependent term of the polarization current $\boldsymbol{J}$ is $\exp(-i\omega t)$, Eq. (6.42) in the time domain is

$$\frac{\partial^2 \boldsymbol{J}}{\partial t^2} + \Gamma \frac{\partial \boldsymbol{J}}{\partial t} = \varepsilon_0 \omega_p^2 \frac{\partial \boldsymbol{E}}{\partial t}. \tag{6.43}$$

Integrating both sides once with respect to t, we obtain

$$\frac{\partial \boldsymbol{J}}{\partial t} + \Gamma \boldsymbol{J} = \varepsilon_0 \omega_p^2 \boldsymbol{E}. \tag{6.44}$$

This is the desired Auxiliary Differential Equation (ADE). Discretizing Eq. (6.44)

$$\frac{\boldsymbol{J}^n - \boldsymbol{J}^{n-1}}{\Delta t} + \Gamma\frac{\boldsymbol{J}^n + \boldsymbol{J}^{n-1}}{2} = \varepsilon_0\omega_p^2\frac{\boldsymbol{E}^n + \boldsymbol{E}^{n-1}}{2}, \tag{6.45}$$

i.e.

$$\boldsymbol{J}^n = \frac{1-\Gamma\Delta t/2}{1+\Gamma\Delta t/2}\boldsymbol{J}^{n-1} + \frac{\varepsilon_0\omega_p^2\Delta t/2}{1+\Gamma\Delta t/2}(\boldsymbol{E}^n + \boldsymbol{E}^{n-1}) \tag{6.46}$$

is obtained. Updating $\boldsymbol{J}^n$ requires $\boldsymbol{J}^{n-1}$, $\boldsymbol{E}^{n-1}$, and $\boldsymbol{E}^n$, but $\boldsymbol{J}^n$ is also required for updating $\boldsymbol{E}^n$, so we have to be creative.

Ampere's equation including the displacement current ($\sigma\boldsymbol{E}$) is

$$\nabla\times\boldsymbol{H} = \varepsilon_0\varepsilon_\infty\frac{\partial\boldsymbol{E}}{\partial t} + \sigma\boldsymbol{E} + \boldsymbol{J}. \tag{6.47}$$

Here, the coefficient of the first term on the right-hand side is not ε_0, but $\varepsilon_0\varepsilon_\infty$. Hereafter, we use this equation. Discretizing, Eq. (6.47), we obtain

$$\nabla\times\boldsymbol{H}^{n-\frac{1}{2}} = \varepsilon_0\varepsilon_\infty\frac{\boldsymbol{E}^n - \boldsymbol{E}^{n-1}}{\Delta t} + \sigma\frac{\boldsymbol{E}^n + \boldsymbol{E}^{n-1}}{2} + \boldsymbol{J}^{n-\frac{1}{2}}, \tag{6.48}$$

where

$$\boldsymbol{J}^{n-\frac{1}{2}} = \frac{1}{2}(\boldsymbol{J}^n + \boldsymbol{J}^{n-1}) \tag{6.49}$$

for matching the time. Substituting Eq. (6.49) into Eq. (6.46), we obtain

$$\begin{aligned}\boldsymbol{J}^{n-\frac{1}{2}} =& \frac{1}{2}\left(1+\frac{1-\Gamma\Delta t/2}{1+\Gamma\Delta t/2}\right)\boldsymbol{J}^{n-1} \\ &+ \frac{1}{2}\left(\frac{\varepsilon_0\omega_p^2\Delta t/2}{1+\Gamma\Delta t/2}\right)(\boldsymbol{E}^n + \boldsymbol{E}^{n-1}).\end{aligned} \tag{6.50}$$

Substituting the Eq. (6.50) into Eq. (6.48),

$$\begin{aligned}\nabla\times\boldsymbol{H}^{n-\frac{1}{2}} =& \frac{\varepsilon_0\varepsilon_\infty}{\Delta t}(\boldsymbol{E}^n - \boldsymbol{E}^{n-1}) + \frac{\sigma}{2}(\boldsymbol{E}^n + \boldsymbol{E}^{n-1}) \\ &+ \frac{1}{2}\left(1+\frac{1-\Gamma\Delta t/2}{1+\Gamma\Delta t/2}\right)\boldsymbol{J}^{n-1} \\ &+ \frac{1}{2}\left(\frac{\varepsilon_0\omega_p^2\Delta t/2}{1+\Gamma\Delta t/2}\right)(\boldsymbol{E}^n + \boldsymbol{E}^{n-1})\end{aligned} \tag{6.51}$$

is obtained. Therefore,

$$\begin{aligned}\boldsymbol{E}^n =& \frac{\left[\frac{\varepsilon_0\varepsilon_\infty}{\Delta t} - \frac{\sigma}{2} - \frac{1}{2}\left(\frac{\varepsilon_0\omega_p^2\Delta t/2}{1+\Gamma\Delta t/2}\right)\right]}{\left[\frac{\varepsilon_0\varepsilon_\infty}{\Delta t} + \frac{\sigma}{2} + \frac{1}{2}\left(\frac{\varepsilon_0\omega_p^2\Delta t/2}{1+\Gamma\Delta t/2}\right)\right]}\boldsymbol{E}^{n-1} \\ &+ \frac{1}{\left[\frac{\varepsilon_0\varepsilon_\infty}{\Delta t} + \frac{\sigma}{2} + \frac{1}{2}\left(\frac{\varepsilon_0\omega_p^2\Delta t/2}{1+\Gamma\Delta t/2}\right)\right]} \\ &\times\left[\nabla\times\boldsymbol{H}^{n-\frac{1}{2}} - \frac{1}{2}\left(1+\frac{1-\Gamma\Delta t/2}{1+\Gamma\Delta t/2}\right)\boldsymbol{J}^{n-1}\right].\end{aligned} \tag{6.52}$$

The calculation procedure is to update $\boldsymbol{E}^n$ using Eq. (6.52) and then update $\boldsymbol{J}^n$ using Eq. (6.46). Writing out the x component of Eq. (6.52),

$$\begin{aligned} E_x\Big|_{i+\frac{1}{2},j,k}^{n} = & \frac{\left[\frac{\varepsilon_0\varepsilon_\infty}{\Delta t} - \frac{\sigma}{2} - \frac{1}{2}\left(\frac{\varepsilon_0\omega_p^2\Delta t/2}{1+\Gamma\Delta t/2}\right)\right]}{\left[\frac{\varepsilon_0\varepsilon_\infty}{\Delta t} + \frac{\sigma}{2} + \frac{1}{2}\left(\frac{\varepsilon_0\omega_p^2\Delta t/2}{1+\Gamma\Delta t/2}\right)\right]} E_x\Big|_{i+\frac{1}{2},j,k}^{n-1} \\ & + \frac{1/\Delta y}{\left[\frac{\varepsilon_0\varepsilon_\infty}{\Delta t} + \frac{\sigma}{2} + \frac{1}{2}\left(\frac{\varepsilon_0\omega_p^2\Delta t/2}{1+\Gamma\Delta t/2}\right)\right]} \\ & \times \left(H_z\Big|_{i+\frac{1}{2},j+\frac{1}{2},k}^{n-\frac{1}{2}} - H_z\Big|_{i+\frac{1}{2},j-\frac{1}{2},k}^{n-\frac{1}{2}}\right) \\ & - \frac{1/\Delta z}{\left[\frac{\varepsilon_0\varepsilon_\infty}{\Delta t} + \frac{\sigma}{2} + \frac{1}{2}\left(\frac{\varepsilon_0\omega_p^2\Delta t/2}{1+\Gamma\Delta t/2}\right)\right]} \\ & \times \left(H_y\Big|_{i+\frac{1}{2},j,k+\frac{1}{2}}^{n-\frac{1}{2}} - H_y\Big|_{i+\frac{1}{2},j,k-\frac{1}{2}}^{n-\frac{1}{2}}\right) \\ & - \frac{1}{2}\frac{1}{\left[\frac{\varepsilon_0\varepsilon_\infty}{\Delta t} + \frac{\sigma}{2} + \frac{1}{2}\left(\frac{\varepsilon_0\omega_p^2\Delta t/2}{1+\Gamma\Delta t/2}\right)\right]} \\ & \times \left(1 + \frac{1-\Gamma\Delta t/2}{1+\Gamma\Delta t/2}\right) J_x\Big|_{i+\frac{1}{2},j,k}^{n-1}. \end{aligned} \tag{6.53}$$

Also, if there is a current source $\boldsymbol{j}$,

$$\begin{aligned} E_x\Big|_{i+\frac{1}{2},j,k}^{n} = & \frac{\left[\frac{\varepsilon_0\varepsilon_\infty}{\Delta t} - \frac{\sigma}{2} - \frac{1}{2}\left(\frac{\varepsilon_0\omega_p^2\Delta t/2}{1+\Gamma\Delta t/2}\right)\right]}{\left[\frac{\varepsilon_0\varepsilon_\infty}{\Delta t} + \frac{\sigma}{2} + \frac{1}{2}\left(\frac{\varepsilon_0\omega_p^2\Delta t/2}{1+\Gamma\Delta t/2}\right)\right]} E_x\Big|_{i+\frac{1}{2},j,k}^{n-1} \\ & + \frac{1/\Delta y}{\left[\frac{\varepsilon_0\varepsilon_\infty}{\Delta t} + \frac{\sigma}{2} + \frac{1}{2}\left(\frac{\varepsilon_0\omega_p^2\Delta t/2}{1+\Gamma\Delta t/2}\right)\right]} \\ & \times \left(H_z\Big|_{i+\frac{1}{2},j+\frac{1}{2},k}^{n-\frac{1}{2}} - H_z\Big|_{i+\frac{1}{2},j-\frac{1}{2},k}^{n-\frac{1}{2}}\right) \\ & - \frac{1/\Delta z}{\left[\frac{\varepsilon_0\varepsilon_\infty}{\Delta t} + \frac{\sigma}{2} + \frac{1}{2}\left(\frac{\varepsilon_0\omega_p^2\Delta t/2}{1+\Gamma\Delta t/2}\right)\right]} \\ & \times \left(H_y\Big|_{i+\frac{1}{2},j,k+\frac{1}{2}}^{n-\frac{1}{2}} - H_y\Big|_{i+\frac{1}{2},j,k-\frac{1}{2}}^{n-\frac{1}{2}}\right) \\ & - \frac{1}{2}\frac{1}{\left[\frac{\varepsilon_0\varepsilon_\infty}{\Delta t} + \frac{\sigma}{2} + \frac{1}{2}\left(\frac{\varepsilon_0\omega_p^2\Delta t/2}{1+\Gamma\Delta t/2}\right)\right]} \\ & \times \left(1 + \frac{1-\Gamma\Delta t/2}{1+\Gamma\Delta t/2}\right) J_x\Big|_{i+\frac{1}{2},j,k}^{n-1} \\ & - \frac{1}{\left[\frac{\varepsilon_0\varepsilon_\infty}{\Delta t} + \frac{\sigma}{2} + \frac{1}{2}\left(\frac{\varepsilon_0\omega_p^2\Delta t/2}{1+\Gamma\Delta t/2}\right)\right]} j_x\Big|_{i+\frac{1}{2},j,k}^{n-\frac{1}{2}}. \end{aligned} \tag{6.54}$$

For actual calculations, the parameters, ε_∞, ω_p, and Γ obtained by fitting the Drude dispersion equation to the experimentally obtained permittivity are used. For example, in the case of gold in the infrared region, there are experimental values obtained by Padalka and Shklyarevskii [39]. These experimental values are the complex permittivity from 1 μm to 11 μm. The parameters obtained by fitting to these experimental values are $\varepsilon_\infty = -16.74$, $\omega_p = 1.034 \times 10^{16}$ Hz, and $\Gamma = 5.384 \times 10^{13}$ Hz. However, the calculation results diverge quickly when using these values. This is because ε_∞ is a negative value. To avoid this, we can fix $\varepsilon_\infty = 1$ and fit remaining parameters. In this case, $\omega_p = 1.038 \times 10^{16}$ Hz and $\Gamma = 5.354 \times 10^{13}$ Hz are obtained.

If we are dealing with an arbitrary medium at a single frequency where the real part of the permittivity is negative, we can treat it as a Drude dispersion medium that will be satisfied at that frequency alone. Suppose that the relative permittivity $\varepsilon = \varepsilon' + i\varepsilon''$ at that frequency ω_0 is given by

$$\varepsilon = \varepsilon' + i\varepsilon'' = 1 - \frac{\omega_p^2}{\omega_0^2 + i\Gamma\omega}. \tag{6.55}$$

From this equation,

$$\varepsilon' = 1 - \frac{\omega_p^2}{\omega_0^2 \Gamma^2}, \tag{6.56}$$

$$\varepsilon'' = \frac{\omega_p^2 \Gamma}{\omega_0(\omega_0^2 + \Gamma^2)}, \tag{6.57}$$

are obtained. Using these, we obtain

$$\omega_p = \sqrt{1 - \varepsilon' + \frac{\varepsilon''^2}{1 - \varepsilon'}}\omega_0, \tag{6.58}$$

$$\Gamma = \frac{\varepsilon''}{1 - \varepsilon'}\omega_0, \tag{6.59}$$

where $\varepsilon_\infty = 1$.

The Drude dispersion can also be used when dealing with ENZ (Epsilon Near Zero) medium, etc.

6.3.2 Lorentz dispersion

Here consider the case of Lorentz dispersion. In this case, the susceptibility in the frequency domain is

$$\chi(\omega) = \frac{\Delta\varepsilon\omega_p^2}{\omega_p^2 - 2i\omega\Gamma - \omega^2}. \tag{6.60}$$

Substituting Eq. (6.60) into Eq. (6.41), we obtain

$$\boldsymbol{J}(\omega) = \varepsilon_0 \frac{\Delta\varepsilon\omega_p^2}{\omega_p^2 - 2i\omega\Gamma - \omega^2}\frac{\partial \boldsymbol{E}}{\partial t} \tag{6.61}$$

$$\omega_p^2 \boldsymbol{J}(\omega) - 2i\omega\Gamma \boldsymbol{J}(\omega) - \omega^2 \boldsymbol{J}(\omega) = \varepsilon_0 \Delta\varepsilon \omega_p^2 \frac{\partial \boldsymbol{E}}{\partial t}. \tag{6.62}$$

Expressing the polarization current $\boldsymbol{J}$ in the time domain, we obtain

$$\omega_p^2 \boldsymbol{J} + 2\Gamma \frac{\partial \boldsymbol{J}}{\partial t} + \frac{\partial^2 \boldsymbol{J}}{\partial t^2} = \varepsilon_0 \Delta\varepsilon \omega_p^2 \frac{\partial \boldsymbol{E}}{\partial t}. \tag{6.63}$$

This is the desired ADE.

Discretizing Eq. (6.63),

$$\begin{aligned} &\omega_p^2 \boldsymbol{J}^{n-1} + 2\Gamma \frac{\boldsymbol{J}^n + \boldsymbol{J}^{n-2}}{2\Delta t} + \frac{\boldsymbol{J}^n - 2\boldsymbol{J}^{n-1} + \boldsymbol{J}^{n-2}}{(\Delta t)^2} \\ &= \varepsilon_0 \Delta\varepsilon \omega_p^2 \frac{\boldsymbol{E}^n - \boldsymbol{E}^{n-2}}{2\Delta t} \end{aligned} \tag{6.64}$$

is obtained. Furthermore, solving with respect to $\boldsymbol{J}^n$,

$$\begin{aligned} \boldsymbol{J}^n =& \frac{2 - \omega_p^2 (\Delta t)^2}{1 + \Gamma \Delta t} \boldsymbol{J}^{n-1} + \frac{\Gamma \Delta t - 1}{1 + \Gamma \Delta t} \boldsymbol{J}^{n-2} \\ &+ \frac{\varepsilon_0 \Delta\varepsilon \omega_p^2 \Delta t}{2 + 2\Gamma \Delta t} (\boldsymbol{E}^n - \boldsymbol{E}^{n-2}) \end{aligned} \tag{6.65}$$

is obtained.

As in the case of Drude dispersion, we use the following approximation as

$$\boldsymbol{J}^{n-\frac{1}{2}} = \frac{1}{2}(\boldsymbol{J}^n + \boldsymbol{J}^{n-1}). \tag{6.66}$$

Substituting Eq. (6.66) into Eq. (6.65), we obtain

$$\begin{aligned} \boldsymbol{J}^{n-\frac{1}{2}} =& \frac{1}{2}\left[1 + \frac{2 - \omega_p^2 (\Delta t)^2}{1 + \Gamma \Delta t}\right] \boldsymbol{J}^{n-1} + \frac{1}{2}\left(\frac{\Gamma \Delta t - 1}{1 + \Gamma \Delta t}\right) \boldsymbol{J}^{n-2} \\ &+ \frac{1}{2}\left(\frac{\varepsilon_0 \Delta\varepsilon \omega_p^2 \Delta t}{2 + 2\Gamma \Delta t}\right) (\boldsymbol{E}^n - \boldsymbol{E}^{n-2}). \end{aligned} \tag{6.67}$$

Furthermore, substituting Eq. (6.67) into Eq. (6.48), we obtain

$$\begin{aligned} \nabla \times \boldsymbol{H}^{n-\frac{1}{2}} =& \frac{\varepsilon_0 \varepsilon_\infty}{\Delta t} (\boldsymbol{E}^n - \boldsymbol{E}^{n-1}) + \frac{\sigma}{2} (\boldsymbol{E}^n + \boldsymbol{E}^{n-1}) \\ &+ \frac{1}{2}\left[1 + \frac{2 - \omega_p^2 (\Delta t)^2}{1 + \Gamma \Delta t}\right] \boldsymbol{J}^{n-1} + \frac{1}{2}\left(\frac{\Gamma \Delta t - 1}{1 + \Gamma \Delta t}\right) \boldsymbol{J}^{n-2} \\ &+ \frac{1}{2}\left(\frac{\varepsilon_0 \Delta\varepsilon \omega_p^2 \Delta t}{2 + 2\Gamma \Delta t}\right) (\boldsymbol{E}^n - \boldsymbol{E}^{n-2}). \end{aligned} \tag{6.68}$$

Therefore,

$$
\begin{aligned}
\boldsymbol{E}^n = & \frac{\left[\frac{\varepsilon_0\varepsilon_\infty}{\Delta t} - \frac{\sigma}{2}\right]}{\left[\frac{\varepsilon_0\tilde{\varepsilon}_\infty}{\Delta t} + \frac{\sigma}{2} + \frac{1}{2}\left(\frac{\varepsilon_0\Delta\varepsilon\omega_p^2\Delta t}{2+2\Gamma\Delta t}\right)\right]}\boldsymbol{E}^{n-1} \\
& - \frac{\frac{1}{2}\left(\frac{\varepsilon_0\Delta\varepsilon\omega_p^2\Delta t}{2+2\Gamma\Delta t}\right)}{\left[\frac{\varepsilon_0\varepsilon_\infty}{\Delta t} + \frac{\sigma}{2} + \frac{1}{2}\left(\frac{\varepsilon_0\Delta\varepsilon\omega_p^2\Delta t}{2+2\Gamma\Delta t}\right)\right]}\boldsymbol{E}^{n-2} \\
& + \frac{1}{\left[\frac{\varepsilon_0\varepsilon_\infty}{\Delta t} + \frac{\sigma}{2} + \frac{1}{2}\left(\frac{\varepsilon_0\Delta\varepsilon\omega_p^2\Delta t}{2+2\Gamma\Delta t}\right)\right]} \\
& \times \left\{\boldsymbol{\nabla}\times\boldsymbol{H}^{n-\frac{1}{2}} - \frac{1}{2}\left[1 + \frac{2-\omega_p^2(\Delta t)^2}{1+\Gamma\Delta t}\right]\boldsymbol{J}^{n-1} - \frac{1}{2}\left(\frac{\Gamma\Delta t - 1}{1+\Gamma\Delta t}\right)\boldsymbol{J}^{n-2}\right\}.
\end{aligned}
\tag{6.69}
$$

The calculation procedure is to update $\boldsymbol{E}^n$ using Eq. (6.69) and then update $\boldsymbol{J}^n$ using Eq. (6.65). Writing out the x component of Eq. (6.69),

$$
\begin{aligned}
E_x\big|_{i+\frac{1}{2},j,k}^{n} = & \frac{\left[\frac{\varepsilon_0\varepsilon_\infty}{\Delta t} - \frac{\sigma}{2}\right]}{\left[\frac{\varepsilon_0\varepsilon_\infty}{\Delta t} + \frac{\sigma}{2} + \frac{1}{2}\left(\frac{\varepsilon_0\Delta\varepsilon\omega_p^2\Delta t}{2+2\Gamma\Delta t}\right)\right]} E_x\big|_{i+\frac{1}{2},j,k}^{n-1} \\
& - \frac{\frac{1}{2}\left(\frac{\varepsilon_0\Delta\varepsilon\omega_p^2\Delta t}{2+2\Gamma\Delta t}\right)}{\left[\frac{\varepsilon_0\varepsilon_\infty}{\Delta t} + \frac{\sigma}{2} + \frac{1}{2}\left(\frac{\varepsilon_0\Delta\varepsilon\omega_p^2\Delta t}{2+2\Gamma\Delta t}\right)\right]} E_x\big|_{i+\frac{1}{2},j,k}^{n-2} \\
& + \frac{1/\Delta y}{\left[\frac{\varepsilon_0\varepsilon_\infty}{\Delta t} + \frac{\sigma}{2} + \frac{1}{2}\left(\frac{\varepsilon_0\Delta\varepsilon\omega_p^2\Delta t}{2+2\Gamma\Delta t}\right)\right]} \\
& \times \left(H_z\big|_{i+\frac{1}{2},j+\frac{1}{2},k}^{n-\frac{1}{2}} - H_z\big|_{i+\frac{1}{2},j-\frac{1}{2},k}^{n-\frac{1}{2}}\right) \\
& - \frac{1/\Delta z}{\left[\frac{\varepsilon_0\varepsilon_\infty}{\Delta t} + \frac{\sigma}{2} + \frac{1}{2}\left(\frac{\varepsilon_0\Delta\varepsilon\omega_p^2\Delta t}{2+2\Gamma\Delta t}\right)\right]} \\
& \times \left(H_y\big|_{i+\frac{1}{2},j,k+\frac{1}{2}}^{n-\frac{1}{2}} - H_y\big|_{i+\frac{1}{2},j,k-\frac{1}{2}}^{n-\frac{1}{2}}\right) \\
& - \frac{1}{2}\frac{\left[1 + \frac{2-\omega_p^2(\Delta t)^2}{1+\Gamma\Delta t}\right]}{\left[\frac{\varepsilon_0\varepsilon_\infty}{\Delta t} + \frac{\sigma}{2} + \frac{1}{2}\left(\frac{\varepsilon_0\Delta\varepsilon\omega_p^2\Delta t}{2+2\Gamma\Delta t}\right)\right]} J_x\big|_{i+\frac{1}{2},j,k}^{n-1} \\
& - \frac{1}{2}\frac{\left(\frac{\Gamma\Delta t-1}{1+\Gamma\Delta t}\right)}{\left[\frac{\varepsilon_0\varepsilon_\infty}{\Delta t} + \frac{\sigma}{2} + \frac{1}{2}\left(\frac{\varepsilon_0\Delta\varepsilon\omega_p^2\Delta t}{2+2\Gamma\Delta t}\right)\right]} J_x\big|_{i+\frac{1}{2},j,k}^{n-2}.
\end{aligned}
\tag{6.70}
$$

As can be seen from this equation and from the Eq. (6.65), the Lorentz dispersion has a second-order pole, so the time evolution requires the values of $\boldsymbol{J}$ and $\boldsymbol{E}$ not only before Δt, but also before $2\Delta t$.

6.4 Perfectly matched layer (PML) absorbing boundary

Since the memory of a computer is finite, the computational domain is finite as well. Therefore, if the periodic boundary condition is not used, the computational domain will have an end face. At this end, one of the points for the calculation of the difference does not exist, so the central difference cannot be performed. Therefore, a perfect conductor must be used for the end face. However, the perfect conductor produces a reflection of 100%, which is a source of large error. Therefore, the electromagnetic field must be sufficiently attenuated before reaching the perfect conductor. Here, we describe the Perfectly Matched Layer (PML) absorption boundary proposed by Berenger [40], which is the most commonly used solution to this problem.

6.4.1 Split field PML

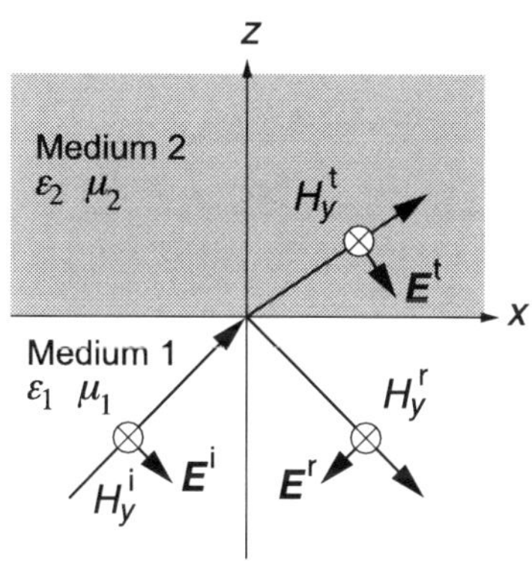

FIGURE 6.8
Incidence of electromagnetic waves from Medium 1 to Medium 2.

Consider the incidence from lossless Medium 1 to lossy Medium 2 (the case where Medium 1 also has losses will be discussed later), as shown in Figure 6.8. The electromagnetic wave in Medium 2 attenuates as it propagates. However, as mentioned above, in order to use this medium as an absorbing boundary, it is important that no reflection occurs at the boundary between Medium 1 and Medium 2. Consider an electromagnetic wave with TE$_y$ polarization that is uniform in the y-direction and whose direction of propagation is in the xz plane (in FDTD, the notation TE$_y$ and TM$_x$ are often used to describe polarization. The former means that the electric field is perpendicular to the y-axis and the latter means that the magnetic field is perpendicular to the x-axis). Suppose the interface is given by $z = z_0$. Maxwell's equations to be satisfied by this electromagnetic wave in Medium 1 are given by

$$\varepsilon_0\varepsilon_1\frac{\partial E_x}{\partial t} = -\frac{\partial H_y}{\partial z}, \tag{6.71}$$

$$\varepsilon_0\varepsilon_1\frac{\partial E_z}{\partial t} = \frac{\partial H_y}{\partial x}, \tag{6.72}$$

$$\mu_0\mu_1\frac{\partial H_y}{\partial t} = -\frac{\partial E_x}{\partial z} + \frac{\partial E_z}{\partial x}. \tag{6.73}$$

On the other hand, Maxwell's equations that this electromagnetic wave must satisfy in Medium 2 are given by the following equations:

$$\varepsilon_0\varepsilon_2\frac{\partial E_x}{\partial t} + \sigma_z E_x = -\frac{\partial H_y}{\partial z}, \tag{6.74}$$

$$\varepsilon_0\varepsilon_2\frac{\partial E_z}{\partial t} + \sigma_x E_z = \frac{\partial H_y}{\partial x}, \tag{6.75}$$

$$\mu_0\mu_2\frac{\partial H_y}{\partial t} + \sigma^* H_y = -\frac{\partial E_x}{\partial z} + \frac{\partial E_z}{\partial x}, \tag{6.76}$$

where σ^* is the magnetic conductivity. Here, let the magnetic field H_y be the sum of the two components H_{yx} and H_{yz},

$$H_y = H_{yx} + H_{yz}. \tag{6.77}$$

Additionally, H_{yx} and H_{yz} are taken to satisfy the following two equations,

$$\mu_0\mu_2\frac{\partial H_{yx}}{\partial t} + \sigma_x^* H_{yx} = \frac{\partial E_z}{\partial x}, \tag{6.78}$$

$$\mu_0\mu_2\frac{\partial H_{yz}}{\partial t} + \sigma_z^* H_{yz} = -\frac{\partial E_x}{\partial z}. \tag{6.79}$$

Substituting Eq. (6.77) into Eqs. (6.74) and (6.75), and considering the time dependency of $\exp(-i\omega t)$, we obtain

$$-i\omega\varepsilon_0\varepsilon_2 E_x + \sigma_z E_x = -\frac{\partial}{\partial z}(H_{yx} + H_{yz}), \tag{6.80}$$

$$-i\omega\varepsilon_0\varepsilon_2 E_z + \sigma_x E_z = \frac{\partial}{\partial x}(H_{yx} + H_{yz}). \tag{6.81}$$

Similarly, from Eqs. (6.78) and (6.79), we obtain

$$-i\omega\mu_0\mu_2 H_{yx} + \sigma_x^* H_{yx} = \frac{\partial E_z}{\partial x}, \tag{6.82}$$

$$-i\omega\mu_0\mu_2 H_{yz} + \sigma_z^* H_{yz} = -\frac{\partial E_x}{\partial z}. \tag{6.83}$$

Here, let

$$s_\nu = 1 - \frac{\sigma_\nu}{i\omega\varepsilon_0\varepsilon_2}, \tag{6.84}$$

$$s_\nu^* = 1 - \frac{\sigma_\nu^*}{i\omega\mu_0\mu_2}, \tag{6.85}$$

where $\nu = x$ or z. Using Eqs. (6.84) and (6.85), Eqs. (6.80)–(6.83) become

$$-i\omega\varepsilon_0\varepsilon_2 s_z E_x = -\frac{\partial}{\partial z}(H_{yx} + H_{yz}), \tag{6.86}$$

$$-i\omega\varepsilon_0\varepsilon_2 s_x E_z = \frac{\partial}{\partial x}(H_{yx} + H_{yz}), \tag{6.87}$$

$$-i\omega\mu_0\mu_2 s_x^* H_{yx} = \frac{\partial E_z}{\partial x}, \tag{6.88}$$

$$-i\omega\mu_0\mu_2 s_z^* H_{yz} = -\frac{\partial E_x}{\partial z}. \tag{6.89}$$

Next, we derive the wave equation from these equations. Partial differentiation of Eq. (6.86) with respect to z and substitution into Eq. (6.89) gives

$$-\omega^2\varepsilon_0\varepsilon_2\mu_0\mu_2 s_z s_z^* H_{yz} = \frac{\partial^2}{\partial z^2}(H_{yx} + H_{yz}). \tag{6.90}$$

Similarly, by partial differentiation of Eq. (6.87) with respect to z and substituting into Eq. (6.88), we obtain

$$-\omega^2\varepsilon_0\varepsilon_2\mu_0\mu_2 s_x s_x^* H_{yx} = \frac{\partial^2}{\partial x^2}(H_{yx} + H_{yz}). \tag{6.91}$$

Summing side by side of Eqs. (6.90) and (6.91), we obtain

$$\begin{aligned} &-\omega^2\varepsilon_0\varepsilon_2\mu_0\mu_2(H_{yx} + H_{yz}) \\ &= \left(\frac{1}{s_z s_z^*}\frac{\partial^2}{\partial z^2} + \frac{1}{s_x s_x^*}\frac{\partial^2}{\partial x^2}\right)(H_{yx} + H_{yz}). \end{aligned} \tag{6.92}$$

Now, once again using Eq. (6.77), we obtain

$$\left(\frac{1}{s_x s_x^*}\frac{\partial^2}{\partial x^2} + \frac{1}{s_z s_z^*}\frac{\partial^2}{\partial z^2} + \omega^2\varepsilon_0\varepsilon_2\mu_0\mu_2\right) H_y = 0. \tag{6.93}$$

This is the wave equation of the electromagnetic wave in Medium 2.

The magnetic field component H_y^t of the plane wave solution satisfying this equation is expressed as

$$H_y^t = tH_y^i \exp(ik_{2x}x + ik_{2z}z), \tag{6.94}$$

where H_y^i is the amplitude of the incident magnetic field and t is the transmission coefficient. Wave numbers k_{2x} and k_{2z} satisfy the dispersion relation for plane waves given by

$$\frac{k_{2x}^2}{s_x s_x^*} + \frac{k_{2z}^2}{s_z s_z^*} = \omega^2\varepsilon_0\varepsilon_2\mu_0\mu_2. \tag{6.95}$$

Using Eqs. (6.77), (6.86), (6.94), and (6.95), the x component of the transmitted electric field is given as

$$E_x^t = \frac{\beta_{2z}}{\omega\varepsilon_0\varepsilon_2}\sqrt{\frac{s_z^*}{s_z}}H_y^t. \tag{6.96}$$

On the other hand, the reflected magnetic field H_y^r in Medium 1 is given as

$$H_y^r = rH_y^i \exp(ik_{1x}x - ik_{1z}z), \tag{6.97}$$

where r is the reflection coefficient. Using the Eq. (6.97), the x component of the reflected electric field is

$$E_x^r = -\frac{k_{1z}}{\omega\varepsilon_0\varepsilon_1}H_y^r. \tag{6.98}$$

From the continuity of H_y at the interface and Eqs. (6.94) and (6.97), we obtain

$$1 + r = t. \tag{6.99}$$

Similarly, from the continuity of E_x and Eqs. (6.96) and (6.98), we obtain

$$\frac{k_{1z}}{\omega\varepsilon_1} - \frac{k_{1z}}{\omega\varepsilon_1}r = \frac{k_{2z}}{\omega\varepsilon_2 s_z}t. \tag{6.100}$$

That is,

$$1 - r = \frac{\varepsilon_1 k_{2z}}{\varepsilon_2 s_z k_{1z}}t \tag{6.101}$$

is obtained. From Eqs. (6.99) and (6.101), the transmission coefficient t and reflection coefficient r are given as

$$t = \frac{2\dfrac{k_{1z}}{\varepsilon_1}}{\dfrac{k_{1z}}{\varepsilon_1} + \dfrac{k_{2z}}{\varepsilon_2 s_z}}, \tag{6.102}$$

$$r = \frac{\dfrac{k_{1z}}{\varepsilon_1} - \dfrac{k_{2z}}{\varepsilon_2 s_z}}{\dfrac{k_{1z}}{\varepsilon_1} + \dfrac{k_{2z}}{\varepsilon_2 s_z}}. \tag{6.103}$$

Next, we derive the condition that the reflection coefficient r is zero. First, let $\varepsilon_2 = \varepsilon_1$ and $\mu_2 = \mu_1$. From Eq. (6.95),

$$\begin{aligned} k_{2z} &= \left(\omega^2\varepsilon_0\varepsilon_2\mu_0\mu_2 - \frac{s_z s_z*}{s_x s_x^*}k_{2x}^2\right)^{1/2} \\ &= \left(\omega^2\varepsilon_0\varepsilon_1\mu_0\mu_1 - \frac{s_z s_z*}{s_x s_x^*}k_{2x}^2\right)^{1/2} \end{aligned} \tag{6.104}$$

is obtained. Also, from the phase matching condition at the interface,

$$k_{2x} = k_{1x} \tag{6.105}$$

must be satisfied. Furthermore, if $\sigma_x = \sigma_x^* = 0$, then $s_x = s_x^* = 1$. In this condition, from Eq. (6.104)

$$k_{2z} = \sqrt{s_z s_z^*}(\omega^2 \varepsilon_0 \varepsilon_1 \mu_0 \mu_1 - k_{1x}^2)^{1/2} = \sqrt{s_z s_z^*} k_{1z} \tag{6.106}$$

is obtained. Using these relationships and Eq. (6.103), the reflection coefficient is

$$r = \frac{1 - \sqrt{\dfrac{s_z^*}{s_z}}}{1 + \sqrt{\dfrac{s_z^*}{s_z}}}. \tag{6.107}$$

Therefore, for the reflection coefficient to be zero,

$$s_z = s_z^* \tag{6.108}$$

must be satisfied. From the Eqs. (6.84) and (6.85), this condition is realized when the following equation is satisfied:

$$\frac{\sigma_z}{\varepsilon_0 \varepsilon_2} = \frac{\sigma_z^*}{\mu_0 \mu_2}. \tag{6.109}$$

In summary, the condition for no reflection at the interface is

$$\varepsilon_2 = \varepsilon_1, \tag{6.110}$$

$$\mu_2 = \mu_1, \tag{6.111}$$

$$\sigma_x = \sigma_x^* = 0, \tag{6.112}$$

$$\frac{\sigma_z}{\varepsilon_0 \varepsilon_2} = \frac{\sigma_z^*}{\mu_0 \mu_2}. \tag{6.113}$$

In this condition, from Eq. (6.106), we obtain

$$k_{z2} = \left(1 - \frac{\sigma_z}{i\omega \varepsilon_0 \varepsilon_1}\right) k_{1z}. \tag{6.114}$$

As a result, the transmitted magnetic field in Medium 2 (PML) is given by

$$H_y^t = H_y^i \exp\left(-\frac{\sigma_z}{\omega \mu_0 \mu_1} k_{1z} z\right) \exp(ik_{1x}x + ik_{1z}z). \tag{6.115}$$

This equation shows that the transmitted wave propagates at the same phase velocity as the incident wave and decays along the z-direction. Thus, Medium 2 is used as the PML.

When the incident wave is propagating light, there is no problem with attenuation, but let us consider what happens when the incident wave is an evanescent one. In this case, k_{z1} is a pure imaginary number. If $k_{1z} = i|k_{1z}|$, Eq. (6.114) is expressed as

$$k_{2z} = -\frac{\sigma_z}{\omega\varepsilon_0}|k_{1z}| + i|k_{1z}|, \tag{6.116}$$

and that the attenuation of the transmitted wave in Medium 2 is no greater than that of the incident evanescent wave in Medium 1. A solution to solve this problem has been proposed by Gedney [41]. He used

$$s_z = \kappa - \frac{\sigma_z}{i\omega\varepsilon_0\varepsilon_1}, \tag{6.117}$$

instead of Equation (6.84). With this s_z, Eq. (6.116) becomes

$$k_{2z} = -\frac{\sigma_z}{\omega\varepsilon_0\varepsilon_1}|k_{1z}| + i\kappa|k_{1z}|. \tag{6.118}$$

That is, the decay of the evanescent wave is accelerated by a factor of κ. However, if κ is too large, it causes side effects for the propagating light.

We will discuss the general case of PML where the system is 3D. All electric and magnetic fields are divided into two components. For the electric field,

$$E_x = E_{xy} + E_{xz}, \tag{6.119}$$

$$E_y = E_{yz} + E_{yx}, \tag{6.120}$$

$$E_z = E_{zx} + E_{zy}, \tag{6.121}$$

and with respect to the magnetic field,

$$H_x = H_{xy} + H_{xz}, \tag{6.122}$$

$$H_y = H_{yz} + H_{yx}, \tag{6.123}$$

$$H_z = H_{zx} + H_{zy}. \tag{6.124}$$

The 12 basic equations of PML are

$$\varepsilon_0\varepsilon_2\frac{\partial E_{xy}}{\partial t} + \sigma_y E_{xy} = \frac{\partial H_z}{\partial y}, \tag{6.125}$$

$$\varepsilon_0\varepsilon_2\frac{\partial E_{xz}}{\partial t} + \sigma_z E_{xz} = -\frac{\partial H_y}{\partial z}, \tag{6.126}$$

$$\varepsilon_0\varepsilon_2\frac{\partial E_{yz}}{\partial t} + \sigma_z E_{yz} = \frac{\partial H_x}{\partial z}, \tag{6.127}$$

$$\varepsilon_0\varepsilon_2\frac{\partial E_{yx}}{\partial t} + \sigma_x E_{yx} = -\frac{\partial H_z}{\partial x}, \tag{6.128}$$

$$\varepsilon_0\varepsilon_2\frac{\partial E_{zx}}{\partial t} + \sigma_x E_{zx} = \frac{\partial H_y}{\partial x}, \tag{6.129}$$

$$\varepsilon_0\varepsilon_2\frac{\partial E_{zy}}{\partial t}+\sigma_y E_{zy}=-\frac{\partial H_x}{\partial y}, \tag{6.130}$$

$$\mu_0\mu_2\frac{\partial H_{xy}}{\partial t}+\sigma_y^* H_{xy}=-\frac{\partial E_z}{\partial y}, \tag{6.131}$$

$$\mu_0\mu_2\frac{\partial H_{xz}}{\partial t}+\sigma_z^* H_{xz}=\frac{\partial E_y}{\partial z}, \tag{6.132}$$

$$\mu_0\mu_2\frac{\partial H_{yz}}{\partial t}+\sigma_z^* H_{yz}=-\frac{\partial E_z}{\partial x}, \tag{6.133}$$

$$\mu_0\mu_2\frac{\partial H_{yx}}{\partial t}+\sigma_x^* H_{yx}=\frac{\partial E_x}{\partial z}, \tag{6.134}$$

$$\mu_0\mu_2\frac{\partial H_{zx}}{\partial t}+\sigma_x^* H_{zx}=-\frac{\partial E_y}{\partial x}, \tag{6.135}$$

$$\mu_0\mu_2\frac{\partial H_{zy}}{\partial t}+\sigma_y^* H_{zy}=\frac{\partial E_x}{\partial y}. \tag{6.136}$$

The conditions for no reflection are $\varepsilon_2=\varepsilon_1$ and $\mu_2=\mu_1$, and in the PML perpendicular to the x-axis,

$$\frac{\sigma_x}{\varepsilon_0\varepsilon_2}=\frac{\sigma_x^*}{\mu_0\mu_2},\quad \sigma_y=\sigma_z=\sigma_y^*=\sigma_z^*=0. \tag{6.137}$$

In the PML perpendicular to the y-axis,

$$\frac{\sigma_y}{\varepsilon_0\varepsilon_2}=\frac{\sigma_y^*}{\mu_0\mu_2},\quad \sigma_z=\sigma_x=\sigma_z^*=\sigma_x^*=0. \tag{6.138}$$

In the PML perpendicular to the z-axis,

$$\frac{\sigma_z}{\varepsilon_0\varepsilon_2}=\frac{\sigma_z^*}{\mu_0\mu_2},\quad \sigma_x=\sigma_y=\sigma_x^*=\sigma_y^*=0. \tag{6.139}$$

In the actual calculation, Eqs. (6.125)–(6.136) must be discretized. As an example, if we discretize Eq. (6.125),

$$\begin{aligned}E_{xy}\big|^{n}_{i+\frac{1}{2},j,k}=&\left(\frac{2\varepsilon_0\varepsilon_2-\sigma_y\Delta t}{2\varepsilon_0\varepsilon_2+\sigma_y\Delta t}\right)E_{xy}\big|^{n-1}_{i+\frac{1}{2},j,k}\\&+\left[\frac{2\Delta t}{(2\varepsilon_0\varepsilon_2+\sigma_y\Delta t)\Delta y}\right]\left(H_z\big|^{n-\frac{1}{2}}_{i+\frac{1}{2},j+\frac{1}{2},k}-H_z\big|^{n-\frac{1}{2}}_{i+\frac{1}{2},j-\frac{1}{2},k}\right).\end{aligned} \tag{6.140}$$

6.4.2 Unsplit PML

Berenger's PML is also known as split-field PML. Although this absorption boundary is very effective, the electromagnetic waves in the PML do not obey Maxwell's equations and are difficult to explain physically. In contrast, Chew and Weedon [42] proposed a PML that does not split the field.

From Eq. (6.85) and the PML conditional Eq. (6.108), Eqs. (6.78) and (6.79) become

$$\mu_0\mu_2\frac{\partial H_{yx}}{\partial t} = \frac{1}{s_x}\frac{\partial E_z}{\partial x} \tag{6.141}$$

$$\mu_0\mu_2\frac{\partial H_{yz}}{\partial t} = -\frac{1}{s_z}\frac{\partial E_x}{\partial z}. \tag{6.142}$$

Adding together Eqs. (6.141) and (6.142) on both sides, we obtain

$$\mu_0\mu_2\frac{\partial H_y}{\partial t} = -\frac{1}{s_z}\frac{\partial E_x}{\partial z} + \frac{1}{s_x}\frac{\partial E_z}{\partial x}. \tag{6.143}$$

The same is true for E,

$$\varepsilon_0\varepsilon_2\frac{\partial E_y}{\partial t} = \frac{1}{s_z}\frac{\partial H_x}{\partial z} - \frac{1}{s_x}\frac{\partial H_z}{\partial x}. \tag{6.144}$$

These are Faraday and Ampere's laws within PML. Since $1/s_z$ and $1/s_x$ on the right-hand side are equivalent to stretching the coordinate system, these expressions are called Stretched-Coordinate Formulation.

These equations can be easily extended to three dimensions. Using Eqs. (6.125)–(6.136), we obtain

$$\varepsilon_0\varepsilon_2\frac{\partial E_x}{\partial t} = \frac{1}{s_y}\frac{\partial H_z}{\partial y} - \frac{1}{s_z}\frac{\partial H_y}{\partial z} \tag{6.145}$$

$$\varepsilon_0\varepsilon_2\frac{\partial E_y}{\partial t} = \frac{1}{s_z}\frac{\partial H_x}{\partial z} - \frac{1}{s_x}\frac{\partial H_z}{\partial x} \tag{6.146}$$

$$\varepsilon_0\varepsilon_2\frac{\partial E_z}{\partial t} = \frac{1}{s_x}\frac{\partial H_y}{\partial x} - \frac{1}{s_y}\frac{\partial H_x}{\partial y} \tag{6.147}$$

$$\mu_0\mu_2\frac{\partial H_x}{\partial t} = -\frac{1}{s_y}\frac{\partial E_z}{\partial y} + \frac{1}{s_z}\frac{\partial E_y}{\partial z} \tag{6.148}$$

$$\mu_0\mu_2\frac{\partial H_y}{\partial t} = -\frac{1}{s_z}\frac{\partial E_x}{\partial z} + \frac{1}{s_x}\frac{\partial E_z}{\partial x} \tag{6.149}$$

$$\mu_0\mu_2\frac{\partial H_z}{\partial t} = -\frac{1}{s_x}\frac{\partial E_y}{\partial x} + \frac{1}{s_y}\frac{\partial E_x}{\partial y}. \tag{6.150}$$

For example, the condition,

$$s_x = s_y = 1, s_z \neq 1 \tag{6.151}$$

is required for a PML layer to be perpendicular to the z-axis.

On the other hand, Gedney [43] showed that by defining the medium in the PML as a medium with uniaxial anisotropy, the effect is similar to Berenger's PML. This PML is called Uniaxial PML (UPML). The UPML is exactly the same as Berenger's PML in the component of the electromagnetic field parallel to the interface, but there is a slight difference between these two in the perpendicular component. Gedney [41] also shows that UPML can be adapted to lossy and/or dispersive media.

6.4.3 Convolutional PML (CPML)

An efficient way to apply Unsplit PML to FDTD for dispersive media is proposed by Roden and Gedney [44]. This method is called Convolutional PML (CPML). To apply Unsplit PML to FDTD, we express Equations (6.145) and (6.148) in the time domain using the convolution theorem:

$$\varepsilon_0\varepsilon_2\frac{\partial E_x}{\partial t} = \bar{s}_y * \frac{\partial H_z}{\partial y} - \bar{s}_z * \frac{\partial H_y}{\partial z} \tag{6.152}$$

$$\mu_0\mu_2\frac{\partial H_x}{\partial t} = -\bar{s}_y * \frac{\partial E_z}{\partial y} + \bar{s}_z\frac{\partial E_y}{\partial z}, \tag{6.153}$$

where $*$ denotes the convolution integral, and

$$\bar{s}_\nu = \mathcal{F}^{-1}\left(\frac{1}{s_\nu}\right) \tag{6.154}$$

where $\mathcal{F}^{-1}$ denotes the inverse Fourier transform. Here we use s_ν as the following generalized form [45] as

$$s_\nu = \kappa_\nu + \frac{\sigma_\nu}{a_\nu - i\omega\varepsilon_0}. \tag{6.155}$$

The PML using this form is called the Complex Frequency Shifted PML (CFS-PML). Parameter a_ν is introduced to prevent the imaginary part of s_ν from diverging in the low frequency region ($\omega \sim 0$). However, the larger a_ν is, the smaller the attenuation in the low-frequency region. Before performing the inverse Fourier transform of $1/s_\nu$, we transform Eq. (6.155) to

$$\frac{1}{s_\nu} = \frac{1}{\kappa_\nu + \frac{\sigma_\nu}{a_\nu - i\omega\varepsilon_0}} = \frac{a_\nu + i\omega\varepsilon_0}{a_\nu\kappa_\nu + \sigma_\nu - i\omega\kappa_\nu\varepsilon_0}. \tag{6.156}$$

Here, we consider the following general form:

$$\begin{aligned}\frac{a - i\omega b}{c - i\omega d} &= \frac{ad - i\omega bd}{d(c - i\omega d)} \\ &= \frac{b(c - i\omega d) + ad - bc}{d(c - i\omega d)} \\ &= \frac{b}{d} + \frac{a/c - b/d}{1 - i\omega d/c}\end{aligned} \tag{6.157}$$

and

$$\mathcal{F}^{-1}\left(\frac{1}{1 - i\omega\tau}\right) = \frac{1}{\tau}\exp\left(-\frac{t}{\tau}\right)u(t), \tag{6.158}$$

where $u(t)$ is a step function,

$$u(t) = \begin{cases} 0 & (t < 0) \\ 1 & (t \geq 0) \end{cases}. \tag{6.159}$$

Substituting Eq. (6.157) into Eq. (6.158),

$$\mathcal{F}^{-1}\left(\frac{b}{d}+\frac{a/c-b/d}{1-i\omega d/c}\right)=\frac{b}{d}\delta(t)+\frac{ad-bc}{d^2}\exp\left(-\frac{ct}{d}\right)u(t), \tag{6.160}$$

where $\delta(t)$ is the Dirac's delta function. Furthermore, substituting $a=a_\nu$, $b=\varepsilon_0$, $c=a_\nu\kappa_\nu+\sigma_\nu$, and $d=\kappa\varepsilon_0$, eventually we obtain

$$\bar{s}_\nu=\frac{1}{\kappa_\nu}\delta(t)-\zeta_\nu(t), \tag{6.161}$$

where

$$\zeta_\nu(t)=-\frac{\sigma_\nu}{\kappa_\nu^2\varepsilon_0}\exp\left[-\left(\frac{a_\nu}{\varepsilon_0}+\frac{\sigma_\nu}{\kappa_\nu\varepsilon_0}\right)t\right]u(t). \tag{6.162}$$

Substituting Eq. (6.161) into Eq. (6.152),

$$\varepsilon_0\varepsilon_2\frac{\partial E_x}{\partial t}=\frac{1}{\kappa_y}\frac{\partial H_z}{\partial y}-\frac{1}{\kappa_z}\frac{\partial H_y}{\partial z}+\zeta_y(t)*\frac{\partial H_z}{\partial y}-\zeta_z(t)*\frac{\partial H_y}{\partial z} \tag{6.163}$$

is obtained. Using the following notation,

$$\Psi_{Exy}(t)=\zeta_y(t)*\frac{\partial H_z}{\partial y}, \tag{6.164}$$

$$\Psi_{Exz}(t)=\zeta_z(t)*\frac{\partial H_y}{\partial z}, \tag{6.165}$$

Eq. (6.163) is written as

$$\varepsilon_0\varepsilon_2\frac{\partial E_x}{\partial t}=\frac{1}{\kappa_y}\frac{\partial H_z}{\partial y}-\frac{1}{\kappa_z}\frac{\partial H_y}{\partial z}+\Psi_{Exy}-\Psi_{Exz}. \tag{6.166}$$

6.4.4 Recursive computation for convolution integrals

Since $\zeta(t)$ is an exponential function, the convolution integral of the above equations can be computed recursively in FDTD. Consider the convolution integral shown in the following equation:

$$G(t)=F(t)*\chi(t)=\chi(t)*F(t)=\int_{-\infty}^{\infty}F(t-\tau)\chi(\tau)d\tau. \tag{6.167}$$

Then, letting $\chi(t)$ be a function of the form

$$\chi(t)=a\exp(-\Gamma t)u(t), \tag{6.168}$$

the integral range of Eq. (6.167) is from zero. Furthermore, discretizing Eq. (6.167) with respect to time t, we obtain

$$G^n = \int_0^{n\Delta t} F(n\Delta t - \tau)\chi(\tau)d\tau. \tag{6.169}$$

Discretizing also for τ, we obtain

$$\begin{aligned} G^n &= \sum_{m=0}^{n-1} F^{n-m}\chi^m \\ &= F^n\chi^0 + \sum_{m=1}^{n-1} F^{n-m}\chi^m \\ &= F^n\chi^0 + \sum_{m=0}^{n-2} F^{n-1-m}\chi^{m+1}, \end{aligned} \tag{6.170}$$

where

$$\chi^m = \int_{m\Delta t}^{(m+1)\Delta t} \chi(\tau)d\tau. \tag{6.171}$$

Substituting Eq. (6.168) into Eq. (6.171), we obtain

$$\begin{aligned} \chi^m &= \int_{m\Delta t}^{(m+1)\Delta t} a\exp(-\Gamma\tau)d\tau \\ &= \frac{a}{\Gamma}[1 - \exp(-\Gamma\Delta t)]\exp(-\Gamma m\Delta t) \end{aligned} \tag{6.172}$$

Therefore,

$$\chi^{m+1} = \exp(-\Gamma\Delta t)\chi^m. \tag{6.173}$$

Using Eq. (6.173), Eq. (6.170) finally becomes

$$\begin{aligned} G^n &= F^n\chi^0 + \sum_{m=0}^{n-2} F^{n-1-m}\chi^{m+1} \\ &= F^n\chi^0 + \exp(-\Gamma\Delta t)\sum_{m=0}^{n-2} F^{n-1-m}\chi^m \\ &= F^n\chi^0 + \exp(-\Gamma\Delta t)G^{n-1}, \end{aligned} \tag{6.174}$$

where

$$\chi^0 = \frac{b}{\Gamma}[1 - \exp(-\Gamma\Delta t)]. \tag{6.175}$$

Using this result, Ψ_{Exy} in Eq. (6.164) can be calculated recursively as follows:

$$\Psi_{Exy}^n = b_y\Psi_{Exy}^{n-1} + c_y \left.\frac{\partial H_z}{\partial y}\right|^n. \tag{6.176}$$

Comparing Eq. (6.162) with Eqs. (6.174), we obtain

$$b_y = \exp\left[-\left(\frac{a_y}{\varepsilon_0} + \frac{\sigma_y}{\kappa_y \varepsilon_0}\right)\Delta t\right] \tag{6.177}$$

and from the Eq. (6.175),

$$\begin{aligned} c_y =& \frac{-\frac{\sigma_y}{\kappa_y^2 \varepsilon_0}}{\frac{a_y}{\varepsilon_0} + \frac{\sigma_y}{\kappa_\nu \varepsilon_0}} \left\{1 - \exp\left[-\left(\frac{a_y}{\varepsilon_0} + \frac{\sigma_y}{\kappa_y \varepsilon_0}\right)\Delta t\right]\right\} \\ =& -\frac{\sigma_y}{\sigma_y \kappa_y + a_y \kappa_y^2}(1 - b_y) \end{aligned} \tag{6.178}$$

and so on.

Using this result to discretize Eq. (6.166), we obtain

$$\begin{aligned} &\frac{\varepsilon_0 \varepsilon_2}{\Delta t}\left(E_x\Big|_{i+\frac{1}{2},j,k}^{n} - E_x\Big|_{i+\frac{1}{2},j,k}^{n-1}\right) \\ &= \frac{1}{\kappa_y \Delta y}\left(H_z\Big|_{i+\frac{1}{2},j+\frac{1}{2},k}^{n-\frac{1}{2}} - H_z\Big|_{i+\frac{1}{2},j-\frac{1}{2},k}^{n-\frac{1}{2}}\right) \\ &- \frac{1}{\kappa_z \Delta z}\left(H_y\Big|_{i+\frac{1}{2},j,k+\frac{1}{2}}^{n-\frac{1}{2}} - H_y\Big|_{i+\frac{1}{2},j,k-\frac{1}{2}}^{n-\frac{1}{2}}\right) \\ &+ \left(\Psi_{Exy}\Big|_{i+\frac{1}{2},j,k}^{n} - \Psi_{Exz}\Big|_{i+\frac{1}{2},j,k}^{n}\right), \end{aligned} \tag{6.179}$$

$$\begin{aligned} E_x\Big|_{i+\frac{1}{2},j,k}^{n} =& E_x\Big|_{i+\frac{1}{2},j,k}^{n-1} \\ &+ \frac{\Delta t}{\varepsilon_0 \varepsilon_2 \kappa_y \Delta y}\left(H_z\Big|_{i+\frac{1}{2},j+\frac{1}{2},k}^{n-\frac{1}{2}} - H_z\Big|_{i+\frac{1}{2},j-\frac{1}{2},k}^{n-\frac{1}{2}}\right) \\ &- \frac{\Delta t}{\varepsilon_0 \varepsilon_2 \kappa_z \Delta z}\left(H_y\Big|_{i+\frac{1}{2},j,k+\frac{1}{2}}^{n-\frac{1}{2}} - H_y\Big|_{i+\frac{1}{2},j,k-\frac{1}{2}}^{n-\frac{1}{2}}\right) \\ &+ \frac{\Delta t}{\varepsilon_0 \varepsilon_2}\left(\Psi_{Exy}\Big|_{i+\frac{1}{2},j,k}^{n} - \Psi_{Exz}\Big|_{i+\frac{1}{2},j,k}^{n}\right), \end{aligned} \tag{6.180}$$

where

$$\begin{aligned} \Psi_{Exy}\Big|_{i+\frac{1}{2},j,k}^{n} =& b_y \Psi_{Exy}\Big|_{i+\frac{1}{2},j,k}^{n-1} \\ &+ \frac{c_y}{\Delta y}\left(H_z\Big|_{i+\frac{1}{2},j+\frac{1}{2},k}^{n-\frac{1}{2}} - H_z\Big|_{i+\frac{1}{2},j-\frac{1}{2},k}^{n-\frac{1}{2}}\right), \end{aligned} \tag{6.181}$$

$$\begin{aligned} \Psi_{Exz}\Big|_{i+\frac{1}{2},j,k}^{n} =& b_z \Psi_{Exz}\Big|_{i+\frac{1}{2},j,k}^{n-1} \\ &+ \frac{c_z}{\Delta z}\left(H_y\Big|_{i+\frac{1}{2},j,k+\frac{1}{2}}^{n-\frac{1}{2}} - H_z\Big|_{i+\frac{1}{2},j,k-\frac{1}{2}}^{n-\frac{1}{2}}\right). \end{aligned} \tag{6.182}$$

6.4.5 Lossy media

The CPML can be applied in the same way for lossy media. In this case, Eq. (6.163) is modified as

$$\varepsilon_0\varepsilon\frac{\partial E_x}{\partial t}+\sigma E_x=\frac{1}{\kappa_y}\frac{\partial H_z}{\partial y}-\frac{1}{\kappa_z}\frac{\partial H_y}{\partial z}+\zeta_y(t)*\frac{\partial H_z}{\partial y}-\zeta_z(t)*\frac{\partial H_y}{\partial z} \tag{6.183}$$

Discretizing Eq. (6.183) with respect to time, we obtain

$$\begin{aligned}&\varepsilon_0\varepsilon\frac{E_x^n-E_x^{n-1}}{\Delta t}+\sigma\frac{E_x^n+E_x^{n-1}}{2}\\&=\frac{1}{\kappa_y}\frac{\partial H_z^{n-\frac{1}{2}}}{\partial y}-\frac{1}{\kappa_z}\frac{\partial H_y^{n-\frac{1}{2}}}{\partial z}+\zeta_y(t)*\frac{\partial H_z^{n-\frac{1}{2}}}{\partial y}-\zeta_z(t)*\frac{\partial H_y^{n-\frac{1}{2}}}{\partial z},\end{aligned} \tag{6.184}$$

and then

$$\begin{aligned}E_x^n=&\left(\frac{2\varepsilon_0\varepsilon-\sigma\Delta t}{2\varepsilon_0\varepsilon+\sigma\Delta t}\right)E_x^{n-1}+\left(\frac{2\Delta t}{2\varepsilon_0\varepsilon+\sigma\Delta t}\right)\\&\times\left[\frac{1}{\kappa_y}\frac{\partial H_z^{n-\frac{1}{2}}}{\partial y}-\frac{1}{\kappa_z}\frac{\partial H_y^{n-\frac{1}{2}}}{\partial z}+\zeta_y(t)*\frac{\partial H_z^{n-\frac{1}{2}}}{\partial y}-\zeta_z(t)*\frac{\partial H_y^{n-\frac{1}{2}}}{\partial z}\right].\end{aligned} \tag{6.185}$$

6.4.5.1 Dispersive media

The CPML is nothing more than replacing the $\nabla\times\boldsymbol{H}$ calculation with a calculation involving convolution integrals. Take the x component as an example,

$$(\nabla\times\boldsymbol{H})_x\rightarrow\frac{1}{\kappa_y}\frac{\partial H_z}{\partial y}-\frac{1}{\kappa_z}\frac{\partial H_y}{\partial z}+\zeta_y(t)*\frac{\partial H_z}{\partial y}-\zeta_z(t)*\frac{\partial H_y}{\partial z}. \tag{6.186}$$

For a Drude dispersive medium using the ADE method, just replace $\nabla\times\boldsymbol{H}$ in Eq. (6.52) (reproduced below) as Eq. (6.186), and we obtain

$$\begin{aligned}\boldsymbol{E}^n=&\frac{\left[\frac{\varepsilon_0\tilde{\varepsilon}_\infty}{\Delta t}-\frac{\sigma}{2}-\frac{1}{2}\left(\frac{\varepsilon_0\omega_p^2\Delta t/2}{1+\Gamma\Delta t/2}\right)\right]}{\left[\frac{\varepsilon_0\tilde{\varepsilon}_\infty}{\Delta t}+\frac{\sigma}{2}+\frac{1}{2}\left(\frac{\varepsilon_0\omega_p^2\Delta t/2}{1+\Gamma\Delta t/2}\right)\right]}\boldsymbol{E}^{n-1}\\&+\frac{1}{\left[\frac{\varepsilon_0\tilde{\varepsilon}_\infty}{\Delta t}+\frac{\sigma}{2}+\frac{1}{2}\left(\frac{\varepsilon_0\omega_p^2\Delta t/2}{1+\Gamma\Delta t/2}\right)\right]}\\&\times\left[\nabla\times\boldsymbol{H}^{n-\frac{1}{2}}-\frac{1}{2}\left(1+\frac{1-\Gamma\Delta t/2}{1+\Gamma\Delta t/2}\right)\boldsymbol{J}^{n-1}\right].\end{aligned} \tag{6.187}$$

If we write out the x component,

$$E_x^n = \frac{\left[\frac{\varepsilon_0\tilde{\varepsilon}_\infty}{\Delta t} - \frac{\sigma}{2} - \frac{1}{2}\left(\frac{\varepsilon_0\omega_p^2\Delta t/2}{1+\Gamma\Delta t/2}\right)\right]}{\left[\frac{\varepsilon_0\tilde{\varepsilon}_\infty}{\Delta t} + \frac{\sigma}{2} + \frac{1}{2}\left(\frac{\varepsilon_0\omega_p^2\Delta t/2}{1+\Gamma\Delta t/2}\right)\right]} E_x^{n-1}$$
$$+ \frac{1}{\left[\frac{\varepsilon_0\tilde{\varepsilon}_\infty}{\Delta t} + \frac{\sigma}{2} + \frac{1}{2}\left(\frac{\varepsilon_0\omega_p^2\Delta t/2}{1+\Gamma\Delta t/2}\right)\right]}$$
$$\times \left[\frac{1}{\kappa_y}\frac{\partial H_z^{n-\frac{1}{2}}}{\partial y} - \frac{1}{\kappa_z}\frac{\partial H_y^{n-\frac{1}{2}}}{\partial z} + \zeta_y(t) * \frac{\partial H_z^{n-\frac{1}{2}}}{\partial y} - \zeta_z(t) * \frac{\partial H_y^{n-\frac{1}{2}}}{\partial z}\right.$$
$$\left. -\frac{1}{2}\left(1 + \frac{1-\Gamma\Delta t/2}{1+\Gamma\Delta t/2}\right) J_x^{n-1}\right]. \tag{6.188}$$

The same is true for the Lorentz dispersion media.

6.4.6 Parameters in PML

The thickness of the PML must be finite, and the outer wall must be terminated with a PEC. To eliminate reflection at the PEC, it is important to gradually increase the loss in the PML. The s_ν used in CFS-PML to achieve this is

$$s_\nu = \kappa_\nu + \frac{\sigma_\nu}{a_\nu - i\omega\varepsilon_0}. \tag{6.189}$$

The following shows how to give σ_ν, κ_ν, and a_ν contained in s_ν.

The conductivity σ_z gives the attenuation of the propagating light. Assuming $z = z_0$ for the coordinates of the PML interface on the $+z$ side and d for the thickness of the PML layer, the commonly used way to give the conductivity σ_z is

$$\sigma_z = \sigma_{\max}\left(\frac{z - z_0}{d}\right)^m, \tag{6.190}$$

where $\sigma_{\max}$ is the conductivity just before PEC. The reflection coefficient of the PML with the conductivity distribution given by this equation is a function of the angle of incidence θ and is given by

$$|r(\theta)| \simeq \exp\left[-\frac{2\sigma_{\max}d}{(m+1)\varepsilon_0 c}\cos\theta\right]. \tag{6.191}$$

The coefficient 2 was caused by the PML round trip. If the maximum value of the reflection coefficient allowed is $|r_{\max}|$, the maximum value of the conductivity $\sigma_{\max}$ is calculated from Eq. (6.191) as

$$\sigma_{\max} = -\frac{(m+1)\varepsilon_0 c}{2d}\ln|r_{\max}|, \tag{6.192}$$

and this value should be used.

The κ_ν gives the multiplier of evanescent wave attenuation, usually similar to $\sigma_{\max}$,

$$\kappa_z = 1 + (\kappa_{\max} - 1)\left(\frac{z - z_0}{d}\right)^m \tag{6.193}$$

is used [46].

On the other hand, a_ν gives the attenuation at low frequencies. The larger a_ν is, the smaller the attenuation is. Therefore, a_ν is set to be maximum at the PML interface and zero at the PEC, contrary to the above two parameters, i.e.

$$a_z = a_{\max}\left(\frac{z_0 + d - z}{d}\right)^m \tag{6.194}$$

is usually used. Figure 6.9 summarizes the above PMA parameters.

As for the multiplier $3 \leq m \leq 4$ is often used for σ_ν and κ_ν. On the other hand, for a_ν, Gedney's book [46] uses $m = 1$ as an example. In this book, $\kappa_{\max} = 15$ and $a_{\max} = 0.2$ are used.

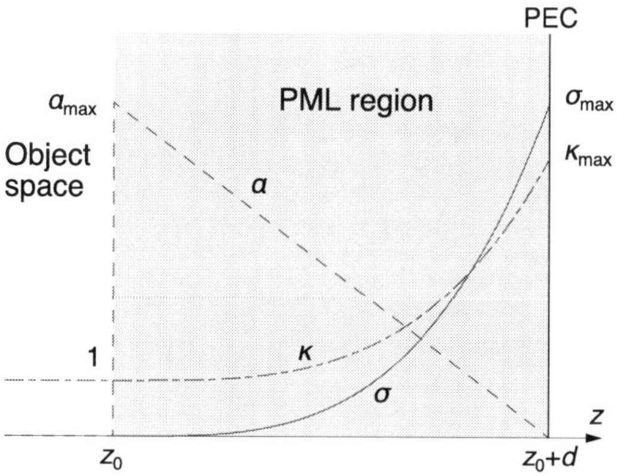

FIGURE 6.9
Parameters in PML.

6.5 Sources

We will discuss the cases where the source is an oscillating electric dipole and a plane wave. For the plane wave, the Total Field/Scattered Field (TF/SF) method is described.

6.5.1 Dipole sources

Consider the case where the source is a micro-oscillating dipole. The dipole moment $\boldsymbol{\mu}(t)$ is given by

$$\boldsymbol{\mu}(t) = \boldsymbol{\mu}_0 \sin \omega t. \tag{6.195}$$

The current $\boldsymbol{I}(t)$ is given by the time derivative of the dipole moment as

$$\boldsymbol{I}(t) = \frac{d\boldsymbol{\mu}(t)}{dt} = \omega \boldsymbol{\mu}_0 \cos \omega t. \tag{6.196}$$

Using the cell size of Yee lattice, the current density $\boldsymbol{j}(t)$ is given as

$$\boldsymbol{j}(t) = \left[\frac{I_x(t)}{\Delta y \Delta z}, \frac{I_y(t)}{\Delta x \Delta z}, \frac{I_z(t)}{\Delta x \Delta y} \right]. \tag{6.197}$$

When a current source is present, Ampere's law is given as

$$\nabla \times \boldsymbol{H} = \varepsilon_0 \varepsilon \frac{\partial \boldsymbol{E}}{\partial t} + \boldsymbol{j}. \tag{6.198}$$

Discretizing this equation as before, we obtain

$$\begin{aligned} E_x|^{n}_{i+\frac{1}{2},j,k} =& E_x|^{n-1}_{i+\frac{1}{2},j,k} + \left(\frac{\Delta t}{\varepsilon_0 \varepsilon \Delta y}\right)\left(H_z|^{n-\frac{1}{2}}_{i+\frac{1}{2},j+\frac{1}{2},k} - H_z|^{n-\frac{1}{2}}_{i+\frac{1}{2},j-\frac{1}{2},k}\right) \\ & - \left(\frac{\Delta t}{\varepsilon_0 \varepsilon \Delta z}\right)\left(H_y|^{n-\frac{1}{2}}_{i+\frac{1}{2},j,k+\frac{1}{2}} - H_y|^{n-\frac{1}{2}}_{i+\frac{1}{2},j,k-\frac{1}{2}}\right) - \left(\frac{\Delta t}{\varepsilon_0 \varepsilon}\right) j_x^{n-\frac{1}{2}}, \end{aligned} \tag{6.199}$$

$$\begin{aligned} E_y|^{n}_{i,j+\frac{1}{2},k} =& E_y|^{n-1}_{i,j+\frac{1}{2},k} + \left(\frac{\Delta t}{\varepsilon_0 \varepsilon \Delta y}\right)\left(H_x|^{n-\frac{1}{2}}_{i,j+\frac{1}{2},k+\frac{1}{2}} - H_x|^{n-\frac{1}{2}}_{i,j-\frac{1}{2},k+\frac{1}{2}}\right) \\ & - \left(\frac{\Delta t}{\varepsilon_0 \varepsilon \Delta x}\right)\left(H_z|^{n-\frac{1}{2}}_{i+\frac{1}{2},j+\frac{1}{2},k} - H_z|^{n-\frac{1}{2}}_{i-\frac{1}{2},j+\frac{1}{2},k}\right) - \left(\frac{\Delta t}{\varepsilon_0 \varepsilon}\right) j_y^{n-\frac{1}{2}}, \end{aligned} \tag{6.200}$$

$$\begin{aligned} E_z|^{n}_{i,j,k+\frac{1}{2}} =& E_z|^{n-1}_{i,j,k+\frac{1}{2}} + \left(\frac{\Delta t}{\varepsilon_0 \varepsilon \Delta x}\right)\left(H_y|^{n-\frac{1}{2}}_{i+\frac{1}{2},j,k+\frac{1}{2}} - H_y|^{n-\frac{1}{2}}_{i-\frac{1}{2},j,k+\frac{1}{2}}\right) \\ & - \left(\frac{\Delta t}{\varepsilon_0 \varepsilon \Delta y}\right)\left(H_x|^{n-\frac{1}{2}}_{i,j+\frac{1}{2},k+\frac{1}{2}} - H_x|^{n-\frac{1}{2}}_{i,j-\frac{1}{2},k+\frac{1}{2}}\right) - \left(\frac{\Delta t}{\varepsilon_0 \varepsilon}\right) j_z^{n-\frac{1}{2}}. \end{aligned} \tag{6.201}$$

As can be seen from these equations the location at which the current source is defined is the same as that of the electric field, and the time is the same as that of the magnetic field. At the grid point where the current source is located, the time evolution can be calculated using Eqs. (6.199), (6.200), and (6.201) instead of the Eqs. (6.7), (6.8), and (6.9). In the presence of Drude dispersion, we can follow Eq. (6.54).

6.5.2 TF/SF method

When a plane wave is incident along the z-axis into a system periodic in the x- and y-directions, the extent of the incident plane wave is virtually infinite and the wave is really a plane wave. The plane wave can be easily introduced by adding the electric field in the $z = z_0$ plane and the magnetic field in the $z = z_0 + \Delta z/2$ plane to the field. However, an isolated system surrounded by PML requires some ingenuity. A commonly used method is the total field/scattered field (TF/SF) method. In this method, the entire electromagnetic field is calculated in the region where the object is located, and only the scattered field is calculated outside the region. As a result, a correction must be made at the boundary between the two to make them consistent.

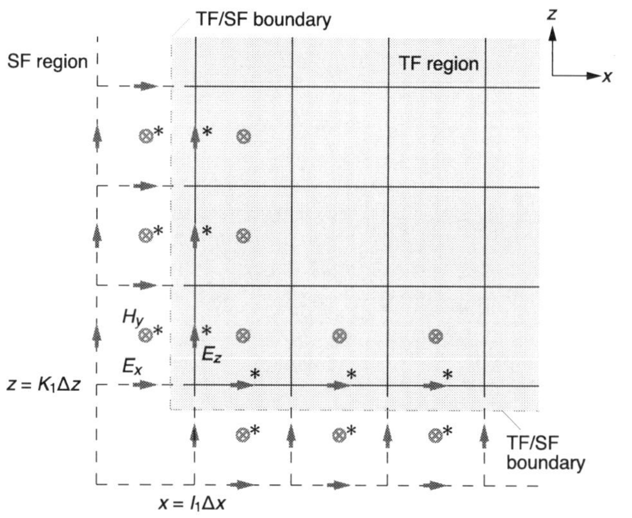

FIGURE 6.10
TF/SF method for the case of TE_y polarized incidence. Corrections are required for the fields marked with $*$.

Consider a plane wave incidence with TEy polarization (the electric field is perpendicular to the y-axis) as shown in Figure 6.10. In this case, non-zero components of incident plane wave are E_x, E_z, and H_y. We treat the total electromagnetic field in the region $I_1\Delta x \le x \le I_2\Delta x$, $J_1\Delta y \le y \le J_2\Delta y$, and $K_1\Delta z \le z \le K_2\Delta z$, and only the scattered field outside this region. When calculating the time evolution of fields adjacent to the boundary between the total electromagnetic field region and the scattered field region (TF/SF boundary), some of the electromagnetic fields required for the calculation of fields in each region are included in different regions. The H_y on $x =$

$(I_1 - 1/2)\Delta x$, which is required for the calculation of the time evolution of E_z on $x = I_1\Delta x$ included in the total electromagnetic field region, is in the scatterred field region. Therefore, it is necessary to add the incident field to make the total electromagnetic field. Namely,

$$H_y^t\Big|_{I_1-\frac{1}{2},j,k+\frac{1}{2}}^{n-\frac{1}{2}} = H_y^s\Big|_{I_1-\frac{1}{2},j,k+\frac{1}{2}}^{n-\frac{1}{2}} + H_y^i\Big|_{I_1-\frac{1}{2},j,k+\frac{1}{2}}^{n-\frac{1}{2}}, \tag{6.202}$$

where the superscripts t, s, and i refer to the total electromagnetic field, the scattered field, and the incident field, respectively. Using this relationship, the time evolution is given as

$$\begin{aligned} E_z^t\Big|_{I_1,j,k+\frac{1}{2}}^{n} &= E_z^t\Big|_{I_1,j,k+\frac{1}{2}}^{n-1} \\ &+ \left(\frac{\Delta t}{\varepsilon_0\varepsilon\Delta x}\right)\left(H_y^t\Big|_{I_1+\frac{1}{2},j,k+\frac{1}{2}}^{n-\frac{1}{2}} - H_y^s\Big|_{I_1-\frac{1}{2},j,k+\frac{1}{2}}^{n-\frac{1}{2}} - H_y^i\Big|_{I_1-\frac{1}{2},j,k+\frac{1}{2}}^{n-\frac{1}{2}}\right) \\ &- \left(\frac{\Delta t}{\varepsilon_0\varepsilon\Delta y}\right)\left(H_x^t\Big|_{I_1,j+\frac{1}{2},k+\frac{1}{2}}^{n-\frac{1}{2}} - H_x^t\Big|_{I_1,j-\frac{1}{2},k+\frac{1}{2}}^{n-\frac{1}{2}}\right). \end{aligned} \tag{6.203}$$

Similarly, since

$$E_z^s\Big|_{I_1,j,k+\frac{1}{2}}^{n} = E_z^t\Big|_{I_1,j,k+\frac{1}{2}}^{n} - E_z^i\Big|_{I_1,j,k+\frac{1}{2}}^{n}, \tag{6.204}$$

the magnetic field in the scattered field region tangential to the TF/SF boundary is

$$\begin{aligned} H_y^s\Big|_{I_1-\frac{1}{2},j,k+\frac{1}{2}}^{n+\frac{1}{2}} &= H_y^s\Big|_{I_1-\frac{1}{2},j,k+\frac{1}{2}}^{n-\frac{1}{2}} \\ &- \left(\frac{\Delta t}{\mu_0\mu\Delta z}\right)\left(E_x^s\Big|_{I_1-\frac{1}{2},j,k+1}^{n} - E_x^s\Big|_{I_1-\frac{1}{2},j,k}^{n}\right) \\ &+ \left(\frac{\Delta t}{\mu_0\mu\Delta x}\right)\left(E_z^t\Big|_{I_1,j,k+\frac{1}{2}}^{n} - E_z^i\Big|_{I_1,j,k+\frac{1}{2}}^{n} - E_z^s\Big|_{I_1-1,j,k+\frac{1}{2}}^{n}\right). \end{aligned} \tag{6.205}$$

Other than the above two components, no correction is necessary because the time evolution calculation does not include a non-zero incident field component. Also, in the $x = I_2\Delta x$ boundary,

$$\begin{aligned} E_z^t\Big|_{I_2,j,k+\frac{1}{2}}^{n} &= E_z^t\Big|_{I_2,j,k+\frac{1}{2}}^{n-1} \\ &+ \left(\frac{\Delta t}{\varepsilon_0\varepsilon\Delta x}\right)\left(H_y^s\Big|_{I_2+\frac{1}{2},j,k+\frac{1}{2}}^{n-\frac{1}{2}} + H_y^i\Big|_{I_2+\frac{1}{2},j,k+\frac{1}{2}}^{n-\frac{1}{2}} - H_y^t\Big|_{I_2-\frac{1}{2},j,k+\frac{1}{2}}^{n-\frac{1}{2}}\right) \\ &- \left(\frac{\Delta t}{\varepsilon_0\varepsilon\Delta y}\right)\left(H_x^t\Big|_{I_2,j+\frac{1}{2},k+\frac{1}{2}}^{n-\frac{1}{2}} - H_x^t\Big|_{I_2,j-\frac{1}{2},k+\frac{1}{2}}^{n-\frac{1}{2}}\right), \end{aligned} \tag{6.206}$$

$$\begin{aligned} H_y^s\Big|_{I_2+\frac{1}{2},j,k+\frac{1}{2}}^{n+\frac{1}{2}} &= H_y^s\Big|_{I_2+\frac{1}{2},j,k+\frac{1}{2}}^{n-\frac{1}{2}} \\ &- \left(\frac{\Delta t}{\mu_0\mu\Delta z}\right)\left(E_x^s\Big|_{I_2+\frac{1}{2},j,k+1}^{n} - E_x^s\Big|_{I_2+\frac{1}{2},j,k}^{n}\right) \\ &+ \left(\frac{\Delta t}{\mu_0\mu\Delta x}\right)\left(E_z^s\Big|_{I_2+1,j,k+\frac{1}{2}}^{n} - E_z^t\Big|_{I_2,j,k+\frac{1}{2}}^{n} + E_z^i\Big|_{I_2,j,k+\frac{1}{2}}^{n}\right). \end{aligned} \tag{6.207}$$

On the other hand, on the $z = K_1 \Delta z$ boundary,

$$E_x^t\Big|_{i+\frac{1}{2},j,K_1}^{n} = E_x^t\Big|_{i+\frac{1}{2},j,K_1}^{n-1} + \left(\frac{\Delta t}{\varepsilon_0 \varepsilon \Delta y}\right)\left(H_z^t\Big|_{i+\frac{1}{2},j+\frac{1}{2},K_1}^{n-\frac{1}{2}} - H_z^t\Big|_{i+\frac{1}{2},j-\frac{1}{2},K_1}^{n-\frac{1}{2}}\right) - \left(\frac{\Delta t}{\varepsilon_0 \varepsilon \Delta z}\right)\left(H_y^t\Big|_{i+\frac{1}{2},j,K_1+\frac{1}{2}}^{n-\frac{1}{2}} - H_y^s\Big|_{i+\frac{1}{2},j,K_1-\frac{1}{2}}^{n-\frac{1}{2}} - H_y^i\Big|_{i+\frac{1}{2},j,K_1-\frac{1}{2}}^{n-\frac{1}{2}}\right) \quad (6.208)$$

$$H_y^s\Big|_{i+\frac{1}{2},j,K_1-\frac{1}{2}}^{n+\frac{1}{2}} = H_y^s\Big|_{i+\frac{1}{2},j,K_1-\frac{1}{2}}^{n-\frac{1}{2}} - \left(\frac{\Delta t}{\mu_0 \mu \Delta z}\right)\left(E_x^t\Big|_{i+\frac{1}{2},j,K_1}^{n} - E_x^i\Big|_{i+\frac{1}{2},j,K_1}^{n} - E_x^s\Big|_{i+\frac{1}{2},j,K_1-1}^{n}\right) + \left(\frac{\Delta t}{\mu_0 \mu \Delta x}\right)\left(E_z^s\Big|_{i+1,j,K_1-\frac{1}{2}}^{n} - E_z^s\Big|_{i,j,K_1-\frac{1}{2}}^{n}\right). \quad (6.209)$$

Also, on the $z = K_2 \Delta z$ boundary,

$$E_x^t\Big|_{i+\frac{1}{2},j,K_2}^{n} = E_x^t\Big|_{i+\frac{1}{2},j,K_2}^{n-1} + \left(\frac{\Delta t}{\varepsilon_0 \varepsilon \Delta y}\right)\left(H_z^t\Big|_{i+\frac{1}{2},j+\frac{1}{2},K_2}^{n-\frac{1}{2}} - H_z^t\Big|_{i+\frac{1}{2},j-\frac{1}{2},K_2}^{n-\frac{1}{2}}\right) - \left(\frac{\Delta t}{\varepsilon_0 \varepsilon \Delta z}\right)\left(H_y^s\Big|_{i+\frac{1}{2},j,K_2+\frac{1}{2}}^{n-\frac{1}{2}} + H_y^i\Big|_{i+\frac{1}{2},j,K_2+\frac{1}{2}}^{n-\frac{1}{2}} - H_y^t\Big|_{i+\frac{1}{2},j,K_2-\frac{1}{2}}^{n-\frac{1}{2}}\right) \quad (6.210)$$

$$H_y^s\Big|_{i+\frac{1}{2},j,K_2+\frac{1}{2}}^{n+\frac{1}{2}} = H_y^s\Big|_{i+\frac{1}{2},j,K_2+\frac{1}{2}}^{n-\frac{1}{2}} - \left(\frac{\Delta t}{\mu_0 \mu \Delta z}\right)\left(E_x^s\Big|_{i+\frac{1}{2},j,K_2+1}^{n} - E_x^t\Big|_{i+\frac{1}{2},j,K_2}^{n} + E_x^i\Big|_{i+\frac{1}{2},j,K_2}^{n}\right) + \left(\frac{\Delta t}{\mu_0 \mu \Delta x}\right)\left(E_z^s\Big|_{i+1,j,K_2+\frac{1}{2}}^{n} - E_z^s\Big|_{i,j,K_2+\frac{1}{2}}^{n}\right). \quad (6.211)$$

Furthermore, on the $y = J_1 \Delta y$ boundary,

$$H_x^s\Big|_{i,J_1-\frac{1}{2},k+\frac{1}{2}}^{n+\frac{1}{2}} = H_x^s\Big|_{i,J_1-\frac{1}{2},k+\frac{1}{2}}^{n-\frac{1}{2}} - \left(\frac{\Delta t}{\mu_0 \mu \Delta y}\right)\left(E_z^t\Big|_{i,J_1,k+\frac{1}{2}}^{n} - E_z^i\Big|_{i,J_1,k+\frac{1}{2}}^{n} - E_z^s\Big|_{i,J_1-1,k+\frac{1}{2}}^{n}\right) + \left(\frac{\Delta t}{\mu_0 \mu \Delta z}\right)\left(E_y^s\Big|_{i,J_1-\frac{1}{2},k+1}^{n} - E_y^s\Big|_{i,J_1-\frac{1}{2},k}^{n}\right), \quad (6.212)$$

$$H_z^s\Big|_{i+\frac{1}{2},J_1-\frac{1}{2},k}^{n+\frac{1}{2}} = H_z^s\Big|_{i+\frac{1}{2},J_1-\frac{1}{2},k}^{n-\frac{1}{2}} - \left(\frac{\Delta t}{\mu_0 \mu \Delta x}\right)\left(E_y^s\Big|_{i+1,J_1-\frac{1}{2},k}^{n} - E_y^s\Big|_{i,J_1-\frac{1}{2},k}^{n}\right) + \left(\frac{\Delta t}{\mu_0 \mu \Delta y}\right)\left(E_x^t\Big|_{i+\frac{1}{2},J_1,k}^{n} - E_x^i\Big|_{i+\frac{1}{2},J_1,k}^{n} - E_x^s\Big|_{i+\frac{1}{2},J_1-1,k}^{n}\right). \quad (6.213)$$

Also, on the $y = J_2\Delta y$ boundary,

$$\begin{aligned} H_x^s\big|_{i,J_2+\frac{1}{2},k+\frac{1}{2}}^{n+\frac{1}{2}} &= H_x^s\big|_{i,J_2+\frac{1}{2},k+\frac{1}{2}}^{n-\frac{1}{2}} \\ &- \left(\frac{\Delta t}{\mu_0\mu\Delta y}\right)\left(E_z^s\big|_{i,J_2+1,k+\frac{1}{2}}^{n} - E_z^t\big|_{i,J_2,k+\frac{1}{2}}^{n} + E_z^i\big|_{i,J_2,k+\frac{1}{2}}^{n}\right) \\ &+ \left(\frac{\Delta t}{\mu_0\mu\Delta z}\right)\left(E_y^s\big|_{i,J_2-\frac{1}{2},k+1}^{n} - E_y^s\big|_{i,J_2-\frac{1}{2},k}^{n}\right), \end{aligned} \tag{6.214}$$

$$\begin{aligned} H_z^s\big|_{i+\frac{1}{2},J_2+\frac{1}{2},k}^{n+\frac{1}{2}} &= H_z^s\big|_{i+\frac{1}{2},J_2+\frac{1}{2},k}^{n-\frac{1}{2}} \\ &- \left(\frac{\Delta t}{\mu_0\mu\Delta x}\right)\left(E_y^s\big|_{i+1,J_2+\frac{1}{2},k}^{n} - E_y^s\big|_{i,J_2+\frac{1}{2},k}^{n}\right) \\ &+ \left(\frac{\Delta t}{\mu_0\mu\Delta y}\right)\left(E_x^s\big|_{i+\frac{1}{2},J_2+1,k+1}^{n} - E_x^t\big|_{i+\frac{1}{2},J_2,k}^{n} + E_x^i\big|_{i+\frac{1}{2},J_2,k}^{n}\right). \end{aligned} \tag{6.215}$$

The only fields that should be stored in the calculation are the total electromagnetic field in the total field region and the scattered field in the scattered field region. After performing the usual time evolution calculations using these fields, we can make corrections for the incident fields.

The same is true for TMy polarization (the magnetic field is perpendicular to the y-axis). In this case, the incident plane wave has only E_y, H_x, and H_z components. On the $x = I_1\Delta x$ boundary,

$$\begin{aligned} E_y^t\big|_{i,j+\frac{1}{2},k}^{n} &= E_y^t\big|_{i,j+\frac{1}{2},k}^{n-1} + \left(\frac{\Delta t}{\varepsilon_0\varepsilon\Delta y}\right)\left(H_x^t\big|_{i,j+\frac{1}{2},k+\frac{1}{2}}^{n-\frac{1}{2}} - H_x^t\big|_{i,j-\frac{1}{2},k+\frac{1}{2}}^{n-\frac{1}{2}}\right) \\ &- \left(\frac{\Delta t}{\varepsilon_0\varepsilon\Delta x}\right)\left(H_z^t\big|_{i+\frac{1}{2},j+\frac{1}{2},k}^{n-\frac{1}{2}} - H_z^s\big|_{i-\frac{1}{2},j+\frac{1}{2},k}^{n-\frac{1}{2}} - H_z^i\big|_{i-\frac{1}{2},j+\frac{1}{2},k}^{n-\frac{1}{2}}\right). \end{aligned} \tag{6.216}$$

Similarly,

$$\begin{aligned} H_z^s\big|_{i-\frac{1}{2},j+\frac{1}{2},k}^{n+\frac{1}{2}} &= H_z^s\big|_{i-\frac{1}{2},j+\frac{1}{2},k}^{n-\frac{1}{2}} \\ &- \left(\frac{\Delta t}{\mu_0\mu\Delta x}\right)\left(E_y^t\big|_{i,j+\frac{1}{2},k}^{n} - E_y^i\big|_{i,j+\frac{1}{2},k}^{n} - E_y^s\big|_{i-1,j+\frac{1}{2},k}^{n}\right) \\ &+ \left(\frac{\Delta t}{\mu_0\mu\Delta y}\right)\left(E_x^s\big|_{i-\frac{1}{2},j+1,k}^{n} - E_x^s\big|_{i-\frac{1}{2},j,k}^{n}\right). \end{aligned} \tag{6.217}$$

On the $x = I_2\Delta x$ boundary,

$$\begin{aligned} E_y^t\big|_{I,j+\frac{1}{2},k}^{n} &= E_y^t\big|_{I,j+\frac{1}{2},k}^{n-1} \\ &+ \left(\frac{\Delta t}{\varepsilon_0\varepsilon\Delta y}\right)\left(H_x^t\big|_{I,j+\frac{1}{2},k+\frac{1}{2}}^{n-\frac{1}{2}} - H_x^t\big|_{I,j-\frac{1}{2},k+\frac{1}{2}}^{n-\frac{1}{2}}\right) \\ &- \left(\frac{\Delta t}{\varepsilon_0\varepsilon\Delta x}\right)\left(H_z^s\big|_{I+\frac{1}{2},j+\frac{1}{2},k}^{n-\frac{1}{2}} + H_z^i\big|_{I+\frac{1}{2},j+\frac{1}{2},k}^{n-\frac{1}{2}} - H_z^t\big|_{I-\frac{1}{2},j+\frac{1}{2},k}^{n-\frac{1}{2}}\right), \end{aligned} \tag{6.218}$$

$$\begin{aligned}H_z^s\Big|_{I+\frac{1}{2},j+\frac{1}{2},k}^{n+\frac{1}{2}} &= H_z^s\Big|_{I+\frac{1}{2},j+\frac{1}{2},k}^{n-\frac{1}{2}}\\ &- \left(\frac{\Delta t}{\mu_0\mu\Delta x}\right)\left(E_y^s\Big|_{I+1,j+\frac{1}{2},k}^{n} - E_y^t\Big|_{I,j+\frac{1}{2},k}^{n} + E_y^i\Big|_{I,j+\frac{1}{2},k}^{n}\right)\\ &+ \left(\frac{\Delta t}{\mu_0\mu\Delta y}\right)\left(E_x^s\Big|_{I+\frac{1}{2},j+1,k}^{n} - E_x^s\Big|_{I+\frac{1}{2},j,k}^{n}\right). \end{aligned} \tag{6.219}$$

On the other hand, on the $y = J_1\Delta y$ boundary,

$$\begin{aligned}E_x^t\Big|_{i+\frac{1}{2},j,k}^{n} &= E_x^t\Big|_{i+\frac{1}{2},j,k}^{n-1}\\ &+ \left(\frac{\Delta t}{\varepsilon_0\varepsilon\Delta y}\right)\left(H_z^t\Big|_{i+\frac{1}{2},j+\frac{1}{2},k}^{n-\frac{1}{2}} - H_z^s\Big|_{i+\frac{1}{2},j-\frac{1}{2},k}^{n-\frac{1}{2}} - H_z^i\Big|_{i+\frac{1}{2},j-\frac{1}{2},k}^{n-\frac{1}{2}}\right)\\ &- \left(\frac{\Delta t}{\varepsilon_0\varepsilon\Delta z}\right)\left(H_y^t\Big|_{i+\frac{1}{2},j,k+\frac{1}{2}}^{n-\frac{1}{2}} - H_y^t\Big|_{i+\frac{1}{2},j,k-\frac{1}{2}}^{n-\frac{1}{2}}\right), \end{aligned} \tag{6.220}$$

$$\begin{aligned}E_z^t\Big|_{i,j,k+\frac{1}{2}}^{n} &= E_z^t\Big|_{i,j,k+\frac{1}{2}}^{n-1}\\ &+ \left(\frac{\Delta t}{\varepsilon_0\varepsilon\Delta x}\right)\left(H_y^t\Big|_{i+\frac{1}{2},j,k+\frac{1}{2}}^{n-\frac{1}{2}} - H_y^t\Big|_{i-\frac{1}{2},j,k+\frac{1}{2}}^{n-\frac{1}{2}}\right)\\ &- \left(\frac{\Delta t}{\varepsilon_0\varepsilon\Delta y}\right)\left(H_x^t\Big|_{i,j+\frac{1}{2},k+\frac{1}{2}}^{n-\frac{1}{2}} - H_x^s\Big|_{i,j-\frac{1}{2},k+\frac{1}{2}}^{n-\frac{1}{2}} - H_x^i\Big|_{i,j-\frac{1}{2},k+\frac{1}{2}}^{n-\frac{1}{2}}\right). \end{aligned} \tag{6.221}$$

Also, on the $y = J_2\Delta y$ boundary,

$$\begin{aligned}E_x^t\Big|_{i+\frac{1}{2},J,k}^{n} &= E_x^t\Big|_{i+\frac{1}{2},J,k}^{n-1}\\ &+ \left(\frac{\Delta t}{\varepsilon_0\varepsilon\Delta y}\right)\left(H_z^s\Big|_{i+\frac{1}{2},J+\frac{1}{2},k}^{n-\frac{1}{2}} + H_z^i\Big|_{i+\frac{1}{2},J+\frac{1}{2},k}^{n-\frac{1}{2}} - H_z^t\Big|_{i+\frac{1}{2},J-\frac{1}{2},k}^{n-\frac{1}{2}}\right)\\ &- \left(\frac{\Delta t}{\varepsilon_0\varepsilon\Delta z}\right)\left(H_y^t\Big|_{i+\frac{1}{2},J,k+\frac{1}{2}}^{n-\frac{1}{2}} - H_y^t\Big|_{i+\frac{1}{2},J,k-\frac{1}{2}}^{n-\frac{1}{2}}\right), \end{aligned} \tag{6.222}$$

$$\begin{aligned}E_z^t\Big|_{i,J,k+\frac{1}{2}}^{n} &= E_z^t\Big|_{i,J,k+\frac{1}{2}}^{n-1}\\ &+ \left(\frac{\Delta t}{\varepsilon_0\varepsilon\Delta x}\right)\left(H_y^t\Big|_{i+\frac{1}{2},J,k+\frac{1}{2}}^{n-\frac{1}{2}} - H_y^t\Big|_{i-\frac{1}{2},J,k+\frac{1}{2}}^{n-\frac{1}{2}}\right)\\ &- \left(\frac{\Delta t}{\varepsilon_0\varepsilon\Delta y}\right)\left(H_x^s\Big|_{i,J+\frac{1}{2},k+\frac{1}{2}}^{n-\frac{1}{2}} + H_x^i\Big|_{i,J+\frac{1}{2},k+\frac{1}{2}}^{n-\frac{1}{2}} - H_x^t\Big|_{i,J-\frac{1}{2},k+\frac{1}{2}}^{n-\frac{1}{2}}\right). \end{aligned} \tag{6.223}$$

Furthermore, on the $z = K_1\Delta z$ boundary,

$$\begin{aligned}E_y^t\Big|_{i,j+\frac{1}{2},k}^{n} &= E_y^t\Big|_{i,j+\frac{1}{2},k}^{n-1}\\ &+ \left(\frac{\Delta t}{\varepsilon_0\varepsilon\Delta z}\right)\left(H_x^t\Big|_{i,j+\frac{1}{2},k+\frac{1}{2}}^{n-\frac{1}{2}} - H_x^s\Big|_{i,j+\frac{1}{2},k-\frac{1}{2}}^{n-\frac{1}{2}} - H_x^i\Big|_{i,j+\frac{1}{2},k-\frac{1}{2}}^{n-\frac{1}{2}}\right)\\ &- \left(\frac{\Delta t}{\varepsilon_0\varepsilon\Delta x}\right)\left(H_z^t\Big|_{i+\frac{1}{2},j+\frac{1}{2},k}^{n-\frac{1}{2}} - H_z^t\Big|_{i-\frac{1}{2},j+\frac{1}{2},k}^{n-\frac{1}{2}}\right), \end{aligned} \tag{6.224}$$

$$
\begin{aligned}
H_x^s\big|_{i,j+\frac{1}{2},k-\frac{1}{2}}^{n+\frac{1}{2}} &= H_x^s\big|_{i,j+\frac{1}{2},k-\frac{1}{2}}^{n-\frac{1}{2}} \\
&- \left(\frac{\Delta t}{\mu_0\mu\Delta y}\right)\left(E_z^s\big|_{i,j+1,k-\frac{1}{2}}^{n} - E_z^s\big|_{i,j,k-\frac{1}{2}}^{n}\right) \\
&+ \left(\frac{\Delta t}{\mu_0\mu\Delta z}\right)\left(E_y^t\big|_{i,j+\frac{1}{2},k}^{n} - E_y^i\big|_{i,j+\frac{1}{2},k}^{n} - E_y^s\big|_{i,j+\frac{1}{2},k-1}^{n}\right). \qquad (6.225)
\end{aligned}
$$

Also, on the $z = K_2\Delta z$ boundary,

$$
\begin{aligned}
E_y^t\big|_{i,j+\frac{1}{2},K}^{n} &= E_y^t\big|_{i,j+\frac{1}{2},K}^{n-1} \\
&+ \left(\frac{\Delta t}{\varepsilon_0\varepsilon\Delta z}\right)\left(H_x^s\big|_{i,j+\frac{1}{2},K+\frac{1}{2}}^{n-\frac{1}{2}} + H_x^i\big|_{i,j+\frac{1}{2},K+\frac{1}{2}}^{n-\frac{1}{2}} - H_x^t\big|_{i,j+\frac{1}{2},K-\frac{1}{2}}^{n-\frac{1}{2}}\right) \\
&- \left(\frac{\Delta t}{\varepsilon_0\varepsilon\Delta x}\right)\left(H_z^t\big|_{i+\frac{1}{2},j+\frac{1}{2},K}^{n-\frac{1}{2}} - H_z^t\big|_{i-\frac{1}{2},j+\frac{1}{2},K}^{n-\frac{1}{2}}\right) \qquad (6.226)
\end{aligned}
$$

$$
\begin{aligned}
H_x^s\big|_{i,j+\frac{1}{2},K+\frac{1}{2}}^{n+\frac{1}{2}} &= H_x^s\big|_{i,j+\frac{1}{2},K+\frac{1}{2}}^{n-\frac{1}{2}} \\
&- \left(\frac{\Delta t}{\mu_0\mu\Delta y}\right)\left(E_z^s\big|_{i,j+1,K+\frac{1}{2}}^{n} - E_z^s\big|_{i,j,K+\frac{1}{2}}^{n}\right) \\
&+ \left(\frac{\Delta t}{\mu_0\mu\Delta z}\right)\left(E_y^s\big|_{i,j+\frac{1}{2},K+1}^{n} - E_y^t\big|_{i,j+\frac{1}{2},K}^{n} + E_y^i\big|_{i,j+\frac{1}{2},K}^{n}\right). \qquad (6.227)
\end{aligned}
$$

6.5.3 Dispersive medium crossing TF/SF boundary

An example of the electric field E_z at $x = I_1\Delta x$ when the medium follows a Drude dispersion and the incident field is TE_y polarized,

$$
\begin{aligned}
E_z^t\big|_{I_1,j,k+\frac{1}{2}}^{n} &= \frac{\left[\frac{\varepsilon_0\varepsilon_\infty}{\Delta t} - \frac{\sigma}{2} - \frac{1}{2}\left(\frac{\varepsilon_0\omega_p^2\Delta t/2}{1+\Gamma\Delta t/2}\right)\right]}{\left[\frac{\varepsilon_0\varepsilon_\infty}{\Delta t} + \frac{\sigma}{2} + \frac{1}{2}\left(\frac{\varepsilon_0\omega_p^2\Delta t/2}{1+\Gamma\Delta t/2}\right)\right]} E_z^t\big|_{I_1,j,k+\frac{1}{2}}^{n-1} \\
&+ \frac{1/\Delta x}{\left[\frac{\varepsilon_0\varepsilon_\infty}{\Delta t} + \frac{\sigma}{2} + \frac{1}{2}\left(\frac{\varepsilon_0\omega_p^2\Delta t/2}{1+\Gamma\Delta t/2}\right)\right]} \\
&\times \left(H_y^t\big|_{I_1+\frac{1}{2},j,k+\frac{1}{2}}^{n-\frac{1}{2}} - H_y^s\big|_{I_1+\frac{1}{2},j,k-\frac{1}{2}}^{n-\frac{1}{2}} - H_y^i\big|_{I_1+\frac{1}{2},j,k-\frac{1}{2}}^{n-\frac{1}{2}}\right) \\
&- \frac{1/\Delta y}{\left[\frac{\varepsilon_0\varepsilon_\infty}{\Delta t} + \frac{\sigma}{2} + \frac{1}{2}\left(\frac{\varepsilon_0\omega_p^2\Delta t/2}{1+\Gamma\Delta t/2}\right)\right]} \\
&\left(H_x^t\big|_{I_1,j+\frac{1}{2},k+\frac{1}{2}}^{n-\frac{1}{2}} - H_x^t\big|_{I_1,j-\frac{1}{2},k+\frac{1}{2}}^{n-\frac{1}{2}}\right) \\
&+ \frac{1}{2}\frac{1}{\left[\frac{\varepsilon_0\varepsilon_\infty}{\Delta t} + \frac{\sigma}{2} + \frac{1}{2}\left(\frac{\varepsilon_0\omega_p^2\Delta t/2}{1+\Gamma\Delta t/2}\right)\right]} \\
&\left(1 + \frac{1-\Gamma\Delta t/2}{1+\Gamma\Delta t/2}\right) J_z^t\big|_{I_1,j,k+\frac{1}{2}}^{n-1}. \qquad (6.228)
\end{aligned}
$$

Therefore, after performing the calculations with the usual ADE, only the correction corresponding to the incident field as

$$-\frac{1/\Delta x}{\left[\frac{\varepsilon_0\varepsilon_\infty}{\Delta t}+\frac{\sigma}{2}+\frac{1}{2}\left(\frac{\varepsilon_0\omega_p^2\Delta t/2}{1+\Gamma\Delta t/2}\right)\right]}H_y^i\Big|_{I_1+\frac{1}{2},j,k-\frac{1}{2}}^{n-\frac{1}{2}} \tag{6.229}$$

is required. In other words, we only need to change the coefficients of the incident field.

6.5.4 Numerical dispersion

If all of the TF/SF boundaries are set in free space with no dispersion, it would seem that no problems arise because the incident electromagnetic field at the boundaries can be calculated analytically. In practice, however, a problem arises. This problem is caused by the numerical dispersion inherent to FDTD. Numerical dispersion is also called grid dispersion.

The wave number $\tilde{k}$ obtained by the FDTD method deviates from the theoretical wave number and is given by

$$\left[\frac{1}{v_p\Delta t}\sin\left(\frac{\omega\Delta t}{2}\right)\right]^2=\left[\frac{1}{\Delta x}\sin\left(\frac{\tilde{k}_x\Delta x}{2}\right)\right]^2+\left[\frac{1}{\Delta y}\sin\left(\frac{\tilde{k}_y\Delta y}{2}\right)\right]^2 +\left[\frac{1}{\Delta z}\sin\left(\frac{\tilde{k}_z\Delta z}{2}\right)\right]^2, \tag{6.230}$$

where v_p is the phase velocity. This relation is the numerical dispersion in FDTD. In the limit of $\Delta x, \Delta y, \Delta z, \Delta t \to 0$, this equation converges to the following dispersion relation for ordinary plane waves,

$$\left(\frac{\omega}{v_p}\right)^2=k_x^2+k_y^2+k_z^2. \tag{6.231}$$

When the analytically determined incident field is applied to the TF/SF boundary, there is a phase difference with the incident field propagating in the TF region as it evolves with time, as shown in Figure 6.11. This is the source of the scattered field that does not actually exist and becomes a source of error.

The TF/SF boundary might be set in a uniform medium, such as plane wave incidence on an isolated object in free space as shown in Figure 6.12(a). However, in real systems, as shown in Figure 6.12(b), the object is often placed on some substrate. In this case, obtaining the incident field analytically is quite complicated, even if numerical dispersion is ignored. In the case of the pulsed light source, the frequency distribution is broad, so the time waveform must be Fourier transformed and expressed in the frequency domain, and the reflection and transmission coefficients for each frequency component must be calculated and transformed back into the time domain.

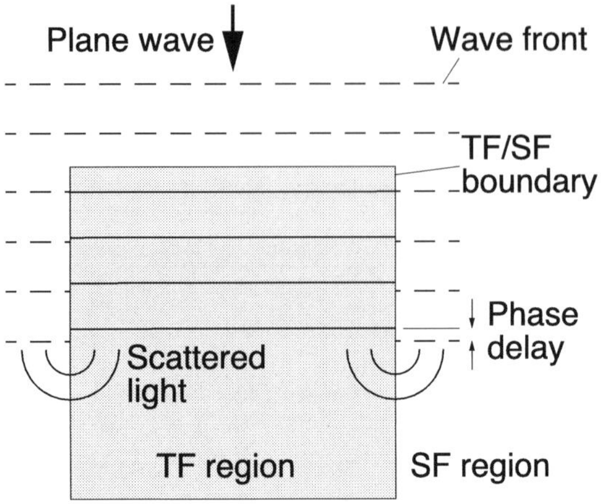

FIGURE 6.11
Scattered light generation due to numerical dispersion. The dashed line is the wavefront obtained analytically and the solid line is the wavefront obtained by the FDTD method.

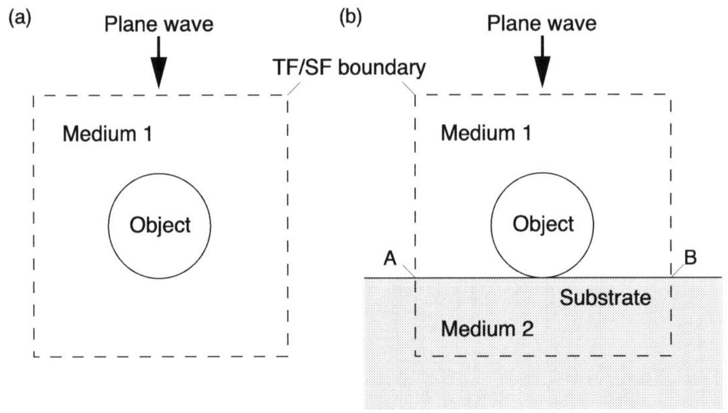

FIGURE 6.12
System with object (a) in free space and (b) on substrate.

A common method to solve these problems is to perform an auxiliary 1D FDTD calculation for the incident field in the propagation direction of the incident field (direction of the wave vector) and use the result as the incident field in the TF/SF method. The advantage of this method is that it can be easily applied even when the computational domain includes substrates or multilayers.

This method is simple if the direction of propagation of the incident field is along the coordinate axes, but if it is inclined to the axes, the position where the electromagnetic field is defined in the 1D FDTD is different from that in the 3D FDTD. Therefore, if one tries to use the results of 1D FDTD as input for 3D FDTD, interpolation and other innovations are required. It also cannot be applied to systems with substrates.

6.5.5 Obliquely incident plane wave

We introduce the method given by Zhang and Seideman [47] for handling an obliquely incident plane wave when the medium contains a substrate or multilayer whose interface is perpendicular to one of the coordinate axes. As an example, consider a system uniform in the y-direction and an incident plane wave with TE$_y$ polarization, where the wave vector $\boldsymbol{k}$ is in the xz plane and the magnetic field points in the y-direction. Assume that the permittivity varies only in the z-direction.

Maxwell's equations in the frequency domain are

$$\frac{\partial E_x}{\partial z} - \frac{\partial E_z}{\partial x} = i\omega\mu_0 H_y, \tag{6.232}$$

$$\frac{\partial H_y}{\partial x} = -i\omega\varepsilon_0\varepsilon(\omega)E_z, \tag{6.233}$$

$$\frac{\partial H_y}{\partial z} = i\omega\varepsilon_0\varepsilon(\omega)E_x, \tag{6.234}$$

where $\varepsilon(\omega)$ is the complex relative permittivity including loss and dispersion, and the relative permeability is assumed to be $\mu = 1$ for all media. First, we have to express these equations as one-dimensional propagation along the z-direction. Differentiating Eq. (6.233) with respect to x and substituting into Eq. (6.232), we obtain

$$\frac{\partial E_x}{\partial z} = i\omega\mu_0 H_y + \frac{1}{-i\omega\varepsilon_0\varepsilon(\omega)}\frac{\partial^2 H_y}{\partial x^2}. \tag{6.235}$$

Due to the phase matching condition, the wave number k_x in the x-direction takes the same value in all layers, and then,

$$\frac{\partial^2 H_y}{\partial x^2} = -k_x^2 H_y \tag{6.236}$$

is obtained. If the first layer on the incident side has no loss, when its relative permittivity is ε_{1r} and the angle of incidence is θ, k_x is given as

$$k_x = \omega\sqrt{\varepsilon_0\varepsilon_{1r}\mu_0}\sin\theta. \tag{6.237}$$

Substituting Eq. (6.237) into Eq. (6.236), and further substituting Eq. (6.236) into Eq. (6.235), we obtain

$$\frac{\partial E_x}{\partial z} = i\omega\mu_0\left[\frac{\varepsilon(\omega) - \varepsilon_{1r}\sin^2\theta}{\varepsilon(\omega)}\right]H_y. \tag{6.238}$$

If the medium is neither lossy nor dispersive, these equations can be easily converted to the time domain and adapted to FDTD. However, for lossy or dispersive media, another effort is needed. This contrivance is made by Jiang et al. [48].

We introduce a new variable H'_z to decompose Eq. (6.238) into two parts as follows:

$$\frac{\partial E_x}{\partial z} = i\omega\mu_0 H'_y, \tag{6.239}$$

$$H'_y = \frac{\varepsilon(\omega) - \varepsilon_{1r}\sin^2\theta}{\varepsilon(\omega)} H_y. \tag{6.240}$$

In the case of a dispersive medium, Eqs. (6.234) and (6.239) can be discretized in the time domain using the usual ADE. Next multiplying both sides of Eq. (6.240) with $\varepsilon_z(\omega)$, we obtain

$$\varepsilon(\omega)H'_y = [\varepsilon(\omega) - \varepsilon_{1r}\sin^2\theta]H_y. \tag{6.241}$$

First, consider the case of a lossy medium, that is,

$$\varepsilon(\omega) = \varepsilon_\infty - \frac{\sigma}{i\omega\varepsilon_0}. \tag{6.242}$$

Substituting Eq. (6.242) into Eq. (6.241), we obtain

$$\left(\varepsilon_\infty - \frac{\sigma}{i\omega\varepsilon_0}\right) H'_y = \left(\varepsilon' - \frac{\sigma}{i\omega\varepsilon_0}\right) H_y, \tag{6.243}$$

$$(-i\omega\varepsilon_0\varepsilon_\infty + \sigma)H'_y = (-i\omega\varepsilon_0\varepsilon' + \sigma)H_y, \tag{6.244}$$

where

$$\varepsilon' = \varepsilon_\infty - \varepsilon_{1r}\sin^2\theta. \tag{6.245}$$

Expressing Eq. (6.244) in the time domain, we obtain

$$\varepsilon_0\varepsilon_\infty \frac{\partial H'_y}{\partial t} + \sigma H'_y = \varepsilon_0\varepsilon' \frac{\partial H_y}{\partial t} + \sigma H_y. \tag{6.246}$$

The time domain Eqs. (6.234), (6.239), and (6.246) are required for the time evolution of the 1D FDTD. These equations can be summarized as

$$\varepsilon_0\varepsilon_\infty \frac{\partial E_x}{\partial t} + \sigma E_x = -\frac{\partial H_y}{\partial z}, \tag{6.247}$$

$$\mu_0 \frac{\partial H'_y}{\partial t} = -\frac{\partial E_x}{\partial z}, \tag{6.248}$$

$$\varepsilon_0\varepsilon_\infty \frac{\partial H'_y}{\partial t} + \sigma H'_y = \varepsilon_0\varepsilon' \frac{\partial H_y}{\partial t} + \sigma H_y. \tag{6.249}$$

Next, we consider the discretization of Eq. (6.249), that is,

$$\begin{aligned}&\varepsilon_0\varepsilon_\infty\frac{H'_y|^{n+\frac{1}{2}}-H'_y|^{n-\frac{1}{2}}}{\Delta t}+\sigma\frac{H'_y|^{n+\frac{1}{2}}+H'_y|^{n-\frac{1}{2}}}{2}\\&=\varepsilon_0\varepsilon'\frac{H_y|^{n+\frac{1}{2}}-H_y|^{n-\frac{1}{2}}}{\Delta t}+\sigma\frac{H_y|^{n+\frac{1}{2}}+H_y|^{n-\frac{1}{2}}}{2}.\end{aligned}\tag{6.250}$$

Therefore, we obtain

$$\begin{aligned}H_y|^{n+\frac{1}{2}}=&\frac{2\varepsilon_0\varepsilon'-\sigma\Delta t}{2\varepsilon_0\varepsilon'+\sigma\Delta t}H_y|^{n-\frac{1}{2}}+\frac{2\varepsilon_0\varepsilon_\infty+\sigma\Delta t}{2\varepsilon_0\varepsilon'+\sigma\Delta t}H'_y|^{n+\frac{1}{2}}\\&-\frac{2\varepsilon_0\varepsilon_\infty-\sigma\Delta t}{2\varepsilon_0\varepsilon'+\sigma\Delta t}H'_y|^{n+\frac{1}{2}}.\end{aligned}\tag{6.251}$$

As an example, let consider the case where the permittivity of the medium obeys the Drude dispersion illustrated as

$$\varepsilon(\omega)=\varepsilon_\infty-\frac{\omega_p^2}{\omega^2+i\Gamma\omega}.\tag{6.252}$$

Substituting Eq. (6.252) into Eq. (6.241), we obtain

$$\left(\varepsilon_\infty-\frac{\omega_p^2}{\omega^2+i\Gamma\omega}\right)H'_y=\left(\varepsilon_\infty-\frac{\omega_p^2}{\omega^2+i\Gamma\omega}-\varepsilon_{1r}\sin^2\theta\right)H_y,\tag{6.253}$$

$$\begin{aligned}&(\varepsilon_\infty\omega^2+i\Gamma\varepsilon_\infty\omega-\omega_p^2)H'_y\\&=[(\varepsilon_\infty-\varepsilon_{1r}\sin^2\theta)\omega^2+i\Gamma(\varepsilon_\infty-\varepsilon_{1r}\sin^2\theta)\omega-\omega_p^2]H_y.\end{aligned}\tag{6.254}$$

In the time domain, it is expressed as

$$\begin{aligned}&\varepsilon_\infty\frac{\partial^2H'_y}{\partial t^2}+\Gamma\varepsilon_\infty\frac{\partial H'_y}{\partial t}+\omega_p^2H'_y\\&=(\varepsilon_\infty-\varepsilon_{1r}\sin^2\theta)\frac{\partial^2H_y}{\partial t^2}+\Gamma(\varepsilon_\infty-\varepsilon_{1r}\sin^2\theta)\frac{\partial H_y}{\partial t}+\omega_p^2H_y.\end{aligned}\tag{6.255}$$

Equations (6.234), (6.239), and (6.255) are required for the time evolution of the 1D FDTD. These equations are expressed in the time domain and summarized as

$$\varepsilon_0\varepsilon_\infty\frac{\partial E_x}{\partial t}+J_x=-\frac{\partial H_y}{\partial z},\tag{6.256}$$

$$\mu_0\frac{\partial H'_y}{\partial t}=-\frac{\partial E_x}{\partial z},\tag{6.257}$$

$$\begin{aligned}&\varepsilon_\infty\frac{\partial^2H'_y}{\partial t^2}+\Gamma\varepsilon_\infty\frac{\partial H'_y}{\partial t}+\omega_p^2H'_y\\&=(\varepsilon_\infty-\varepsilon_{1r}\sin^2\theta)\frac{\partial^2H_y}{\partial t^2}+\Gamma(\varepsilon_\infty-\varepsilon_{1r}\sin^2\theta)\frac{\partial H_y}{\partial t}+\omega_p^2H_y.\end{aligned}\tag{6.258}$$

Next, we consider the discretization of Eq. (6.258). Here we need to deal with the second-order derivative with respect to time. Discretizing this second-order derivative in the same way as before, we obtain

$$\frac{\partial^2 H}{\partial t^2} = \frac{H|^{n+\frac{1}{2}} - 2H|^{n-\frac{1}{2}} + H|^{n-\frac{3}{2}}}{\Delta t^2}. \tag{6.259}$$

The first-order and zeroth-order derivatives usually used are

$$\frac{\partial H}{\partial t} = \frac{H|^{n+\frac{1}{2}} - H|^{n-\frac{1}{2}}}{\Delta t}, \tag{6.260}$$

$$H = \frac{H|^{n+\frac{1}{2}} + H|^{n-\frac{1}{2}}}{2}. \tag{6.261}$$

Equation (6.259) is the field at time $t = (n - 1/2)\Delta t$, but Eqs. (6.260) and (6.261) are at time $t = n\Delta t$, so the times are not identical. Therefore, to match the time to $t = (n - 1/2)\Delta t$,

$$\frac{\partial H}{\partial t} = \frac{H|^{n+\frac{1}{2}} - H|^{n-\frac{3}{2}}}{2\Delta t} \tag{6.262}$$

$$H = H|^{n-\frac{1}{2}} \tag{6.263}$$

must be used. As an alternative, Zhang et al. [47] use the following equation instead of the above equation:

$$H = \frac{H|^{n+\frac{1}{2}} + H|^{n-\frac{3}{2}}}{2}. \tag{6.264}$$

Substituting Eqs. (6.259), (6.262), and (6.263) into Eq. (6.258), we obtain

$$\begin{aligned}
&\varepsilon_\infty \frac{H'_y|^{n+\frac{1}{2}} - 2H'_y|^{n-\frac{1}{2}} + H'_y|^{n-\frac{3}{2}}}{\Delta t^2} + \Gamma\varepsilon_\infty \frac{H'_y|^{n+\frac{1}{2}} - H'_y|^{n-\frac{3}{2}}}{2\Delta t} + \omega_p^2 H'_y|^{n-\frac{1}{2}} \\
=&\varepsilon' \frac{H_y|^{n+\frac{1}{2}} - 2H_y|^{n-\frac{1}{2}} + H_y|^{n-\frac{3}{2}}}{\Delta t^2} + \Gamma\varepsilon' \frac{H_y|^{n+\frac{1}{2}} - H_y|^{n-\frac{3}{2}}}{2\Delta t} + \omega_p^2 H_y|^{n-\frac{1}{2}}.
\end{aligned} \tag{6.265}$$

Therefore,

$$\begin{aligned}
H_y|^{n+\frac{1}{2}} =& \frac{4\varepsilon' - 2\omega_p^2\Delta t^2}{\varepsilon'(2+\Gamma\Delta t)} H_y|^{n-\frac{1}{2}} - \frac{\varepsilon'(2-\Gamma\Delta t)}{\varepsilon'(2+\Gamma\Delta t)} H_y|^{n-\frac{3}{2}} \qquad (6.266) \\
&+ \frac{\varepsilon_\infty(2+\Gamma\Delta t)}{\varepsilon'(2+\Gamma\Delta t)} H'_y|^{n+\frac{1}{2}} - \frac{4\varepsilon_\infty - 2\omega_p^2\Delta t^2}{\varepsilon'(2+\Gamma\Delta t)} H'_y|^{n-\frac{1}{2}} \\
&+ \frac{\varepsilon_\infty(2-\Gamma\Delta t)}{\varepsilon'(2+\Gamma\Delta t)} H'_y|^{n-\frac{3}{2}}.
\end{aligned} \tag{6.267}$$

For 1D FDTDs, the usual CPML can be applied without modification [48].

In the CPML domain, Eqs. (6.256) and (6.257) can be rewritten as follows, respectively,

$$\varepsilon_0 \varepsilon_\infty \frac{\partial E_x}{\partial t} + \sigma E_x + J_x = -\frac{1}{\kappa_z}\frac{\partial H_y}{\partial z} - \zeta_z * \frac{\partial H_y}{\partial z}, \tag{6.268}$$

$$\mu_0 \frac{\partial H'_y}{\partial t} = -\frac{1}{\kappa_z}\frac{\partial E_x}{\partial z} - \zeta_z * \frac{\partial E_x}{\partial z}. \tag{6.269}$$

The Courant condition must be taken care of when actually performing 1D FDTD in the z-direction. The phase velocity v_z in the z-direction is

$$v_z = \frac{c}{\sqrt{\varepsilon_{1r}} \cos\theta}. \tag{6.270}$$

Hence, the time step must satisfy the following condition:

$$\Delta t_{1\mathrm{D}} < \frac{\Delta z}{v_z} = \frac{\sqrt{\varepsilon_{1r}} \cos\theta}{c} \Delta z. \tag{6.271}$$

As the incident angle θ increases, v_z becomes faster than the speed of light in vacuum. Therefore, $\Delta t_{1\mathrm{D}}$ must be reduced accordingly. On the other hand, if the time step $\Delta t_{3\mathrm{D}}$ in the 3D FDTD is set to the same value as $\Delta t_{1\mathrm{D}}$ in the 1D FDTD, the computation time becomes longer. In the 3D FDTD, however, we can employ the usual Courant condition for the time step, $\Delta t_{3\mathrm{D}}$. A method to solve this problem has been proposed by Çapoğlu and Smith [49]. The time step of 1D FDTD satisfying $\Delta t_{1\mathrm{D}} = \Delta t_{3\mathrm{D}}/k, (k = 3, 5, 7, ...)$. Then, the method uses the 1D FDTD results as the incident field for the 3D FDTD by thinning out the 1D FDTD results.

In the 1D FDTD calculation, corrections at the TF/SF boundary must be made for E_x and H'_y, as can be seen from Eqs. (6.256) and (6.257). If the incident field is TE$_y$ polarization, the fields that require compensation at the TF/SF boundary perpendicular to the x-axis are H_y on line A and E_z on line B in Figure 6.13. On the TF/SF boundary perpendicular to the z-axis, H_y on line D and E_x on line E are required for the compensation. On the TF/SF boundary perpendicular to the y-axis, compensation are required for E_x and E_z. The calculation procedure is as follows. First, perform 1D FDTD on line A to obtain H_y. Next, E_z on line B is obtained, which cannot be obtained directly. Therefore, first, H_y on line C is obtained by applying a time delay to H_y on line A. Next, from H_y on lines A and C, E_z on line B is obtained. The H_y on line F and E_x on line G are obtained by applying a time delay to H_y and E_x on line A, respectively.

The time delay τ in the field at positions with the same z-position and distance in the x-direction $I\Delta x$ can be calculated as follows. The phase velocity in the x-direction is given by

$$v_x = \frac{c}{k_x} = \frac{c}{\sqrt{\varepsilon_{1r}} \sin\theta}. \tag{6.272}$$

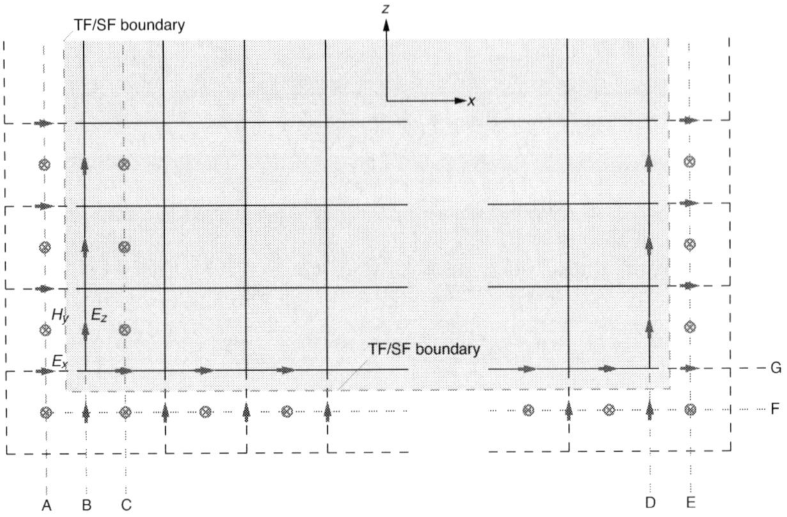

FIGURE 6.13
TF/SF boundary.

Thus, the time delay is

$$\tau = \frac{I\Delta x}{v_x} = \frac{\sqrt{\varepsilon_{1r}}\sin\theta}{c} I\Delta x. \tag{6.273}$$

When the delay time is represented by $\tau = (l + w)\Delta t$ (l is an integer, $0 \leq w < 1$), for example, take E_x can be obtained by interpolation as

$$E_x^i\big|_{i+I+\frac{1}{2},j,k}^{n} = (1-w)E_x^i\big|_{i+\frac{1}{2},j,k}^{n-l} + wE_x^i\big|_{i+\frac{1}{2},j,k}^{n-l+1}. \tag{6.274}$$

6.5.6 Source waveform

When determining the temporal waveform of a wave source, several things need to be taken into account. One is the presence or absence of a DC component, which does not propagate, so it stays in place forever. Another is the inclusion of high-frequency components when the source is excited by continuous waves (CW) with a single frequency. If a sinusoidal wave whose amplitude varies stepwise at time $t = 0$ is used, a high-frequency component with a large amplitude is generated. To avoid this, it is necessary to use a sine wave whose amplitude varies slowly. One example of this is the waveform given by the following equation:

$$j(t) = \begin{cases} 0 & (t < 0) \\ \frac{1}{2}(1 - \cos a\omega t)\sin\omega t & (0 \leq t < \pi/a\omega) \\ \sin\omega t & (\pi/a\omega \leq t) \end{cases}. \tag{6.275}$$

Figure 6.14(a) is the waveform when $a = 1/4$. The dashed line is the electric

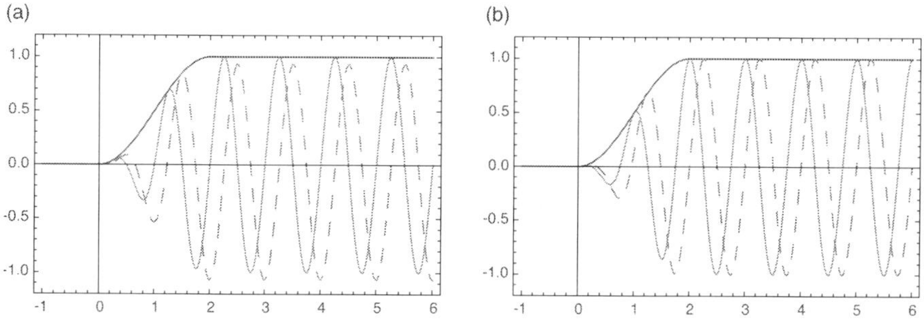

FIGURE 6.14
Wave source waveform given by (a) Eq. (6.275) and (b) Eq. (6.277). Solid curves represent $j(t)$ and dashed curves represent $Q(t)$.

charge $Q(t)$ calculated with

$$Q(t) = \int_0^t j(t')dt'. \tag{6.276}$$

As can be seen from this figure, the average value of the steady-state charge is not zero, but shifts to the negative side, indicating that a steady-state charge (DC component) is generated. The solution to this is to use a cosine wave instead of a sine wave as the carrier wave. Namely,

$$j(t) = \begin{cases} 0 & (t < 0) \\ \frac{1}{2}(1 - \cos a\omega t)\cos\omega t & (0 \le t < \pi/a\omega) \\ \cos\omega t & (\pi/a\omega \le t) \end{cases}. \tag{6.277}$$

Figure 6.14(b) shows its waveform. It can be seen that the average (DC component) of the integral is zero. This is always true when $a = 1/m$ (m is an integer).

6.5.7 Frequency analysis

One approach is to use a continuous wave with a single frequency as the incident light and calculate the steady state of the system, which is performed sequentially at different frequencies. This method is not appropriate because it takes time to obtain a steady-state solution with the FDTD method, unlike frequency-domain methods such as RCWA, where a steady-state solution is directly obtained from the beginning. However, since the FDTD method provides a time-domain solution, the frequency response can be obtained by using short-pulse light as input. The frequency component of the short pulse is in a Fourier transform relationship with the time waveform of the short pulse. Therefore, the frequency response of the system can be obtained by Fourier transforming the time waveform of the output and dividing it by that of the input. This method has the advantage that frequency analysis can be performed with a single calculation of the time evolution. When performing the Fourier transform, it is necessary to store the values of the time series of the electric and magnetic fields at the desired observation location. However, it is sufficient to store the values after appropriate thinning. By thinning out the values, the memory required for storage can be significantly reduced.

Sinusoidally modulated Gaussian waveforms are often used as pulses to determine the frequency response. The sinusoidally modulated Gaussian pulse is given by

$$E(t) = \exp\left[-\left(\frac{t}{\tau}\right)^2\right] \sin\omega_0 t. \tag{6.278}$$

The frequency spectrum $\hat{E}(\omega)$ of this waveform is obtained by Fourier transform as

$$\begin{aligned}\hat{E}(\omega) &= \int_{-\infty}^{\infty} E(t)\exp(-i\omega t)dt \\ &= \frac{\sqrt{\pi}\tau}{2i}\left\{\exp\left[-\left(\frac{\tau}{2}\right)^2(\omega+\omega_0)^2\right] - \exp\left[-\left(\frac{\tau}{2}\right)^2(\omega-\omega_0)^2\right]\right\}.\end{aligned} \tag{6.279}$$

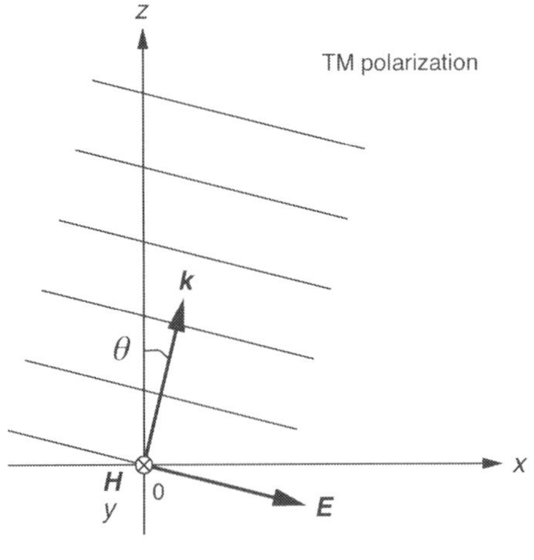

FIGURE 6.15
TM polarized plane wave.

As an example of an incident field propagating in vacuum, consider a TE_y wave incident at an angle of incidence θ as shown in Figure 6.15. For the sinusoidally modulated Gaussian pulse described above, the electric and magnetic fields are given by

$$E_x = \exp\left[-\left(\frac{t - t_0 - \frac{x}{c}\sin\theta - \frac{z}{c}\cos\theta}{\tau}\right)^2\right] \times \sin\left[\omega_0\left(t - t_0 - \frac{x}{c}\sin\theta - \frac{z}{c}\cos\theta\right)\right]\cos\theta, \tag{6.280}$$

$$E_z = -\exp\left[-\left(\frac{t - t_0 - \frac{x}{c}\sin\theta - \frac{z}{c}\cos\theta}{\tau}\right)^2\right] \times \sin\left[\omega_0\left(t - t_0 - \frac{x}{c}\sin\theta - \frac{z}{c}\cos\theta\right)\right]\sin\theta, \tag{6.281}$$

$$H_y = \frac{1}{z_0}\exp\left[-\left(\frac{t - t_0 - \frac{x}{c}\sin\theta - \frac{z}{c}\cos\theta}{\tau}\right)^2\right] \times \sin\left[\omega_0\left(t - t_0 - \frac{x}{c}\sin\theta - \frac{z}{c}\cos\theta\right)\right]. \tag{6.282}$$

The problem here is how much t_0 should be taken. If t_0 is small, a residual charge will be generated. In our experience, it is sufficient to set $t_0 \geq 5\tau$, but of course, it depends on the required accuracy.

6.5.8 Oblique incidence under periodic boundaries

When calculating the optical response of spatially periodic objects, there is no problem at all when the incidence is perpendicular to the direction of the period. However, for oblique incidence, difficulties arise in setting the periodic boundary conditions.

Various methods have been proposed to introduce obliquly incident light for periodic boundaries. Their features are shown in Figure 6.16. The Sin-Cosine method [51] allows only one point in $k_x - \omega$ space to be computed simultaneously where k_x is the in-plane wave vector. The most sophisticated is the split-field method developed by Roden et al. [52]. This method provides a broadband frequency response for a single angle of incidence. In contrast, a simpler method, named Spectral FDTD, was proposed by Aminian and Rahmat-Samii [53]. This method allows a one-time computation of the frequency response on the line of $k_x = const.$ in $k_x - \omega$ space. The problem with this method arises from the fact that the incident pulse plane wave contains evanescent waves. Since evanescent waves propagate parallel to the period and PML has no effect on them, the evanescent waves diverge where some resonances exist. A solution to this problem has been proposed by Yang et al. [50]. Schurig [54] used the fact that the periodic boundary condition is directly applicable when an integer multiple of the unit cell is equal to an integer multiple

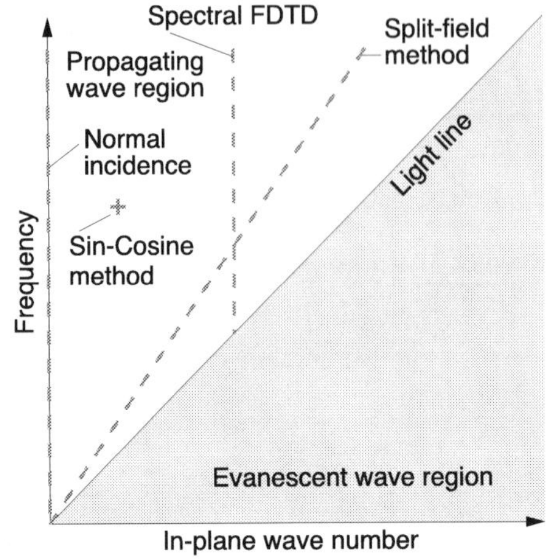

FIGURE 6.16
Combination of wavenumbers and frequencies that can be obtained at once by various methods in $k - \omega$ space [50].

of the wavelength in that direction. However, this method is only applicable to the points on discrete curves in space spanned by the angle of incidence and frequency. Therefore, Schurig also describes an interpolation method between them.

6.6 Transformation from near field to far field

It is not practical to calculate the far field directly by FDTD, when calculating the scattered field by scatterers. Therefore, a method to calculate the far field from the near field is required. A method that calculates virtual electric current and magnetic current sources on a closed surface surrounding the scatterer and then calculates the far field has been proposed (surface integral method) [55]. On the other hand, a volume integral method has also been proposed. This method calculates the energy dissipation from the electric field in the scatterers. Only the electric field in the scatterers is used. This method can also be applied when there is absorption in the medium.

Zhai et al. [56] compare the superiority of the surface and volume integration methods in terms of the computer resources required. They show that the surface integration method is advantageous when the refractive index of the scatterers is high or when the size parameter ($= \pi nd/\lambda$; n and d are the refractive index and the size of the scatterer, respectively, and λ is the wavelength in vacuum) is large. The reason for this is that the higher the refractive index, the shorter the wavelength inside the particle, and the larger the size parameter, the steeper the change in the field inside the particle, which requires a finer grid. Here, we introduce the surface integration method proposed by Luebbers et al. [57].

If electric and magnetic current sources are confined in closed space V, the far field can be calculated from their distributions $\boldsymbol{J}(\boldsymbol{r})$ and $\boldsymbol{M}(\boldsymbol{r})$. This is illustrated as

$$\boldsymbol{E}(\boldsymbol{r}) = i\omega\mu_0\mu\boldsymbol{A}(\boldsymbol{r}) - \frac{1}{i\omega\varepsilon_0\varepsilon}\nabla\nabla\cdot\boldsymbol{A}(\boldsymbol{r}) - \nabla\times\boldsymbol{F}(\boldsymbol{r}), \tag{6.283}$$

$$\boldsymbol{H}(\boldsymbol{r}) = i\omega\varepsilon_0\varepsilon\boldsymbol{F}(\boldsymbol{r}) - \frac{1}{i\omega\mu_0\mu}\nabla\nabla\cdot\boldsymbol{F}(\boldsymbol{r}) + \nabla\times\boldsymbol{A}(\boldsymbol{r}), \tag{6.284}$$

where $\boldsymbol{r}$ is the position vector, $\boldsymbol{A}$ is the magnetic vector potential, and $\boldsymbol{F}$ is the electric vector potential, each given by

$$\boldsymbol{A}(\boldsymbol{r}) = \int_V \boldsymbol{J}(\boldsymbol{r}')\frac{e^{ikR}}{4\pi R}d^3\boldsymbol{r}', \tag{6.285}$$

$$\boldsymbol{F}(\boldsymbol{r}) = \int_V \boldsymbol{M}(\boldsymbol{r}')\frac{e^{ikR}}{4\pi R}d^3\boldsymbol{r}', \tag{6.286}$$

where $R = |\boldsymbol{r} - \boldsymbol{r}'|$. Using these equations, the far field can be calculated, but it is necessary to perform calculations for all electric and magnetic current sources in the closed space, which requires much computation time. Therefore, we consider replacing the volume integral with a surface integral.

Consider an arbitrary closed surface S surrounding electric and magnetic current sources. Then, consider a virtual electric current source $\boldsymbol{J}_s$ and a virtual magnetic current source $\boldsymbol{M}_s$ on this surface. These electric and magnetic current sources can be calculated from the electric field $\boldsymbol{E}(\boldsymbol{r}_s)$ and magnetic field $\boldsymbol{H}(\boldsymbol{r}_s)$ on the closed surface and are expressed as follows:

$$\boldsymbol{J}_s(\boldsymbol{r}_s) = \hat{\boldsymbol{n}}\times\boldsymbol{H}(\boldsymbol{r}_s), \tag{6.287}$$

$$\boldsymbol{M}_s(\boldsymbol{r}_s) = \boldsymbol{E}(\boldsymbol{r}_s)\times\hat{\boldsymbol{n}}, \tag{6.288}$$

where $\hat{\boldsymbol{n}}$ is the unit normal vector toward the outside of the closed surface. Using Green's theorem, the radiation to the outside of the closed surface can be expressed in terms of radiation by surface electric and magnetic currents on the closed surface. As a result, the Eqs. (6.285) and (6.286) are given as

$$\boldsymbol{A}(\boldsymbol{r}) = \int_S \boldsymbol{J}_s(\boldsymbol{r}')\frac{e^{ikR}}{4\pi R}d^2\boldsymbol{r}', \tag{6.289}$$

$$\boldsymbol{F}(\boldsymbol{r}) = \int_S \boldsymbol{M}_s(\boldsymbol{r}')\frac{e^{ikR}}{4\pi R}d^2\boldsymbol{r}'. \tag{6.290}$$

Next, we describe the specific calculation method. First, we approximate $e^{-jkR}/4\pi R$ at the far end:

$$R = [r^2 + r'^2 - 2rr'(\hat{\boldsymbol{r}} \cdot \hat{\boldsymbol{r}}')]^{1/2}, \tag{6.291}$$

where $r = |\boldsymbol{r}|$, $r' = |\boldsymbol{r}'|$, $\hat{\boldsymbol{r}} = \boldsymbol{r}/r$, and $\hat{\boldsymbol{r}}' = \boldsymbol{r}'/r'$. Since $r \gg r'$, it can be approximated as $R \simeq r$, but this approximation is not sufficient when considering the phase. Therefore,

$$R \simeq r - r'(\hat{\boldsymbol{r}} \cdot \hat{\boldsymbol{r}}') \tag{6.292}$$

should be used. In the far distance if these approximations are used, we obtain

$$\frac{e^{ikR}}{4\pi R} \simeq \frac{e^{ikr} e^{-ikr'(\hat{\boldsymbol{r}} \cdot \hat{\boldsymbol{r}}')}}{4\pi r}. \tag{6.293}$$

Using Eqs. (6.283), (6.284), (6.289), (6.290), and (6.293), the far field is approximated as

$$E_r \simeq 0, \tag{6.294}$$

$$E_\theta \simeq ik \frac{e^{ikr}}{4\pi r}(L_\phi + ZN_\theta), \tag{6.295}$$

$$E_\phi \simeq -ik \frac{e^{ikr}}{4\pi r}(L_\theta - ZN_\phi), \tag{6.296}$$

$$H_r \simeq 0, \tag{6.297}$$

$$H_\theta \simeq -ik \frac{e^{ikr}}{4\pi r}\left(N_\phi - \frac{1}{Z}L_\theta\right) = -\frac{1}{Z}E_\phi, \tag{6.298}$$

$$H_\phi \simeq ik \frac{e^{ikr}}{4\pi r}\left(N_\theta + \frac{1}{Z}L_\phi\right) = \frac{1}{Z}E_\theta, \tag{6.299}$$

where Z is the wave impedance and N_θ, N_ϕ, L_θ, and L_ϕ are the components of the following vectors:

$$\boldsymbol{N}(\theta, \phi) = \int_S \boldsymbol{J}_s e^{-i\boldsymbol{k}\cdot\boldsymbol{r}'} d^2\boldsymbol{r}', \tag{6.300}$$

$$\boldsymbol{L}(\theta, \phi) = \int_S \boldsymbol{M}_s e^{-i\boldsymbol{k}\cdot\boldsymbol{r}'} d^2\boldsymbol{r}', \tag{6.301}$$

where (θ, ϕ) is the direction of $\boldsymbol{r}$, i.e., the observation angle (θ is the polar angle and ϕ is the azimuthal angle), and the vector $\boldsymbol{k}$ is the wave vector toward this direction,

$$\boldsymbol{k} = k\hat{\boldsymbol{r}} = k(\sin\theta\cos\phi\hat{\boldsymbol{e}}_x + \sin\theta\sin\phi\hat{\boldsymbol{e}}_y + \cos\theta\hat{\boldsymbol{e}}_z), \tag{6.302}$$

where $\hat{\boldsymbol{r}}$, $\hat{\boldsymbol{e}}_x$, $\hat{\boldsymbol{e}}_y$, and $\hat{\boldsymbol{e}}_z$ are unit vectors. Using Eqs. (6.294) and (6.299), the Poynting vector has only components in the r direction remaining,

$$S_r = E_\theta H_\phi^* - E_\phi H_\theta^* = \frac{k^2}{16\pi^2 r^2 Z}\left(|L_\phi + ZN_\theta|^2 + |L_\theta - ZN_\phi|^2\right). \tag{6.303}$$

Next, we will discuss how to obtain the surface electric and magnetic currents. It is most usual to take a closed surface S to be a rectangular solid consisting of a set of E-cells. Let $(i_{s1}\Delta x, j_{s1}\Delta y, k_{s1}\Delta z)$ and $(i_{s2}\Delta x, j_{s2}\Delta y, k_{s2}\Delta z)$ denote the diagonal coordinates of the rectangular solid, respectively. The relation between the magnetic current $\boldsymbol{M}$ and the electric field $\boldsymbol{E}$ is given by

$$\boldsymbol{M} = \boldsymbol{E} \times \boldsymbol{n}. \tag{6.304}$$

As an example, consider the magnetic current on $x = i_{s1}\Delta x$ surface of a rectangular S,

$$\hat{\boldsymbol{e}}_z M_z|^n_{i_{s1},j+\frac{1}{2},k} = -\hat{\boldsymbol{e}}_y E_y|^n_{i_{s1},j+\frac{1}{2},k} \times \hat{\boldsymbol{e}}_x = \hat{\boldsymbol{e}}_z E_y|^n_{i_{s1},j+\frac{1}{2},k}. \tag{6.305}$$

Similarly,

$$\hat{\boldsymbol{e}}_y M_y|^n_{i_{s1},j,k+\frac{1}{2}} = -\hat{\boldsymbol{e}}_y E_z|^n_{i_{s1},j,k+\frac{1}{2}}. \tag{6.306}$$

Thus,

$$M_z|^n_{i_{s1},j+\frac{1}{2},k} = E_y|^n_{i_{s1},j+\frac{1}{2},k}, \tag{6.307}$$

$$M_y|^n_{i_{s1},j,k+\frac{1}{2}} = -E_z|^n_{i_{s1},j,k+\frac{1}{2}}. \tag{6.308}$$

Similarly, on $x = i_{s2}\Delta x$ surface,

$$M_z|^n_{i_{s2},j+\frac{1}{2},k} = -E_y|^n_{i_{s2},j+\frac{1}{2},k}, \tag{6.309}$$

$$M_y|^n_{i_{s2},j,k+\frac{1}{2}} = E_z|^n_{i_{s2},j,k+\frac{1}{2}}. \tag{6.310}$$

On $y = j_{s1}\Delta y$ surface,

$$M_x|^n_{i,j_{s1},k+\frac{1}{2}} = E_z|^n_{i,j_{s1},k+\frac{1}{2}}, \tag{6.311}$$

$$M_z|^n_{i+\frac{1}{2},j_{s1},k} = -E_x|^n_{i+\frac{1}{2},j_{s1},k}. \tag{6.312}$$

On $y = j_{s2}\Delta y$ surface,

$$M_x|^n_{i,j_{s2},k+\frac{1}{2}} = -E_z|^n_{i,j_{s2},k+\frac{1}{2}}, \tag{6.313}$$

$$M_z|^n_{i+\frac{1}{2},j_{s2},k} = E_x|^n_{i+\frac{1}{2},j_{s2},k}. \tag{6.314}$$

Similarly, on $z = k_{s1}\Delta z$ surface,

$$M_y|^n_{i+\frac{1}{2},j,k_{s1}} = E_x|^n_{i+\frac{1}{2},j,k_{s1}}, \tag{6.315}$$

$$M_x|^n_{i,j+\frac{1}{2},k_{s1}} = -E_y|^n_{i,j+\frac{1}{2},k_{s1}}. \tag{6.316}$$

On the $z = k_{s2}\Delta z$ surface,

$$M_y|^n_{i+\frac{1}{2},j,k_{s2}} = -E_x|^n_{i+\frac{1}{2},j,k_{s2}}, \tag{6.317}$$

$$M_x|^n_{i,j+\frac{1}{2}j,k_{s2}} = E_y|^n_{i,j+\frac{1}{2},k_{s2}}. \tag{6.318}$$

For example, $L_x(\theta,\phi)$ at $z = k_{s2}\Delta z$ surface can be expressed by using Eq. (6.318) as follows:

$$L_x(\theta,\phi)|^n_{k_{s2}} \simeq \sum_{j=j_{s1}}^{j_{s2}-1} \sum_{i=i_{s1}}^{i_{s2}} \nu^i_{i_{s1},i_{s2}} M_x|^n_{i,j+\frac{1}{2},k_{s2}} \exp(-i\boldsymbol{k}\cdot\boldsymbol{r}'_{i,j+\frac{1}{2},k_{s2}})\Delta x\Delta y$$
$$= \sum_{j=j_{s1}}^{j_{s2}-1} \sum_{i=i_{s1}}^{i_{s2}} \nu^i_{i_{s1},i_{s2}} E_y|^n_{i,j+\frac{1}{2},k_{s2}} \exp(-i\boldsymbol{k}\cdot\boldsymbol{r}'_{i,j+\frac{1}{2},k_{s2}})\Delta x\Delta y, \tag{6.319}$$

where

$$\nu^i_{i_{s1},i_{s2}} = \begin{cases} \frac{1}{2} & \text{if } i = i_{s1} \text{ or } i = i_{s2} \\ 1 & \text{else} \end{cases}. \tag{6.320}$$

Considering the contributions from all surfaces of rectangular S, we obtain

$$\begin{aligned} L_x(\theta,\phi)|^n \simeq &- \sum_{j=j_{s1}}^{j_{s2}-1} \sum_{i=i_{s1}}^{i_{s2}} \nu^i_{i_{s1},i_{s2}} E_y|^n_{i,j+\frac{1}{2},k_{s1}} \exp(-i\boldsymbol{k}\cdot\boldsymbol{r}'_{i,j+\frac{1}{2},k_{s1}})\Delta x\Delta y \\ &+ \sum_{j=j_{s1}}^{j_{s2}-1} \sum_{i=i_{s1}}^{i_{s2}} \nu^i_{i_{s1},i_{s2}} E_y|^n_{i,j+\frac{1}{2},k_{s2}} \exp(-i\boldsymbol{k}\cdot\boldsymbol{r}'_{i,j+\frac{1}{2},k_{s2}})\Delta x\Delta y \\ &+ \sum_{k=k_{s1}}^{k_{s2}-1} \sum_{i=i_{s1}}^{i_{s2}} \nu^i_{i_{s1},i_{s2}} E_z|^n_{i,j_{s1},k+\frac{1}{2}} \exp(-i\boldsymbol{k}\cdot\boldsymbol{r}'_{i,j_{s1},k+\frac{1}{2}})\Delta z\Delta x \\ &- \sum_{k=k_{s1}}^{k_{s2}-1} \sum_{i=i_{s1}}^{i_{s2}} \nu^i_{i_{s1},i_{s2}} E_z|^n_{i,j_{s2},k+\frac{1}{2}} \exp(-i\boldsymbol{k}\cdot\boldsymbol{r}'_{i,j_{s2},k+\frac{1}{2}})\Delta z\Delta x. \end{aligned} \tag{6.321}$$

There is no contribution from the surface perpendicular to x-axis. Similarly, we obtain

$$\begin{aligned} L_y(\theta,\phi)|^n \simeq &- \sum_{k=k_{s1}}^{k_{s2}-1} \sum_{j=j_{s1}}^{j_{s2}} \nu^j_{j_{s1},j_{s2}} E_z|^n_{i_{s1},j,k+\frac{1}{2}} \exp(-i\boldsymbol{k}\cdot\boldsymbol{r}'_{i_{s1},j,k+\frac{1}{2}})\Delta y\Delta z \\ &+ \sum_{k=k_{s1}}^{k_{s2}-1} \sum_{j=j_{s1}}^{j_{s2}} \nu^j_{j_{s1},j_{s2}} E_z|^n_{i_{s2},j,k+\frac{1}{2}} \exp(-i\boldsymbol{k}\cdot\boldsymbol{r}'_{i_{s2},j,k+\frac{1}{2}})\Delta y\Delta z \\ &+ \sum_{i=i_{s1}}^{i_{s2}-1} \sum_{j=j_{s1}}^{j_{s2}} \nu^j_{j_{s1},j_{s2}} E_x|^n_{i+\frac{1}{2},j,k_{s1}} \exp(-i\boldsymbol{k}\cdot\boldsymbol{r}'_{i+\frac{1}{2},j,k_{s1}})\Delta x\Delta y \\ &- \sum_{i=i_{s1}}^{i_{s2}-1} \sum_{j=j_{s1}}^{j_{s2}} \nu^j_{j_{s1},j_{s2}} E_x|^n_{i+\frac{1}{2},j,k_{s2}} \exp(-i\boldsymbol{k}\cdot\boldsymbol{r}'_{i+\frac{1}{2},j,k_{s2}})\Delta x\Delta y, \end{aligned} \tag{6.322}$$

$$
\begin{aligned}
L_z(\theta,\phi)|^n \simeq &-\sum_{i=i_{s1}}^{i_{s2}-1}\sum_{k=k_{s1}}^{k_{s2}} \nu^k_{k_{s1},k_{s2}} E_x|^n_{i+\frac{1}{2},j_{s1},k} \exp(-i\boldsymbol{k}\cdot\boldsymbol{r}'_{i+\frac{1}{2},j_{s1},k})\Delta z\Delta x \\
&+\sum_{i=i_{s1}}^{i_{s2}-1}\sum_{k=k_{s1}}^{k_{s2}} \nu^k_{k_{s1},k_{s2}} E_x|^n_{i+\frac{1}{2},j_{s2},k} \exp(-i\boldsymbol{k}\cdot\boldsymbol{r}'_{i+\frac{1}{2},j_{s2},k})\Delta z\Delta x \\
&+\sum_{j=j_{s1}}^{j_{s2}-1}\sum_{k=k_{s1}}^{k_{s2}} \nu^k_{k_{s1},k_{s2}} E_y|^n_{i_{s1},j+\frac{1}{2},k} \exp(-i\boldsymbol{k}\cdot\boldsymbol{r}'_{i_{s1},j+\frac{1}{2},k})\Delta y\Delta z \\
&-\sum_{j=j_{s1}}^{j_{s2}-1}\sum_{k=k_{s1}}^{k_{s2}} \nu^k_{k_{s1},k_{s2}} E_y|^n_{i_{s2},j+\frac{1}{2},k} \exp(-i\boldsymbol{k}\cdot\boldsymbol{r}'_{i_{s2},j+\frac{1}{2},k})\Delta y\Delta z.
\end{aligned}
\tag{6.323}
$$

The tangential components of the magnetic field are necessary to obtain the surface electric currents, but since these are defined 1/2 cell away from the closed surface composed with E-cells, a little ingenuity is required. Simply take the average of two values that are only $\pm 1/2$ cells apart across the closed surface. Since the defined time of the magnetic field also deviates from that of the electric field by $\Delta t/2$, it is necessary to take the average of the time as well. That is, on the surface $z = k_{s2}\Delta z$,

$$
\begin{aligned}
J_y|^n_{i,j+\frac{1}{2},k_{s2}} =& \frac{1}{4}\left(H_x|^{n+\frac{1}{2}}_{i,j+\frac{1}{2},k_{s2}-\frac{1}{2}} + H_x|^{n+\frac{1}{2}}_{i,j+\frac{1}{2},k_{s2}+\frac{1}{2}}\right. \\
&\left.+H_x|^{n-\frac{1}{2}}_{i,j+\frac{1}{2},k_{s2}-\frac{1}{2}} + H_x|^{n-\frac{1}{2}}_{i,j+\frac{1}{2},k_{s2}+\frac{1}{2}}\right),
\end{aligned}
\tag{6.324}
$$

$$
\begin{aligned}
J_x|^n_{i+\frac{1}{2},j,k_{s2}} =& -\frac{1}{4}\left(H_y|^{n+\frac{1}{2}}_{i+\frac{1}{2},j,k_{s2}-\frac{1}{2}} + H_y|^{n+\frac{1}{2}}_{i+\frac{1}{2},j,k_{s2}+\frac{1}{2}}\right. \\
&\left.+H_y|^{n-\frac{1}{2}}_{i+\frac{1}{2},j,k_{s2}-\frac{1}{2}} + H_y|^{n-\frac{1}{2}}_{i+\frac{1}{2},j,k_{s2}+\frac{1}{2}}\right).
\end{aligned}
\tag{6.325}
$$

Similarly on $x = i_{s2}\Delta x$ surface,

$$
\begin{aligned}
J_z|^n_{i_{s2},j,k+\frac{1}{2}} =& \frac{1}{4}\left(H_y|^{n+\frac{1}{2}}_{i_{s2}-\frac{1}{2},j,k+\frac{1}{2}} + H_y|^{n+\frac{1}{2}}_{i_{s2}+\frac{1}{2},j,k+\frac{1}{2}}\right. \\
&\left.+H_y|^{n-\frac{1}{2}}_{i_{s2}-\frac{1}{2},j,k+\frac{1}{2}} + H_y|^{n-\frac{1}{2}}_{i_{s2}+\frac{1}{2},j,k+\frac{1}{2}}\right),
\end{aligned}
\tag{6.326}
$$

$$
\begin{aligned}
J_y|^n_{i_{s2},j+\frac{1}{2},k} =& -\frac{1}{4}\left(H_z|^{n+\frac{1}{2}}_{i_{s2}-\frac{1}{2},j+\frac{1}{2},k} + H_z|^{n+\frac{1}{2}}_{i_{s2}+\frac{1}{2},j+\frac{1}{2},k}\right. \\
&\left.+H_z|^{n-\frac{1}{2}}_{i_{s2}-\frac{1}{2},j+\frac{1}{2},k} + H_z|^{n-\frac{1}{2}}_{i_{s2}+\frac{1}{2},j+\frac{1}{2},k}\right).
\end{aligned}
\tag{6.327}
$$

On $y = j_{s2}\Delta y$ surface,

$$
\begin{aligned}
J_x|^n_{i+\frac{1}{2},j,k_{s2}} =& \frac{1}{4}\left(H_z|^{n+\frac{1}{2}}_{i+\frac{1}{2},j_{s2}-\frac{1}{2},k} + H_z|^{n+\frac{1}{2}}_{i+\frac{1}{2},j_{s2}+\frac{1}{2},k}\right. \\
&\left.+H_z|^{n-\frac{1}{2}}_{i+\frac{1}{2},j_{s2}-\frac{1}{2},k} + H_z|^{n-\frac{1}{2}}_{i+\frac{1}{2},j_{s2}+\frac{1}{2},k}\right),
\end{aligned}
\tag{6.328}
$$

$$J_z|^n_{i,j_{s2},k+\frac{1}{2}} = -\frac{1}{4}\left(H_x|^{n+\frac{1}{2}}_{i,j_{s2}-\frac{1}{2},k+\frac{1}{2}} + H_x|^{n+\frac{1}{2}}_{i,j_{s2}+\frac{1}{2},k+\frac{1}{2}} + H_x|^{n-\frac{1}{2}}_{i,j_{s2}-\frac{1}{2},k+\frac{1}{2}} + H_x|^{n-\frac{1}{2}}_{i,j_{s2}+\frac{1}{2},k+\frac{1}{2}}\right). \tag{6.329}$$

It is known that employing the geometric mean instead of the arithmetic mean improves the accuracy [58].

The contribution of $N_x(\theta,\phi)$ from the $z = k_{s2}\Delta z$ surface can be expressed by using Eq. (6.325) as

$$\begin{aligned} &N_x(\theta,\phi)|^n_{k_{s2}} \\ &\simeq \sum_{i=i_{s1}}^{i_{s2}-1}\sum_{j=j_{s1}}^{j_{s2}} \nu^j_{j_{s1},j_{s2}} J_x|^n_{i+\frac{1}{2},j,k_{s2}} \exp(j\boldsymbol{k}\cdot\boldsymbol{r}'_{i+\frac{1}{2},j,k_{s2}})\Delta x\Delta y \\ &= \frac{1}{4}\sum_{i=i_{s1}}^{i_{s2}-1}\sum_{j=j_{s1}}^{j_{s2}} \nu^j_{j_{s1},j_{s2}}\left(H_y|^{n+\frac{1}{2}}_{i+\frac{1}{2},j,k_{s2}-\frac{1}{2}} + H_y|^{n+\frac{1}{2}}_{i+\frac{1}{2},j,k_{s2}+\frac{1}{2}}\right. \\ &\left.+H_y|^{n-\frac{1}{2}}_{i+\frac{1}{2},j,k_{s2}-\frac{1}{2}} + H_y|^{n-\frac{1}{2}}_{i+\frac{1}{2},j,k_{s2}+\frac{1}{2}}\right)\exp(j\boldsymbol{k}\cdot\boldsymbol{r}'_{i+\frac{1}{2},j,k_{s2}})\Delta x\Delta y. \end{aligned} \tag{6.330}$$

Considering the contributions from all surfaces, we obtain

$$\begin{aligned} &N_x(\theta,\phi)|^n \simeq \\ &+\frac{1}{4}\sum_{i=i_{s1}}^{i_{s2}-1}\sum_{j=j_{s1}}^{j_{s2}} \nu^j_{j_{s1},j_{s2}}\left(H_y|^{n+\frac{1}{2}}_{i+\frac{1}{2},j,k_{s1}-\frac{1}{2}} + H_y|^{n+\frac{1}{2}}_{i+\frac{1}{2},j,k_1+\frac{1}{2}}\right. \\ &\left.+H_y|^{n-\frac{1}{2}}_{i+\frac{1}{2},j,k_{s1}-\frac{1}{2}} + H_y|^{n-\frac{1}{2}}_{i+\frac{1}{2},j,k_1+\frac{1}{2}}\right)\exp(-i\boldsymbol{k}\cdot\boldsymbol{r}'_{i+\frac{1}{2},j,k_{s1}})\Delta x\Delta y \\ &-\frac{1}{4}\sum_{i=i_{s1}}^{i_{s2}-1}\sum_{j=j_{s1}}^{j_{s2}} \nu^j_{j_{s1},j_{s2}}\left(H_y|^{n+\frac{1}{2}}_{i+\frac{1}{2},j,k_{s2}-\frac{1}{2}} + H_y|^{n+\frac{1}{2}}_{i+\frac{1}{2},j,k_{s2}+\frac{1}{2}}\right. \\ &\left.+H_y|^{n-\frac{1}{2}}_{i+\frac{1}{2},j,k_{s2}-\frac{1}{2}} + H_y|^{n-\frac{1}{2}}_{i+\frac{1}{2},j,k_{s2}+\frac{1}{2}}\right)\exp(-i\boldsymbol{k}\cdot\boldsymbol{r}'_{i+\frac{1}{2},j,k_{s2}})\Delta x\Delta y \\ &-\frac{1}{4}\sum_{i=i_{s1}}^{i_{s2}-1}\sum_{k=k_{s1}}^{k_{s2}} \nu^k_{k_{s1},k_{s2}}\left(H_z|^{n+\frac{1}{2}}_{i+\frac{1}{2},j_{s1}-\frac{1}{2},k} + H_z|^{n+\frac{1}{2}}_{i+\frac{1}{2},j_{s1}+\frac{1}{2},k}\right. \\ &\left.+H_z|^{n-\frac{1}{2}}_{i+\frac{1}{2},j_{s1}-\frac{1}{2},k} + H_z|^{n-\frac{1}{2}}_{i+\frac{1}{2},j_{s1}+\frac{1}{2},k}\right)\exp(-i\boldsymbol{k}\cdot\boldsymbol{r}'_{i+\frac{1}{2},j_{s1},k})\Delta z\Delta x \\ &+\frac{1}{4}\sum_{i=i_{s1}}^{i_{s2}-1}\sum_{k=k_{s1}}^{k_{s2}} \nu^k_{k_{s1},k_{s2}}\left(H_z|^{n+\frac{1}{2}}_{i+\frac{1}{2},j_{s2}-\frac{1}{2},k} + H_z|^{n+\frac{1}{2}}_{i+\frac{1}{2},j_{s2}+\frac{1}{2},k}\right. \\ &\left.+H_z|^{n-\frac{1}{2}}_{i+\frac{1}{2},j_{s2}-\frac{1}{2},k} + H_z|^{n-\frac{1}{2}}_{i+\frac{1}{2},j_{s2}+\frac{1}{2},k}\right)\exp(-i\boldsymbol{k}\cdot\boldsymbol{r}'_{i+\frac{1}{2},j_{s2},k})\Delta z\Delta x, \end{aligned} \tag{6.331}$$

$$
\begin{aligned}
&N_y(\theta,\phi)|^n \simeq \\
&+\frac{1}{4}\sum_{j=j_{s1}}^{j_{s2}-1}\sum_{k=k_{s1}}^{k_{s2}} \nu^k_{k_{s1},k_{s2}} \left(H_z|^{n+\frac{1}{2}}_{i_{s1}-\frac{1}{2},j+\frac{1}{2},k} + H_z|^{n+\frac{1}{2}}_{i_{s1}+\frac{1}{2},j+\frac{1}{2},k}\right. \\
&\left.+H_z|^{n-\frac{1}{2}}_{i_{s1}-\frac{1}{2},j+\frac{1}{2},k} + H_z|^{n-\frac{1}{2}}_{i_{s1}+\frac{1}{2},j+\frac{1}{2},k}\right) \exp(-i\boldsymbol{k}\cdot\boldsymbol{r}'_{i_{s1},j+\frac{1}{2},k})\Delta y\Delta z \\
&-\frac{1}{4}\sum_{j=j_{s1}}^{j_{s2}-1}\sum_{k=k_{s1}}^{k_{s2}} \nu^k_{k_{s1},k_{s2}} \left(H_z|^{n+\frac{1}{2}}_{i_{s2}-\frac{1}{2},j+\frac{1}{2},k} + H_z|^{n+\frac{1}{2}}_{i_{s2}+\frac{1}{2},j+\frac{1}{2},k}\right. \\
&\left.+H_z|^{n-\frac{1}{2}}_{i_{s2}-\frac{1}{2},j+\frac{1}{2},k} + H_z|^{n-\frac{1}{2}}_{i_{s2}+\frac{1}{2},j+\frac{1}{2},k}\right) \exp(-i\boldsymbol{k}\cdot\boldsymbol{r}'_{i_{s2},j+\frac{1}{2},k})\Delta y\Delta z \\
&-\frac{1}{4}\sum_{j=j_{s1}}^{j_{s2}-1}\sum_{i=i_{s1}}^{i_{s2}} \nu^i_{i_{s1},i_{s2}} \left(H_x|^{n+\frac{1}{2}}_{i,j+\frac{1}{2},k_{s1}-\frac{1}{2}} + H_x|^{n+\frac{1}{2}}_{i,j+\frac{1}{2},k_{s1}+\frac{1}{2}}\right. \\
&\left.+H_x|^{n-\frac{1}{2}}_{i,j+\frac{1}{2},k_{s1}-\frac{1}{2}} + H_x|^{n-\frac{1}{2}}_{i,j+\frac{1}{2},k_{s1}+\frac{1}{2}}\right) \exp(-i\boldsymbol{k}\cdot\boldsymbol{r}'_{i,j+\frac{1}{2},k_{s1}})\Delta x\Delta y \\
&+\frac{1}{4}\sum_{j=j_{s1}}^{j_{s2}-1}\sum_{i=i_{s1}}^{i_{s2}} \nu^i_{i_{s1},i_{s2}} \left(H_x|^{n+\frac{1}{2}}_{i,j+\frac{1}{2},k_{s2}-\frac{1}{2}} + H_x|^{n+\frac{1}{2}}_{i,j+\frac{1}{2},k_{s2}+\frac{1}{2}}\right. \\
&\left.+H_x|^{n-\frac{1}{2}}_{i,j+\frac{1}{2},k_{s2}-\frac{1}{2}} + H_x|^{n-\frac{1}{2}}_{i,j+\frac{1}{2},k_{s2}+\frac{1}{2}}\right) \exp(-i\boldsymbol{k}\cdot\boldsymbol{r}'_{i,j+\frac{1}{2},k_{s2}})\Delta x\Delta y, \qquad (6.332)
\end{aligned}
$$

$$
\begin{aligned}
&N_z(\theta,\phi)|^n \simeq \\
&+\frac{1}{4}\sum_{k=k_{s1}}^{k_{s2}-1}\sum_{i=i_{s1}}^{i_{s2}} \nu^i_{i_{s1},i_{s2}} \left(H_x|^{n+\frac{1}{2}}_{i,j_{s1}-\frac{1}{2},k+\frac{1}{2}} + H_x|^{n+\frac{1}{2}}_{i,j_{s1}+\frac{1}{2},k+\frac{1}{2}}\right. \\
&\left.+H_x|^{n-\frac{1}{2}}_{i,j_{s1}-\frac{1}{2},k+\frac{1}{2}} + H_x|^{n-\frac{1}{2}}_{i,j_{s1}+\frac{1}{2},k+\frac{1}{2}}\right) \exp(-i\boldsymbol{k}\cdot\boldsymbol{r}'_{i,j_{s1},k+\frac{1}{2}})\Delta z\Delta x \\
&-\frac{1}{4}\sum_{k=k_{s1}}^{k_{s2}-1}\sum_{i=i_{s1}}^{i_{s2}} \nu^i_{i_{s1},i_{s2}} \left(H_x|^{n+\frac{1}{2}}_{i,j_{s2}-\frac{1}{2},k+\frac{1}{2}} + H_x|^{n+\frac{1}{2}}_{i,j_{s2}+\frac{1}{2},k+\frac{1}{2}}\right. \\
&\left.+H_x|^{n-\frac{1}{2}}_{i,j_{s2}-\frac{1}{2},k+\frac{1}{2}} + H_x|^{n-\frac{1}{2}}_{i,j_{s2}+\frac{1}{2},k+\frac{1}{2}}\right) \exp(-i\boldsymbol{k}\cdot\boldsymbol{r}'_{i,j_{s2},k+\frac{1}{2}})\Delta z\Delta x \\
&-\frac{1}{4}\sum_{k=k_{s1}}^{k_{s2}-1}\sum_{j=j_{s1}}^{j_{s2}} \nu^j_{j_{s1},j_{s2}} \left(H_y|^{n+\frac{1}{2}}_{i_{s1}-\frac{1}{2},j,k+\frac{1}{2}} + H_y|^{n+\frac{1}{2}}_{i_{s1}+\frac{1}{2},j,k+\frac{1}{2}}\right. \\
&\left.+H_y|^{n-\frac{1}{2}}_{i_{s1}-\frac{1}{2},j,k+\frac{1}{2}} + H_y|^{n-\frac{1}{2}}_{i_{s1}+\frac{1}{2},j,k+\frac{1}{2}}\right) \exp(-i\boldsymbol{k}\cdot\boldsymbol{r}'_{i_{s1},j,k+\frac{1}{2}})\Delta y\Delta z \\
&+\frac{1}{4}\sum_{k=k_{s1}}^{k_{s2}-1}\sum_{j=j_{s1}}^{j_{s2}} \nu^j_{j_{s1},j_{s2}} \left(H_y|^{n+\frac{1}{2}}_{i_{s2}-\frac{1}{2},j,k+\frac{1}{2}} + H_y|^{n+\frac{1}{2}}_{i_{s2}+\frac{1}{2},j,k+\frac{1}{2}}\right. \\
&\left.+H_y|^{n-\frac{1}{2}}_{i_{s2}-\frac{1}{2},j,k+\frac{1}{2}} + H_y|^{n-\frac{1}{2}}_{i_{s2}+\frac{1}{2},j,k+\frac{1}{2}}\right) \exp(-i\boldsymbol{k}\cdot\boldsymbol{r}'_{i_{s2},j,k+\frac{1}{2}})\Delta y\Delta z. \qquad (6.333)
\end{aligned}
$$

If $\boldsymbol{L}(\theta,\phi)$ and $\boldsymbol{N}(\theta,\phi)$ obtained above are expressed in the polar coordinate system, we obtain

$$L_\theta(\theta,\phi) = L_x \cos\theta\cos\phi + L_y \cos\theta\sin\phi - L_z \sin\theta \tag{6.334}$$

$$L_\phi(\theta,\phi) = -L_x \sin\phi + L_y \cos\phi, \tag{6.335}$$

$$N_\theta(\theta,\phi) = N_x \cos\theta\cos\phi + N_y \cos\theta\sin\phi - N_z \sin\theta \tag{6.336}$$

$$N_\phi(\theta,\phi) = -N_x \sin\phi + N_y \cos\phi. \tag{6.337}$$

Substituting Eqs. (6.334)–(6.337) into Eqs. (6.294)–(6.299), the electromagnetic field in the far field is obtained.

6.7 Postprocess

6.7.1 Scattering, absorption, extinction cross-section

The scattering cross-section of a scatterer can be calculated from the sum of the power of the scattered waves leaving the closed surface that completely surrounds the scatterer. Letting $\boldsymbol{E}_{\text{sca}}$ and $\boldsymbol{H}_{\text{sca}}$ be the scattered electric and magnetic fields, respectively, the Poynting vector of the scatterer $\boldsymbol{S}_{\text{sca}}$ is given by

$$\boldsymbol{S}_{\text{sca}} = \frac{1}{2}\text{Re}\left(\tilde{\boldsymbol{E}}_{\text{sca}} \times \tilde{\boldsymbol{H}}^*_{\text{sca}}\right), \tag{6.338}$$

where * denotes complex conjugation. The electric and magnetic fields used in this equation are those in the frequency domain. In the FDTD method, the electric and magnetic fields are usually expressed in real numbers. The values in the frequency domain are obtained by Fourier transforming the response to a short pulse wave source as described previously. Therefore, the electromagnetic field in the frequency domain is generally a complex number with a phase term.

The total scattered power W_{sca} is given as

$$W_{\text{sca}} = \int_S \boldsymbol{S}_{\text{sca}} \cdot \hat{\boldsymbol{n}} dS. \tag{6.339}$$

In the FDTD method, it is common to take this closed surface as a rectangular solid. Furthermore, by using the TF/SF method and setting all the faces of this rectangular solid to be in the scattering field region, the power of only the scattered waves can be easily calculated.

Assuming that the rectangular solid is made up of a collection of E-cells, the energy flow through one face of the E-cell is calculated. The Poynting vector at this face is represented by the value at the centre of the E-cell face. As an example, consider a face of an E-cell perpendicular to the z-axis. To calculate the Poynting vector at the centre of this face, we need the values of E_x and H_y and E_y and H_x in the frequency domain. However, since only H_z is defined at this location, a little ingenuity is required for other fields. Furthermore, the definition times of E and H differ by $\Delta t/2$, which must also be taken into account. When obtaining the spectrum, it is necessary to store the time series data of the electric and magnetic fields, and it is convenient to correct these positions and times at every time evolution sequence. Here, the time is to be aligned with the time at which E is defined. That is, at the centre of the face perpendicular to the z-axis of the E-cell:

$$E_x|^n_{i+\frac{1}{2},j+\frac{1}{2},k} = \frac{1}{2}\left(E_x|^n_{i+\frac{1}{2},j,k} + E_x|^n_{i+\frac{1}{2},j+1,k}\right), \tag{6.340}$$

$$\begin{aligned} H_y|^n_{i+\frac{1}{2},j+\frac{1}{2},k} =& \frac{1}{8}\left(H_y|^{n-\frac{1}{2}}_{i+\frac{1}{2},j,k-\frac{1}{2}} + H_y|^{n-\frac{1}{2}}_{i+\frac{1}{2},j+1,k-\frac{1}{2}} + H_y|^{n-\frac{1}{2}}_{i+\frac{1}{2},j,k+\frac{1}{2}}\right. \\ &+ H_y|^{n-\frac{1}{2}}_{i+\frac{1}{2},j+1,k+\frac{1}{2}} + H_y|^{n+\frac{1}{2}}_{i+\frac{1}{2},j,k-\frac{1}{2}} + H_y|^{n+\frac{1}{2}}_{i+\frac{1}{2},j+1,k-\frac{1}{2}} \\ &\left.+ H_y|^{n+\frac{1}{2}}_{i+\frac{1}{2},j,k+\frac{1}{2}} + H_y|^{n+\frac{1}{2}}_{i+\frac{1}{2},j+1,k+\frac{1}{2}}\right). \end{aligned} \tag{6.341}$$

The E_y and H_x can be calculated in the same way. The power W_z flowing out of this face in the $+z$-direction is given as

$$\begin{aligned} W_z|^n_{i+\frac{1}{2},j+\frac{1}{2},k} =& \Delta x \Delta y \left(\tilde{E}_x|_{i+\frac{1}{2},j+\frac{1}{2},k} \tilde{H}^*_y|_{i+\frac{1}{2},j+\frac{1}{2},k}\right. \\ &\left. - \tilde{E}_y|_{i+\frac{1}{2},j+\frac{1}{2},k} \tilde{H}^*_x|_{i+\frac{1}{2},j+\frac{1}{2},k}\right). \end{aligned} \tag{6.342}$$

Similar calculations are performed for the six faces of the rectangular solid and by summing them, the total power of the scattered light is obtained. The scattering cross-section is obtained by dividing this total scattered power by the incident light intensity (incident power per unit area).

Next, we will discuss how to obtain the absorption cross section and extinction cross section. The Poynting vector $\boldsymbol{S}$ in the medium surrounding the scatterer is given by the sum of the three terms as [8],

$$\boldsymbol{S} = \frac{1}{2}\mathrm{Re}(\tilde{\boldsymbol{E}}_{\mathrm{tot}} \times \tilde{\boldsymbol{H}}^*_{\mathrm{tot}}) = \boldsymbol{S}_{\mathrm{inc}} + \boldsymbol{S}_{\mathrm{sca}} + \boldsymbol{S}_{\mathrm{ext}}, \tag{6.343}$$

where $\boldsymbol{S}_{\mathrm{inc}}$ is the Poynting vector of the incident field, and $\boldsymbol{S}_{\mathrm{sca}}$ and $\boldsymbol{S}_{\mathrm{ext}}$ are the Poynting vectors outgoing and incoming to the scatterer, respectively. The $\boldsymbol{S}_{\mathrm{sca}}$ and $\boldsymbol{S}_{\mathrm{ext}}$ are expressed as follows, respectively

$$\boldsymbol{S}_{\mathrm{inc}} = \frac{1}{2}\mathrm{Re}(\tilde{\boldsymbol{E}}_{\mathrm{inc}} \times \tilde{\boldsymbol{H}}^*_{\mathrm{inc}}), \tag{6.344}$$

$$\boldsymbol{S}_{\mathrm{sca}} = \frac{1}{2}\mathrm{Re}(\tilde{\boldsymbol{E}}_{\mathrm{sca}} \times \tilde{\boldsymbol{H}}^*_{\mathrm{sca}}). \tag{6.345}$$

Since $\tilde{\boldsymbol{E}}_{\text{tot}} = \tilde{\boldsymbol{E}}_{\text{inc}} + \tilde{\boldsymbol{E}}_{\text{sca}}$ in the medium, from Eq. (6.343), we obtain

$$\begin{aligned}\boldsymbol{S} &= \frac{1}{2}\text{Re}[(\tilde{\boldsymbol{E}}_{\text{inc}} + \tilde{\boldsymbol{E}}_{\text{sca}}) \times (\tilde{\boldsymbol{H}}^*_{\text{inc}} + \tilde{\boldsymbol{H}}^*_{\text{sca}})] \\ &= \frac{1}{2}\text{Re}(\tilde{\boldsymbol{E}}_{\text{inc}} \times \tilde{\boldsymbol{H}}^*_{\text{inc}} + \tilde{\boldsymbol{E}}_{\text{sca}} \times \tilde{\boldsymbol{H}}^*_{\text{sca}} + \tilde{\boldsymbol{E}}_{\text{inc}} \times \tilde{\boldsymbol{H}}^*_{\text{sca}} + \tilde{\boldsymbol{E}}_{\text{sca}} \times \tilde{\boldsymbol{H}}^*_{\text{inc}}).\end{aligned} \tag{6.346}$$

Substituting Eqs. (6.344) and (6.345) into Eq. (6.346), we obtain

$$\boldsymbol{S} = \boldsymbol{S}_{\text{inc}} + \boldsymbol{S}_{\text{sca}} + \frac{1}{2}\text{Re}(\tilde{\boldsymbol{E}}_{\text{inc}} \times \tilde{\boldsymbol{H}}^*_{\text{sca}} + \tilde{\boldsymbol{E}}_{\text{sca}} \times \tilde{\boldsymbol{H}}^*_{\text{inc}}). \tag{6.347}$$

Comparing this equation with Eq. (6.343),

$$\boldsymbol{S}_{\text{ext}} = \frac{1}{2}\text{Re}(\tilde{\boldsymbol{E}}_{\text{inc}} \times \tilde{\boldsymbol{H}}^*_{\text{sca}} + \tilde{\boldsymbol{E}}_{\text{sca}} \times \tilde{\boldsymbol{H}}^*_{\text{inc}}) \tag{6.348}$$

is obtained. The extinction power W_{ext} due to the scatterer is given by

$$W_{\text{ext}} = -\int_S \boldsymbol{S}_{\text{ext}} \cdot \hat{\boldsymbol{n}} dS. \tag{6.349}$$

As above, taking a closed surface to the surface of a rectangular solid and using the incident and scattered fields at this surface. On the $-x$-side surface of the rectangular solid, we obtain

$$\begin{aligned}(\boldsymbol{S}_{\text{ext}} \cdot \hat{\boldsymbol{n}})_{-x} =& \frac{1}{2}\left(\tilde{E}'_{iy}\tilde{H}'_{sz} + \tilde{E}''_{iy}\tilde{H}''_{sz} - \tilde{E}'_{iz}\tilde{H}'_{sy} - \tilde{E}''_{iz}\tilde{H}''_{sy}\right. \\ &\left. + \tilde{E}'_{sy}\tilde{H}'_{iz} + \tilde{E}''_{sy}\tilde{H}''_{iz} - \tilde{E}'_{sz}\tilde{H}'_{iy} - \tilde{E}''_{sz}\tilde{H}''_{iy}\right),\end{aligned} \tag{6.350}$$

where $'$ denotes the real part and $''$ the imaginary part. On the $-y$-side surface,

$$\begin{aligned}(\boldsymbol{S}_{\text{ext}} \cdot \hat{\boldsymbol{n}})_{-y} =& \frac{1}{2}\left(-\tilde{E}'_{ix}\tilde{H}'_{sz} - \tilde{E}''_{ix}\tilde{H}''_{sz} + \tilde{E}'_{iz}\tilde{H}'_{sx} + \tilde{E}''_{iz}\tilde{H}''_{sx}\right. \\ &\left. - \tilde{E}'_{sx}\tilde{H}'_{iz} - \tilde{E}''_{sx}\tilde{H}''_{iz} + \tilde{E}'_{sz}\tilde{H}'_{ix} + \tilde{E}''_{sz}\tilde{H}''_{ix}\right).\end{aligned} \tag{6.351}$$

On the $-z$-side surface,

$$\begin{aligned}(\boldsymbol{S}_{\text{ext}} \cdot \hat{\boldsymbol{n}})_{-z} =& \frac{1}{2}\left(\tilde{E}'_{ix}\tilde{H}'_{sy} + \tilde{E}''_{ix}\tilde{H}''_{sy} - \tilde{E}'_{iy}\tilde{H}'_{sx} - \tilde{E}''_{iy}\tilde{H}''_{sx}\right. \\ &\left. + \tilde{E}'_{sx}\tilde{H}'_{iy} + \tilde{E}''_{sx}\tilde{H}''_{iy} - \tilde{E}'_{sy}\tilde{H}'_{ix} - \tilde{E}''_{sy}\tilde{H}''_{ix}\right).\end{aligned} \tag{6.352}$$

At the + side surface, the signs of each term are all reversed. These calculations are performed on all surfaces of the rectangular solid, and the extinction cross-section is obtained by dividing the sum by the incident light intensity.

For example, if the incident field is z propagating x polarization, $\tilde{E}_{iy} = \tilde{E}_{iz} = \tilde{H}_{ix} = \tilde{H}_{iz} = 0$ at $-x$-side surface, Eq. (6.350) becomes

$$(\boldsymbol{S}_{\text{ext}} \cdot \hat{\boldsymbol{n}})_{-x} = -\frac{1}{2}(\tilde{E}'_{sz}\tilde{H}'_{iy} + \tilde{E}''_{sz}\tilde{H}''_{iy}). \tag{6.353}$$

On the $-y$-side surface, Eq. (6.351) becomes

$$(\boldsymbol{S}_{\text{ext}} \cdot \hat{\boldsymbol{n}})_{-y} = -\frac{1}{2}(\tilde{E}'_{ix}\tilde{H}'_{sz} + \tilde{E}''_{ix}\tilde{H}''_{sz}). \tag{6.354}$$

On the $-z$-side surface, Eq. (6.352) becomes

$$(\boldsymbol{S}_{\text{ext}} \cdot \hat{\boldsymbol{n}})_{-z} = \frac{1}{2}(\tilde{E}'_{ix}\tilde{H}'_{sy} + \tilde{E}''_{ix}\tilde{H}''_{sy} + \tilde{E}'_{sx}\tilde{H}'_{iy} + \tilde{E}''_{sx}\tilde{H}''_{iy}). \tag{6.355}$$

Also, if the incident field is z propagating y polarization, $\tilde{E}_{ix} = \tilde{E}_{iz} = \tilde{H}_{iy} = \tilde{H}_{iz} = 0$ on the $-x$-side surface, Eqs. (6.350)–(6.352) become

$$(\boldsymbol{S}_{\text{ext}} \cdot \hat{\boldsymbol{n}})_{-x} = \frac{1}{2}(\tilde{E}'_{iy}\tilde{H}'_{sz} + \tilde{E}''_{iy}\tilde{H}''_{sz}), \tag{6.356}$$

$$(\boldsymbol{S}_{\text{ext}} \cdot \hat{\boldsymbol{n}})_{-y} = \frac{1}{2}(\tilde{E}'_{sz}\tilde{H}'_{ix} + \tilde{E}''_{sz}\tilde{H}''_{ix}), \tag{6.357}$$

$$(\boldsymbol{S}_{\text{ext}} \cdot \hat{\boldsymbol{n}})_{-z} = -\frac{1}{2}(\tilde{E}'_{iy}\tilde{H}'_{sx} - \tilde{E}''_{iy}\tilde{H}''_{sx} + E'_{sy}\tilde{H}'_{ix} - \tilde{E}''_{sy}\tilde{H}''_{ix}). \tag{6.358}$$

The absorption cross section, C_{abs} can be calculated from the following definition:

$$C_{\text{abs}} = C_{\text{ext}} - C_{\text{sca}}. \tag{6.359}$$

6.7.2 Absorption distribution

The absorbed power per unit cell, ΔP can be calculated from Poynting's theorem,

$$\Delta P = \frac{1}{2}\sigma|\boldsymbol{E}|^2 \Delta x \Delta y \Delta z. \tag{6.360}$$

In the optics field, complex permittivity is often used instead of conductivity σ. The relationship between the complex relative permittivity ε^* and σ is given by

$$\varepsilon^* = \varepsilon + \frac{i\sigma}{\varepsilon_0 \omega}. \tag{6.361}$$

Therefore, in the case of Drude dispersion, the conductivity is given by

$$\sigma = \frac{\varepsilon_0 \omega_p^2 \Gamma}{\omega^2 + \Gamma^2}. \tag{6.362}$$

Also, in the case of Lorenz dispersion,

$$\sigma = \frac{2\varepsilon_0 \Delta\varepsilon_p \omega_p^2 \omega^2 \Gamma}{(\omega_p^2 - \omega^2)^2 + 4\omega^2\Gamma^2}. \tag{6.363}$$

6.7.3 Charge density distribution

The charge density $\rho(\boldsymbol{r})$ is obtained from Gauss's law,

$$\nabla \cdot \boldsymbol{D}(\boldsymbol{r}) = \rho(\boldsymbol{r}). \tag{6.364}$$

It is easy to obtain the value at the vertex of E-cell (the centre of H-cell), thus:

$$\begin{aligned}\rho|_{i,j,k} =& \frac{\varepsilon_x|_{i+\frac{1}{2},j,k}E_x|_{i+\frac{1}{2},j,k} - \varepsilon_x|_{i-\frac{1}{2},j,k}E_x|_{i-\frac{1}{2},j,k}}{\Delta x} \\ &+ \frac{\varepsilon_y|_{i,j+\frac{1}{2},k}E_y|_{i,j+\frac{1}{2},k} - \varepsilon_y|_{i,j-\frac{1}{2},k}E_y|_{i,j-\frac{1}{2},k}}{\Delta y} \\ &+ \frac{\varepsilon_z|_{i,j,k+\frac{1}{2}}E_z|_{i,j,k+\frac{1}{2}} - \varepsilon_z|_{i,j,k-\frac{1}{2}}E_z|_{i,j,k-\frac{1}{2}}}{\Delta z}.\end{aligned} \tag{6.365}$$

6.7.4 Amplitude and phase of damping harmonic oscillation

When using a continuous wave light source, there are many cases where we want to obtain not only the electric field at a certain time, but also the amplitude of the electric field including the phase. In addition, it is necessary to obtain the amplitude of the damping harmonic oscillation when one wants to obtain the mode pattern of a localized surface plasmon due to dipole excitation, for example. In this case, the unknowns are the damping constant (inverse of the time constant) Γ, the amplitude a, the phase ϕ, and the bias component b that cannot be completely removed. Furthermore, if the sampling interval ΔT, which is constant, is another unknown, then there are five unknowns in total. Therefore, to obtain these five unknowns, we need the values of the electromagnetic field sampled at least five times at a constant interval ΔT. The electric fields obtained by sampling five times are

$$\begin{aligned}E_1 =& E[t_0 - (3/2)\Delta T] \\ =& b + a\exp[(3/2)\Gamma\Delta T]\cos\{\omega[t_0 - (3/2)\Delta T] + \phi'\},\end{aligned} \tag{6.366}$$

$$\begin{aligned}E_2 =& E[t_0 - (1/2)\Delta T] \\ =& b + a\exp[(1/2)\Gamma\Delta T]\cos\{\omega[t_0 - (1/2)\Delta T] + \phi'\},\end{aligned} \tag{6.367}$$

$$\begin{aligned}E_3 =& E[t_0 + (1/2)\Delta T] \\ =& b + a\exp[-(1/2)\Gamma\Delta T]\cos\{\omega[t_0 + (1/2)\Delta T] + \phi'\},\end{aligned} \tag{6.368}$$

$$\begin{aligned}E_4 =& E[t_0 + (3/2)\Delta T] \\ =& b + a\exp[-(3/2)\Gamma\Delta T]\cos\{\omega[t_0 + (3/2)\Delta T] + \phi'\},\end{aligned} \tag{6.369}$$

$$E_5 = E[t_0 + (5/2)\Delta T]$$
$$= b + a\exp[-(5/2)\Gamma\Delta T]\cos\{\omega[t_0 + (5/2)\Delta T] + \phi'\}. \quad (6.370)$$

The bias component b is obtained from the above five equations as

$$b = (E_3^2 - 2E_2E_3E_4 + E_1E_4^2 + E_2^2E_5 - E_1E_3E_5)$$
$$/(E_2^2 - E_1E_3 - 2E_2E_3 + 3E_3^2 + 2E_1E_4 - 2E_2E_4$$
$$- 2E_3E_4 + E_4^2 - E_1E_5 + 2E_2E_5 - E_3E_5). \quad (6.371)$$

By subtracting this bias component from Eqs. (6.366)–(6.366), we obtain

$$E_1' = a\exp[(3/2)\gamma]\cos(\phi - 3\Phi), \quad (6.372)$$

$$E_2' = a\exp[(1/2)\gamma]\cos(\phi - \Phi), \quad (6.373)$$

$$E_3' = a\exp[-(1/2)\gamma]\cos(\phi + \Phi), \quad (6.374)$$

$$E_4' = a\exp[-(3/2)\gamma]\cos(\phi + 3\Phi), \quad (6.375)$$

where $\gamma = \Gamma\Delta T$, $\phi = \omega t_0 + \phi'$, and $\Phi = (1/2)\omega\Delta T$. By taking the product or square of these equations,

$$E_1'E_3' = \frac{1}{2}a^2\exp(\gamma)\left[\cos(2\phi - 2\Phi) + \cos 2\Phi\right], \quad (6.376)$$

$$E_2'^2 = \frac{1}{2}a^2\exp(\gamma)\left[\cos(2\phi - 2\Phi) + 1\right], \quad (6.377)$$

$$E_2'E_4' = \frac{1}{2}a^2\exp(-\gamma)\left[\cos(2\phi + 2\Phi) + \cos 2\Phi\right], \quad (6.378)$$

$$E_3'^2 = \frac{1}{2}a^2\exp(-\gamma)\left[\cos(2\phi + 2\Phi) + 1\right], \quad (6.379)$$

are obtained. Using Eqs. (6.376) and (6.377),

$$E_2'^2 - E_1'E_3' = \frac{1}{2}a^2\exp(\gamma)\left[1 - \cos 2\Phi\right], \quad (6.380)$$

$$E_3'^2 - E_2'E_4' = \frac{1}{2}a^2\exp(-\gamma)\left[1 - \cos 2\Phi\right], \quad (6.381)$$

are obtained. Dividing Eq. (6.380) by the Eq. (6.377) on each side and taking the square root, we obtain

$$\exp(-\gamma) = \sqrt{\frac{E_3'^2 - E_2'E_4'}{E_2'^2 - E_1'E_3'}}. \quad (6.382)$$

Next, we use the electric field divided by the above damping term, we obtain

$$E_1'' = a\cos(\phi - 3\Phi), \quad (6.383)$$

$$E_2'' = a\cos(\phi - \Phi), \tag{6.384}$$

$$E_3'' = a\cos(\phi + \Phi), \tag{6.385}$$

$$E_4'' = a\cos(\phi + 3\Phi). \tag{6.386}$$

Adding and subtracting Eqs. (6.383) and (6.386) as well as Eqs. (6.384) and (6.385) on each side, we obtain

$$E_1'' + E_4'' = 2a\cos\phi\cos 3\Phi, \tag{6.387}$$

$$E_1'' - E_4'' = 2a\sin\phi\sin 3\Phi, \tag{6.388}$$

$$E_2'' + E_3'' = a\cos\phi\cos\Phi, \tag{6.389}$$

$$E_2'' - E_3'' = a\sin\phi\sin\Phi. \tag{6.390}$$

From Eqs. (6.387) to (6.390), we obtain

$$\frac{E_1'' + E_4''}{E_2'' + E_3''} = \frac{\cos 3\Phi}{\cos\Phi} = 4\cos^2\Phi - 3, \tag{6.391}$$

$$\frac{E_1'' - E_4''}{E_2'' - E_3''} = \frac{\sin 3\Phi}{\sin\Phi} = -4\sin^2\Phi + 3. \tag{6.392}$$

From these equations, we obtain

$$\cos^2\Phi = \frac{1}{4}\left(3 + \frac{E_1'' + E_4''}{E_2'' + E_3''}\right), \tag{6.393}$$

$$\sin^2\Phi = \frac{1}{4}\left(3 - \frac{E_1'' - E_4''}{E_2'' - E_3''}\right). \tag{6.394}$$

On the other hand, if we square the sides of the Eqs. (6.389) to (6.390),

$$(E_2'' + E_3'')^2 = a^2\cos^2\phi\cos^2\Phi, \tag{6.395}$$

$$(E_2'' - E_3'')^2 = a^2\sin^2\phi\sin^2\Phi, \tag{6.396}$$

are obtained. From Eqs. (6.393) and (6.395) as well as Eqs. (6.394) and (6.396),

$$\frac{(E_2'' + E_3'')^2}{\left(3 + \frac{E_1''+E_4''}{E_2''+E_3''}\right)} = a^2\cos^2\phi, \tag{6.397}$$

$$\frac{(E_2'' - E_3'')^2}{\left(3 - \frac{E_1''-E_4''}{E_2''-E_3''}\right)} = a^2\sin^2\phi, \tag{6.398}$$

are obtained. Adding up each sides of Eqs. (6.397) and (6.398), we obtain the amplitude

$$a^2 = \frac{(E_2'' + E_3'')^3}{(E_1'' + E_4'') + 3(E_2'' + E_3'')} - \frac{(E_2'' - E_3'')^3}{(E_1'' - E_4'') - 3(E_2'' - E_3'')}. \tag{6.399}$$

Next, we calculate the phase. From Eqs. (6.389) and (6.390),

$$\tan\phi = \frac{\cos\Phi}{\sin\Phi}\frac{E_2'' - E_3''}{E_2'' + E_3''} \tag{6.400}$$

is obtained. Using this equation and Eqs. (6.393) and (6.394), we obtain phase, ϕ. Considering the signs in the denominator and numerator of Eq. (6.400), ϕ is uniquely determined in the range $(-\pi, \pi]$. Also, if we take $\Phi \sim \pi/4$, i.e. $\Delta T \sim \pi/2\omega$, the signs of $\cos\Phi$ and $\sin\Phi$ are both positive.

In actual calculations, the right side of Eqs. (6.393) and (6.394) may become negative due to rounding errors when the amplitude is small. In this case, $\cos\Phi$ and $\sin\Phi$ become imaginary numbers. To avoid this, in an actual calculation, for example, when $\cos\Phi$ is calculated, it is better to use

$$\cos\Phi = \frac{1}{2}\sqrt{3 + \left|\frac{E_1'' + E_4''}{E_2'' + E_3''}\right|}. \tag{6.401}$$

Once the amplitude and phase of the electric and magnetic fields have been obtained, the distribution of the electric and magnetic fields at any time in the steady state can be obtained. The question arises as to which time (phase) the fields should be displayed. If we want to display enhanced fields such as localized surface plasmon resonance, it would be better to use the phase of the field at the position where the amplitude takes the maximum value as a reference. If this phase is ϕ_0, for example, the amplitude the electric field E_x to be displayed is

$$E_x(x, y, z) = |E_x(x, y, z)| \sin\left[\phi_{Ex}(x, y, z) - \phi_0 + \pi/2\right]. \tag{6.402}$$

It is important to note that at resonant frequencies, the phase of the resonant mode oscillation is delayed by $\pi/2$ from the phase of the incident field. Therefore, when the field is calculated according to the above equation, the amplitude of the incident field is almost zero. On the other hand, if $\pi/2$ in the hook brackets in the above equation is set to zero, only the incident field is obtained and the enhanced field due to resonance is almost zero.

6.8 Example of localized surface plasmon resonance calculation

As an example of FDTD calculation, we describe the analysis of localized surface plasmon resonance in metallic nanoparticles. Localized surface plasmon resonance is a phenomenon in which a collection of free electrons in metal nanoparticles resonantly oscillates in response to an incident field [59]. Here, we investigate the localized surface plasmon resonance in a gold disk

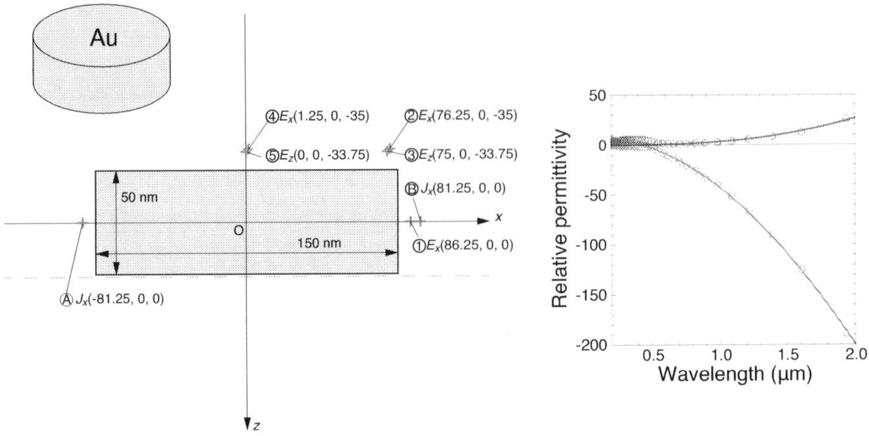

FIGURE 6.17
Gold disk and coordinate system; A and B are dipole positions; positions 1–5 are the observation points. The right figure shows the relative permittivity of gold; experimental values (circled) [12] and values fitted with the Drude model (solid line).

with a 150 nm diameter and 50 nm thickness placed in vacuum as shown in Figure 6.17. The disk is placed in the centre of the coordinate system so that its central axis coincides with the z-axis; the size of the Yee cell is $2.5 \times 2.5 \times 2.5$ nm^3 and the object space is $500 \times 500 \times 500$ nm^3. The entire object space was terminated with eight layers of PML. The relative permittivity of gold was obtained by fitting with the Drude dispersion formula using literature values [12] of in a wavelength region of 0.5–2.0 μm. The parameters obtained from the fitting are $\varepsilon_\infty = 10.38$, $\omega_p = 1.375 \times 10^{16}$ Hz, and $\Gamma = 1.181 \times 10^{14}$ Hz. The resultant relative permittivity of gold using these values is also shown in Figure 6.17. The deviation of the experimental values from the Drude model in the short wavelength range is due to the interband transitions of gold. In order to express the permittivity of gold more faithfully, it is necessary to express the permittivity as the sum of the Drude dispersion and the Lorentz dispersion.

First, an x-polarized plane wave propagating in the $+z$-direction is incident using the TS/FS method to obtain absorption, scattering, and extinction spectra of the disk. For this purpose, a scattering field region is set up for three cells thick outside the object region. The pulse waveform is a sinusoidally modulated Gaussian pulse with a centre frequency of 600 nm in terms of wavelength in vacuum. The standard deviation of the Gaussian waveform was set to half of the period of the modulated wave. The obtained scattering, absorption and extinction spectra are shown in Figure 6.18(a). A large peak

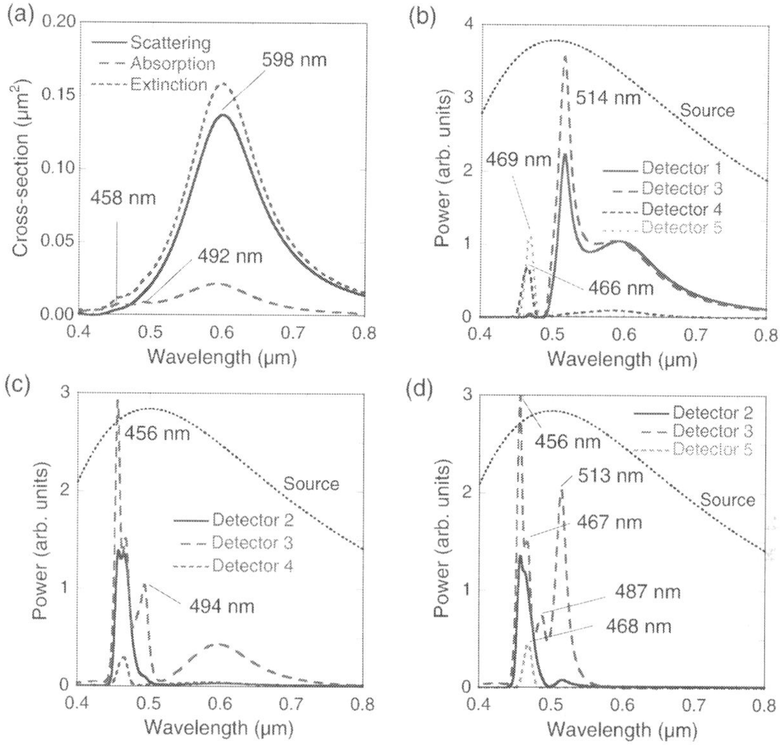

FIGURE 6.18
(a) Absorption, scattering, and extinction cross-section spectra of the gold disk. (b) Power spectra of the electric field at each detector position when a single dipole is placed on the side of the metallic disk (point A) and excited with a Gaussian pulse. (c) Power spectrum of the electric field at each detector position when two dipoles oriented in the x-direction are placed on both sides of the metal disk (points A and B) and excited with a Gaussian pulse. (d) Same as (c) but with the dipoles oscillating symmetrically (excited in opposite phase).

at a wavelength of 598 nm and two smaller peaks at wavelengths of 458 nm and 491 nm are observed.

Next, we visualize the electromagnetic field distribution of the resonance mode in order to investigate what kind of resonance mode of localized surface plasmon these peaks correspond to. First, we investigate the resonance mode at a wavelength of 598 nm. For this purpose, a continuous monochromatic plane wave of a wavelength of 598 nm is incident, and the electromagnetic field distribution in the steady state is calculated. The results are shown in Figures 6.19(a) and (b), which are plots of the electric field E_x at $z = 0$ and

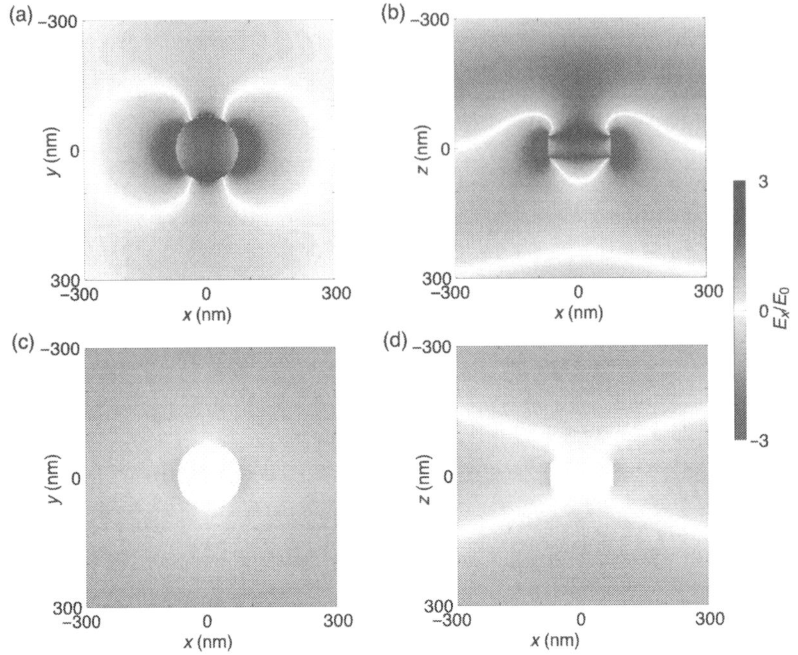

FIGURE 6.19
Electric field (E_x) distribution at steady state for an incident continuous wave of wavelength 598 nm. (a) and (b) are for at the time when the maximum electric field is obtained, and (c) and (d) are the electric field distributions at the time shifted by a quarter cycle from (a) and (b). Distributions (a) and (b) are at the cross-section $z = 0$ and (b) and (d) are at the cross-section $y = 0$.

$y = 0$ cross-sections, respectively. The electric field amplitude and phase at each location were calculated, and these were used to show the electric field at the time giving its maximum. As can be seen from Figure 6.19(b), the electric field distribution is not only that of the resonant mode, since the electric field enhanced by localized surface plasmon resonance and the incident electric field are superimposed. On the other hand, Figures 6.19(c) and (d) show the electric field distribution at the time shifted by a quarter cycle from Figures 6.19(a) and (b), where the enhanced field is barely visible and almost only the incident field is shown.

To remove the incident field, excitation by the incident field should be terminated in the middle of the excitation. Since resonance modes have a long lifetime, they continue to oscillate for a while after the excitation is terminated. Thus, the amplitude of the remaining oscillation gives only that of resonance mode. Here, we must be careful about the spectral waveform of the incident field to be used. The incident field is expressed as the product of a sine wave and a function representing an envelope with a finite time width (envelope function). The spectrum of this waveform is represented by the convolution of the delta function and the Fourier transform of the envelope function. The Fourier transform of the envelope function generally results in a peak with a certain width. In addition, non-negligible side lobes appear on both sides of the peak, and these side lobes decay away from the peak, but continue indefinitely. Therefore, if there is another resonance mode with a high Q value in the vicinity of the desired resonance mode, this side lobe excites this high Q-value mode at the same time. Since the peak frequency of the incident light is matched to the resonance frequency of the desired low Q-value mode, this mode is excited with a large amplitude. At the same time, however, a nearby mode with a high Q value is also excited, albeit with a smaller amplitude. Since the mode with the higher Q value has a longer lifetime, this mode is excited more strongly if the excitation is continued. In addition, the oscillation continues for a long time after the excitation is terminated. As a result, if the amplitude distribution is detected after all the incident light has left the calculation area, the amplitude distribution of a nearby mode with a high Q value may be detected instead of the desired mode with a low Q value.

To avoid this, the shape of the envelope function of the incident field should be devised. This is consistent with the window function problem that has been studied in frequency analysis. The envelope function is the window function itself. The shapes of the peaks and side lobes in the spectrum of the incident field depend on the shape of the window function. In this case, the width of the peak and the magnitude of the sidelobe are in a contradictory relationship. When the peak width is reduced, the side lobe becomes larger, but when the side lobe is reduced, the peak width becomes larger. One function that satisfies above desired condition is the Nuttall window [60] given by

$$w(t) = \frac{1}{L}\sum_{k=0}^{K} a_k \cos(2\pi kt/L) \qquad \text{for} \quad |t| \le L/2, \tag{6.403}$$

and several combination of a_k have been proposed so far. One of them is called "4-Term with Continuous First Derivative" as

$$\begin{aligned} a_0 &= 0.355768, \\ a_1 &= 0.487396, \\ a_2 &= 0.144232, \\ a_3 &= 0.012604. \end{aligned} \tag{6.404}$$

This set has the excellent characteristics of a maximum lobe intensity of −93.32 dB and a lobe attenuation of 18 dB/octave.

The distribution of resonance modes at a wavelength of 598 nm obtained in this way is shown in Figure 6.20. A highly symmetric electric field distribution with the incident field removed is obtained. This figure shows that this mode is a dipole mode.

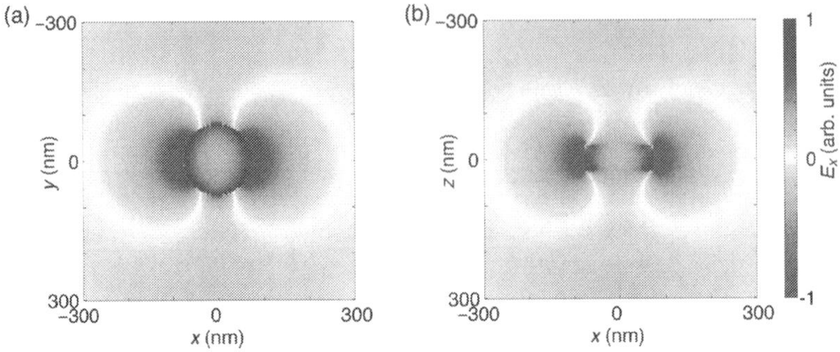

FIGURE 6.20
Distribution of the electric field E_x after the excitation with a monochromatic plane wave of a wavelength of 598 nm and terminated it.

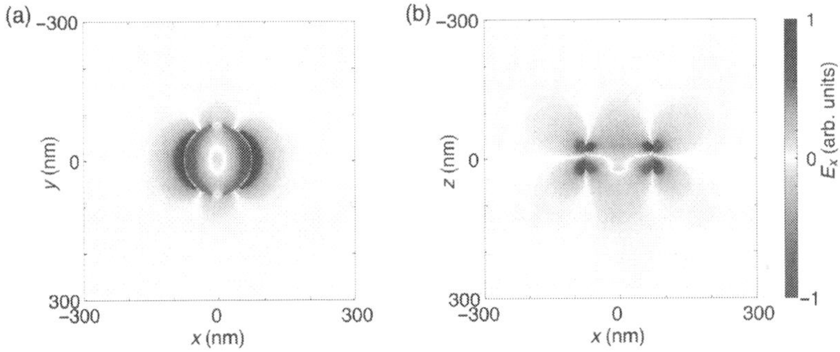

FIGURE 6.21
Distribution of electric field E_x at steady state for an incident monochromatic plane wave of a wavelength of 492 nm. (a) It is in the $z = 26.25$ nm plane and (b) is in the $y = 0$ plane. Both are field distributions at the time when the maximum electric field is obtained.

Next, we examine the resonance peak at a wavelength of 492 nm. As in the case of the dipole mode at a wavelength of 598 nm, the electric field distribution after a while of continuous excitation with a monochromatic plane wave at a wavelength of 492 nm and after the excitation has been terminated is shown in Figure 6.21. This distribution is less symmetric than that for the 598 nm wavelength mode. One possible reason for this is that this electric field distribution does not represent that of only one mode, but a superposition of multiple modes. As can be seen from the spectrum shown in Figure 6.18(a), the hem of the resonance peak at 598 nm extends to this wavelength, suggesting that this mode is superimposed. In such a case, no matter how narrow the width of the excitation spectrum, the influence of other modes with overlapping resonance peak hem cannot be eliminated.

Thus, modes that cannot be completely separated by the spectrum obtained by plane wave excitation must be separated by using sources that excite only the mode to be observed, instead of excitation by plane waves. One of the methods is excitation by a set of dipoles. Figure 6.18(b) shows the power spectrum of the electric field at four positions when a single dipole oscillating in the x-direction is placed at 6.25 nm away from the side of the disk on the x-axis and excited with a pulse. The position of the detector and the orientation of the detected electric field are shown in Figure 6.17. New resonance peaks appeared at 514 nm and 468 nm, but the two peaks on the short wavelength side of Figure 6.18(a) disappeared. Therefore, the resonant mode at 492 nm cannot be observed with this dipole configuration.

In order to reproduce the electric field distribution more similar to plane wave incidence, two dipoles oriented in the x-direction (antisymmetric direction with respect to the plane of $x = 0$) were placed at symmetric positions on both sides of the disk for excitation. The power spectrum of the detected electric field is shown in Figure 6.18(c). It can be seen that the resonance peak on the short wavelength side observed in Figure 6.18(a) is strongly excited compared to the dipole mode peak. Furthermore, the peak on the shortest wavelength side, which appeared to be one peak in Fig. 6.18(a), shows that two resonance peaks were overlapped. The result of plotting the electric field distribution after the dipoles are excited and terminated at the frequency corresponding to a wavelength of 494 nm with this dipole arrangement is shown in Figure 6.22. Since it is difficult to understand the mode only with the E_x display, the amplitude distribution of E_z on the plane perpendicular to the z-axis 1.25 nm down from the lower surface of the disk is shown in Figure 6.22(c). From this figure, it is clear that this mode is a sextupole mode. The two modes at the shortest wavelengths cannot be well separated by the set of positions of these dipoles. To separate the two modes, we need to further improve the arrangement of the dipoles.

Next, we obtain the electric field distribution of the 514 nm wavelength mode shown in Figure 6.18(b). Since this mode overlaps with the large hem of the dipole mode at a wavelength of 598 nm, it is difficult to excite only this mode by the excitation with one dipole. Therefore, the following arrangement of dipoles is used. The excitation is the same as in the case of Figure 6.18(c) up to the point where two dipoles are placed on both sides of the disk, but the phases of the oscillation of the dipoles are shifted by π with each other so that their oscillations are symmetric. The result is shown in Figure 6.18(d). With this excitation method, the dipole mode is completely obscured, and at the same time, the intensity of the peak on the short wavelength side becomes larger and clearer. Figure 6.23 shows the distribution of E_z obtained after being excited and terminated by the oscillation of two dipoles at frequencies corresponding to wavelengths of 513 nm and 487 nm. It can be seen that they are quadrupole and octupole modes, respectively. Further work on the dipole configuration is needed to identify resonance peaks at shorter wavelengths.

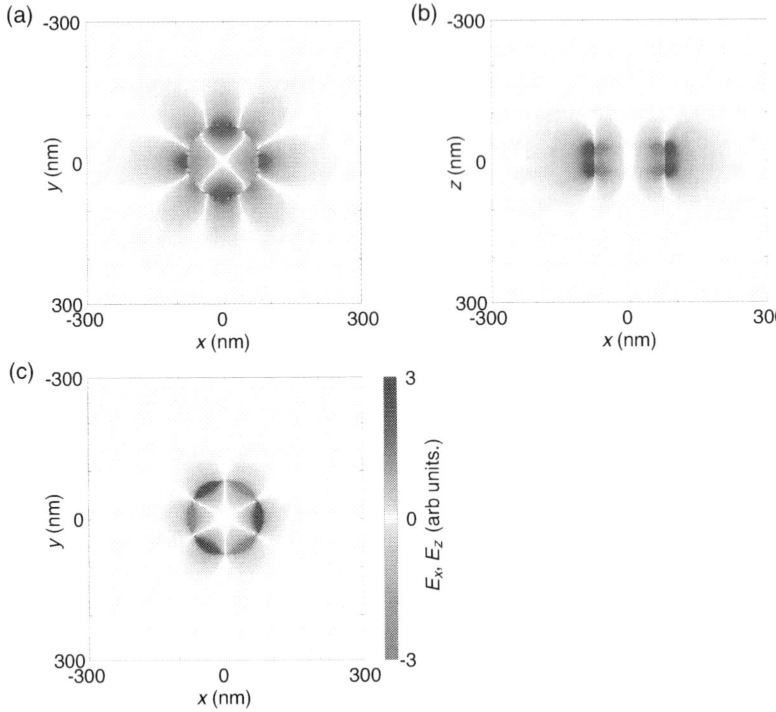

FIGURE 6.22
Distribution of the electric field after excitation and termination of the oscillation of the dipoles at a frequency corresponding to a wavelength of 494 nm, (a) and (b) are the electric field E_x in the x-direction in the plane containing the origin, and (c) is the electric field E_z in the z-direction in the plane at $z = 26.25$ nm.

6.9 Sample program

Sample programs for the FDTD method are shown in the Appendix (A.6.1 (runfdtd.py), A.6.2 (fdtd.py), and A.6.3 (preprocess.py). The outside of the object space is terminated by PML. The sources correspond to the plane wave of x-polarized z propagation or dipoles. The plane waves are introduced using the TF/SF method. All units are in the SI unit system.

`regionx`, `regiony` and `regionz` are the spatial sizes of object space and `dx`, `dy` and `dz` are the cell sizes. The `source` specifies whether the source is a plane wave (`'plane'`) or dipoles (`'dipole'`). `pulse` specifies whether the source is a continuous wave (`'cw'`) or a Gaussian pulse modulated with a sinusoidal wave (`'pulse'`). `lambda0` is the centre wavelength of the source. `mt` is the

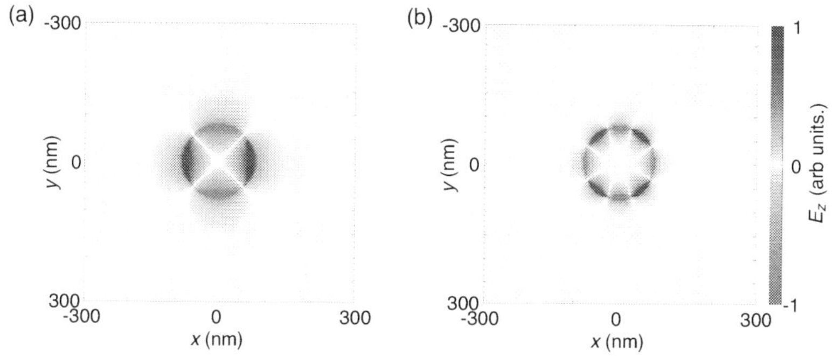

FIGURE 6.23
Distribution of the electric field E_z in the z-direction in the $z = 26.25$ nm plane after pulse excitation by the dipoles at frequencies corresponding to wavelengths of (a) 513 nm and (b) 487 nm.

number of time evolutions, `mfft` is the sampling number of the calculated waveforms and `extrapol` is the multiple of zero filling to extend the duration of the waveforms for the spectrum calculation. This multiple is equal to the density of interpolation of the sampling points in the resulting spectrum. The `msf` gives the width of the scattering region in units of cell size and the `mpml` gives the number of PML layers. Also, `kappamax`, `amax` and `mpow` are $\kappa_{\max}$, $a_{\max}$, and multiplier m of PML parameters.

`objs` specifies the object to be placed in the object space. The only object shapes incorporated in this program are spheres and flat substrate. The installed media are vacuum (`'vacuum'`), silica (`'SiO2'`), gold (`'Au'`) and silver (`'Ag'`). `dipoles` specifies the polarization, phase (only 0 or π) and position of the dipoles as sources. `fieldmons` specifies the parameters for preserving the electric or magnetic field distribution in the cross-section perpendicular to the coordinate axes at constant time intervals. `epsmons` specifies the location of the distribution of the media filling the object space in the cross-section perpendicular to the coordinate axes to be preserved. `detectors` specifies the electric or magnetic fields and their locations for preserving time evolving waveforms and spectra.

The values of the indices of the arrays storing the electric and magnetic fields correspond to the coordinates in units of cell size as shown in Figure 6.24.

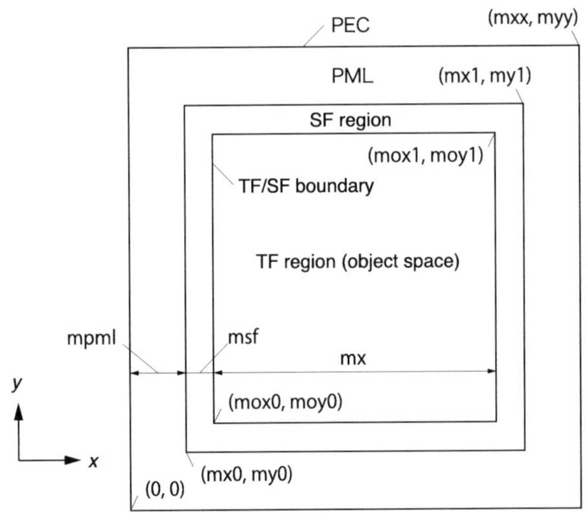

FIGURE 6.24
The relationship between the indices of the array of electric and magnetic fields in the computer and the spatial coordinates in units of cell size. The same applies to the z-direction.

7

Discrete Dipole Approximation

There are various methods for electromagnetic field analysis, one of which is the discrete dipole approximation (DDA). The optical response is derived by approximating the structure of interest as an assembly of dipoles induced by the incident optical electric field and the dipole–dipole interaction. It is suitable for estimating the optical response of an isolated structure, especially scattering and absorption. This method initially predicted interstellar matter's scattering, absorption, and extinction spectra. The software developed by Draine and Flatau (DDSCAT) is available. The user guide describes how to use this software in detail. This chapter introduces some programs for processing data output using DDSCAT.

7.1 DDA principle

In the DDA, the optical response is derived by approximating the structure of interest as an assembly of dipoles induced by the incident optical electric field and the dipole-dipole interaction, as shown in Figure 7.1 [61,62]. The dipole $\boldsymbol{p}_j$ arising at position $\boldsymbol{r}_j$ can be written using the local electric field $\boldsymbol{E}_{j,\mathrm{loc}}$ and the polarization tensor $\tilde{\alpha}$ as follows:

$$\boldsymbol{p}_j = \tilde{\alpha}\boldsymbol{E}_{j,\mathrm{loc}} \tag{7.1}$$

The local electric field $\boldsymbol{E}_{j,\mathrm{loc}}$ is the sum of the externally incident electric field of light $\boldsymbol{E}_{j,\mathrm{ext}}$ at position $\boldsymbol{r}_j$ and the electric field $\boldsymbol{E}_{j,\mathrm{dip}}$ created by other dipoles at position $\boldsymbol{r}_j$.

$$\boldsymbol{E}_{j,\mathrm{loc}} = \boldsymbol{E}_{j,\mathrm{ext}} + \boldsymbol{E}_{j,\mathrm{dip}} \tag{7.2}$$

The $\boldsymbol{E}_{j,\mathrm{dip}}$ can be written using the dipole $\boldsymbol{p}_k$ at position $\boldsymbol{r}_k$, including retardation effects due to propagation, as follows:

$$\begin{aligned}\boldsymbol{E}_{j,\mathrm{dip}} &= \frac{\exp(ikr_{jk})}{r_{jk}}\Big(k^2(\boldsymbol{r}_{jk}\times(\boldsymbol{r}_{jk}\times\boldsymbol{p}_k)\\ &+ \frac{1-ikr_{jk}}{r_{jk}^2}(r_{jk}^2\boldsymbol{p}_k - 3\boldsymbol{r}_{jk}(\boldsymbol{r}_{jk}\cdot\boldsymbol{p}_k))\Big)\end{aligned} \tag{7.3}$$

Here, k is the wavenumber in vacuum, $\boldsymbol{r}_{jk} = \boldsymbol{r}_k - \boldsymbol{r}_j$, and $r_{jk} = |\boldsymbol{r}_{jk}|$. The dipole number N is the number of dipoles considered. Summarizing Equations

DOI: 10.1201/9781003357674-7

(7.2) and (7.3), we have

$$\boldsymbol{E}_{j,\text{loc}} = \boldsymbol{E}_{j,\text{ext}} + \sum_{j \neq k}^{N} \tilde{A}_{jk} \boldsymbol{p}_k. \tag{7.4}$$

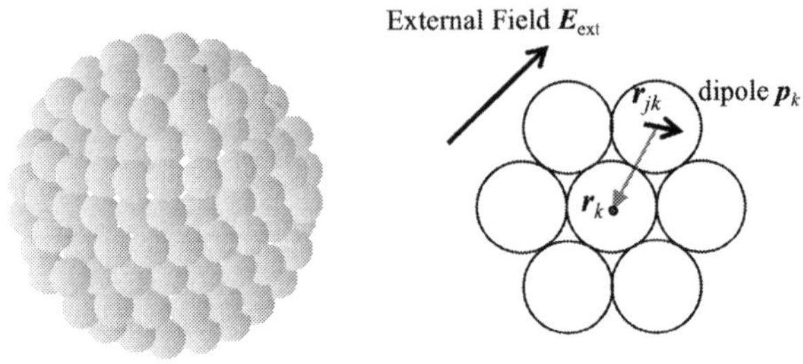

FIGURE 7.1
Discrete dipole approximation.

The $\tilde{A}_{jk}$ can be obtained by component calculations, and $\boldsymbol{p}_k$ is a dipole induced by the local electric field $\boldsymbol{E}_{k,\text{loc}}$ at position $\boldsymbol{r}_k$. Solving this, $\boldsymbol{E}_{j,\text{ext}}$ is given as follows:

$$\boldsymbol{E}_{j,\text{ext}} = \tilde{\alpha}^{-1} \boldsymbol{p}_j + \sum_{j \neq k}^{N} \tilde{A}_{jk} \boldsymbol{p}_k. \tag{7.5}$$

To consider all dipoles, the electric field and dipoles at each position are lined up vertically. Writing the corresponding tensor $\tilde{A}_{jk}$ in the form of a matrix, we obtain

$$\begin{pmatrix} \boldsymbol{E}_{1,\text{ext}} \\ \vdots \\ \boldsymbol{E}_{j,\text{ext}} \\ \vdots \\ \boldsymbol{E}_{N,\text{ext}} \end{pmatrix} = \begin{pmatrix} \tilde{A}_{11} & \dots & \tilde{A}_{1j} & \dots & \tilde{A}_{1N} \\ \vdots & \ddots & & & \vdots \\ \tilde{A}_{j1} & \dots & \tilde{A}_{jj} & \dots & \tilde{A}_{jN} \\ \vdots & & & \ddots & \vdots \\ \tilde{A}_{N1} & \dots & \tilde{A}_{Nj} & \dots & \tilde{A}_{NN} \end{pmatrix} \begin{pmatrix} \boldsymbol{p}_1 \\ \vdots \\ \boldsymbol{p}_j \\ \vdots \\ \boldsymbol{p}_N \end{pmatrix}. \tag{7.6}$$

The vector $\boldsymbol{E}_{\text{ext}}$, which is a vertical vector of external electric fields, and $\tilde{A}_{jk}$ are known. The vector $\boldsymbol{P}$ of dipoles arranged vertically is unknown. It seems that we can solve a simultaneous equation with the components of $\boldsymbol{P}$ as unknowns, but it is difficult to find the inverse matrix considering that $\tilde{A}$ has $3N \times 3N$ components and $N = 10^3 - 10^7$. Therefore, algorithms using non-stationary iterative methods, such as conjugate gradient methods, converge the solution; DDSCAT 7.3 implements five different algorithms, which can be switched as required.

The polarizability α is obtained from the refractive index of a substance. The well-known relationship between a substance's refractive index n and its polarizability α is the Clausius-Mosotti relation. It is

$$\alpha = \frac{3V}{4\pi}\frac{n^2 - 1}{n^2 + 2}, \tag{7.7}$$

in the cgs unit. V is the volume occupied by one dipole, and $V = d^3$ for a cubic lattice, where d is the dipole spacing. The Clausius-Mosotti relation approximates zero frequency and is not necessarily a good approximation for optical frequencies. Several more advanced approximations considering frequency dependence have been proposed [63] and implemented.

7.2 Actual use of DDSCAT

DDSCAT is distributed in Fortran source code, which can be used by compiling; executable files are available for Windows. The source specified in the user guide must be mentioned if the calculated results are published. The execution conditions are described in the file named ddscat.par. If they are described in a file with a name other than this, specify that file name as an argument. The range of wavelengths and sizes to be calculated, as well as the direction of polarization and rotation of the structure, are also described in this file. Typical structures such as spheres, ellipsoids, right-angle prisms, and aggregates can be calculated by specifying keywords. The keywords are listed in ddscat.par. Files containing some materials' refractive indices and dielectric constants (Au, graphite, etc.) are stored in a diel directory. If one wishes to work with other substances, one may prepare them. The format can be found in the file.

To calculate shapes other than the prepared ones, specify "FROM_FILE" as a keyword and place a file named shape.dat in the same directory that lists the coordinates of the dipole and the material. Details on how to list dipole coordinates, etc., are described in the User's Guide. The coordinates are normalized by the dipole spacing d in the sample coordinate system.

While other calculation methods, such as FDTD, directly specify the cell size, the DDSCAT calculation does not directly specify d. Instead, the number of dipoles of the structure to be calculated is specified as N and the radius of the sphere with a volume equal to the volume of the structure as the radius of execution a_{eff}. The relationship between them is

$$d = a_{\text{eff}}\left(\frac{3N}{4\pi}\right)^{-1/3}. \tag{7.8}$$

If you specify a provided shape, N is automatically calculated. The d is listed in the file of calculation results, but if you want to know d before the calculation, calculate N from the shape parameters and obtain d from Eq. (7.8). The calculation of N depends on the geometry, so refer to the user guide. "FROM_FILE" is specified, and the value of N is entered directly into shape.dat.

The calculation results are stored in several files: mtable lists the wavelengths, refractive indices, and dielectric constants used in the calculations; qtable lists the extinction efficiency Q_{ext}, absorption efficiency Q_{abs}, scattering efficiency Q_{sca}, differential scattering efficiency Q_{bk} for each wavelength and each effective radius a_{eff}, differential scattering efficiency Q_{bk}, and so on. The following equation calculates the extinction efficiency Q_{ext}.

$$Q_{\text{ext}} = \frac{C_{\text{ext}}}{\pi a_{\text{eff}}^2} = \frac{1}{\pi a_{\text{eff}}^2} \frac{4\pi k}{|E_{\text{ext}}|^2} \sum_{j=1}^{N} \text{Im}(E_{\text{ext}}^* \cdot \boldsymbol{p}_j). \tag{7.9}$$

Here, C_{ext} is the extinction cross-section, $\boldsymbol{p}_j$ is the dipole moment, and k is the wavenumber. The absorption efficiency Q_{abs} is calculated by the following equation:

$$Q_{\text{abs}} = \frac{C_{\text{abs}}}{\pi a_{\text{eff}}^2} = \frac{1}{\pi a_{\text{eff}}^2} \frac{4\pi k}{|E_{\text{ext}}|^2} \sum_{j=1}^{N} \left(\text{Im}(\boldsymbol{p}_j \cdot (\tilde{\alpha}_j^{-1})^* \cdot \boldsymbol{p}_j^*) - \frac{2}{3} k^3 |\boldsymbol{p}_j|^2| \right), \tag{7.10}$$

where C_{abs} is the absorption cross-section. The scattering efficiency Q_{sca} is obtained from $Q_{\text{sca}} = Q_{\text{ext}} - Q_{\text{abs}}$.

The phase lag efficiency Q_{pha}, polarization efficiency index Q_{pol}, and circular polarization efficiency index Q_{cpol} are listed in qtable2. The polarization efficiency index Q_{pol} is the difference of extinction efficiency Q_{ext} in two orthogonal polarizations. The circular polarization efficiency index Q_{cpol} is defined as $Q_{\text{cpol}} = Q_{\text{pol}} Q_{\text{pha}}$. The phase lag efficiency Q_{pha} is a parameter that indicates the degree of phase delay that occurs when light passes through the structure and is useful for calculating the phase delay of light passing through dust, etc. The phase shift efficiency is given by

$$Q_{\text{pha}} = \frac{C_{\text{pha}}}{\pi a_{\text{eff}}^2} = \frac{1}{\pi a_{\text{eff}}^2} \frac{2\pi k}{|\boldsymbol{E}_{\text{ext}}|^2} \sum_{j=1}^{N} \text{Re}(\boldsymbol{E}_{\text{ext}}^* \cdot \boldsymbol{p}_j). \tag{7.11}$$

Here, C_{pha} is the phase lag cross-section.

The file wXXXrYYYY.avg contains the calculation results at the XXXXth wavelength and YYYYth effective radius. The file target.out, generated when the structure is created using the keyword, contains the coordinates of the dipole, corresponding to the shape.dat file used when FROM_FILE is used.

7.3 Programs for DDSCAT

In DDSCAT, representative shapes can be calculated by specifying keywords. For example, in the ddscat.par file, specify "ELLIPSOID" in the "Target

Geometry and Composition" field on Line 11 and "shape parameters" on Line 12 as, for example, "30 30 30", and in the case of gold particles, specify. "/diel/Au_evap" in the case of gold particles. The size of the sphere is specified in effective radii on Line 31. Each wavelength's extinction, absorption, and scattering efficiencies are stored in qtable. Program 7.1, which directly reads the data stored in the qtable and plots them, is shown below.

In Line 6, f = open("qtable", "r") opens the file and reads its contents into the variable dat, which is separated using the split function. Then, assign it to Qext and plot it. Use the float command to convert from string to value.

Program 7.1

```
1  import matplotlib.pyplot as plt
2  import numpy as np
3  from scipy import zeros
4  from matplotlib.pyplot import plot,show,xlabel,ylabel,title,
       legend,grid,axis,tight_layout
5
6  f = open("qtable","r")  # open "qtable" file
7  dat = f.read()   # Read all strings
8  f.close()      # close "qtable" file
9
10 dat = dat.split("\n")   # split character variable dat
11 datLEN=len(dat)-15   # Find the number of lines excluding the
       header section
12
13 WLx=zeros(datLEN)
14 Qext=zeros(datLEN)
15 Qabs=zeros(datLEN)
16 Qsca=zeros(datLEN)
17
18 DDSversion=dat[0] # Line 0 of "qtable" DDSCAT version
19 Target=dat[1]  # Line 1 "qtable" Keyword (target type)
20 Shape=dat[4]  # Line 4 of "qtable"  Shape of target
21 NumDipole=dat[5]  # Line 5 of "qtable"  number of dipoles
22 aEff=dat[15][1:11]  # Line 15 of "qtable"  effective radius
23
24 i=14
25 j=0
26 while j <= datLEN-1:
27     WLx[j]=float(dat[i][12:22])*1000      # read wavelength in nm
28     Qext[j]=float(dat[i][23:33])   #  read extinction cross-
           section in nm
29     Qabs[j]=float(dat[i][34:44])   #  read absorption cross-
           section in nm
30     Qsca[j]=float(dat[i][45:55])   #  read scattering cross-
           section in nm
31     i=i+1
32     j=j+1
33
34 plot(WLx,Qsca, label=r"$Q_{\rm sca}$",linewidth = 3.0, color='
       black')
```

```
plot(WLx,Qext, label=r"$Q_{\rm ext}$",linewidth = 3.0, color='
    gray')
plot(WLx,Qabs, label=r"$Q_{\rm abs}$",linewidth = 3.0, color='
    black',linestyle='dashed')

xlabel("wavelength (nm)",fontsize=22)
ylabel("Efficiency",fontsize=22)
title("Efficiency",fontsize=22)
grid(True)
axis([400,800,0,15])
plt.tick_params(labelsize=20)
legend(fontsize=20,loc='upper right')
tight_layout()
show()
```

Next, we introduce Program 7.2, which uses the FROM_FILE keyword to output the shape.dat file needed to calculate the next arbitrary structure. shape.dat contains the number of dipoles, the coordinates of each dipole, and information about the material. First, specify the range of coordinates in xmin and xmax. The actual size is specified by effective radii in Line 31 of the ddscat.par file. The dipole spacing d is determined by the effective radii $a_{\rm eff}$ and the number of dipoles from Equation (7.8). The shape of the structure is determined in Lines 25–36. That is, a loop is turned from xmin to xmax in the x-, y-, and z-directions, and 1 is assigned to p[x,y,z] if the expression described in Line 28 applies (true), and 0 if it does not (false). Finally, the coordinates of p[x,y,z]=1 are output to a file. Rewriting Line 28, one can output a shape.dat file for any shape. Program A.7 (ShapePlot) shown in the Appendix can be used to check if the conditions described in Lines 25–36 are the desired structure.

Program 7.2

```
1  import matplotlib.pyplot as plt
2  import numpy as np
3  from scipy import zeros, array
4  from matplotlib.pyplot import plot,show,xlabel,ylabel,title,
       legend,grid, axis,subplot
5
6  xmin = -100      # Calculation range setting
7  xmax = 100
8  ymin = -100
9  ymax = 100
10 zmin = -100
11 zmax = 100
12
13 numx = xmax-xmin+1  #  Number of points to calculate in x-
       direction
```

```
numy = ymax-ymin+1  #  Number of points to calculate in y-direction
numz = zmax-zmin+1  #  Number of points to calculate in z-direction
num = numx*numy*numz  #  Number of points to calculate in all directions

p = np.zeros([numx,numy,numz],dtype=int)   #   initialization of flag p(x,y,z)

iii=0
xorigin=0  #     initialization of gravity center in x-direction
yorigin=0  #     initialization of gravity center in y-direction
zorigin=0  #     initialization of gravity center in z-direction

for z in range(zmin, zmax):
    for y in range(ymin, ymax):
        for x in range(xmin, xmax):
            if x**2 + y**2 + z**2 <= 10**2:   #  determine whether the coordinates constitute a shape
                p[x-xmin,y-ymin,z-zmin] = 1   # p=1 for the coordinates that make up the shape
                    #  Since the array of p is an integer greater than or equal to 0, it is shifted by xmin
                xorigin=xorigin+(x-xmin)   #    Sum the x-coordinates to find the gravity center
                yorigin=yorigin+(y-ymin)   #    Sum the y-coordinates to find the gravity center
                zorigin=zorigin+(z-zmin)   #    Sum the z-coordinates to find the gravity center
                iii+=1
            else:
                p[x-xmin,y-ymin,z-zmin] = 0     #       p=0 if the coordinates do not constitute a shape

Xorigin=xorigin/iii   # the gravity center x component
Yorigin=yorigin/iii   # the gravity center y component
Zorigin=zorigin/iii   # the gravity center z component

l1="--- ddscat calc for FROM_FILE ---"

l3="1.000   0.000   0.000"   # a1 vector
l4="1.000   1.000   0.000"   # a2 vector
l5="1.      1.      1.   "   # d_x/d  d_y/d  d_z/d  (normally 1 1 1)
l7="J     JX      JY      JZ     ICOMPX    ICOMPY   ICOMPZ"

f = open("shape.dat","w")  # open "shape.dat" in write mode
f.write(l1+"\n")   #  write 1st line
f.write(str(iii)+"\n")   #  write 2nd line  (number of dipoles)
f.write(l3+"\n")    # write 3rd line (a1 vector)
f.write(l4+"\n")    # write 4th line (a2 vector)
f.write(l5+"\n")    # write 5th line
f.write(str(Xorigin)+"   "+str(Yorigin)+"   "+str(Zorigin)+"\n")    #   write 6th line
f.write(l7+"\n")    #  write 7th line
```

```

ii=1
for z in range(zmin, zmax):
    for y in range(ymin, ymax):
        for x in range(xmin, xmax):
            if p[x-xmin,y-ymin,z-zmin] == 1:
                f.write(str(ii)+"    "+str(x-xmin)+"    "+str(y-ymin)+"    "+str(z-zmin)+"    1      1      1"+"\n")
                ii+=1
f.close()
```

8

Appendix

8.1 Program of surface plasmon resonance

Program A.1(plannerSPR.py)

```
import scipy as sp
import matplotlib as mpl
import matplotlib.pyplot as plt

from scipy import pi,sin,cos,tan,arcsin,exp,linspace,arrange,sqrt,zeros,array,matrix,asmatrix
from matplotlib.pyplot import plot,show,xlabel,ylabel,title,legend,grid,axis,tight_layout

def mMATs(n1z,n2z):
    return (1/(2*n1z))*matrix([[n1z+n2z,n1z-n2z],[n1z-n2z,n1z+n2z]])
                                        # s-pol Mij matrix
def mMATp(n1z,n2z,n1,n2):
    return (1/(2*n1*n2*n1z))*\
           matrix([[n1**2*n2z+n2**2*n1z,n1**2*n2z-n2**2*n1z],\
                   [n1**2*n2z-n2**2*n1z,n1**2*n2z+n2**2*n1z]])
                                         # p-pol Mij matrix
def matFAI(n1z,d1,k0):
    return matrix([[exp(1j*n1z*k0*d1), 0],[0,exp(-1j*n1z*k0*d1)]])
                                  #  Phi matrix

n1=1.86           # medium 1 (prism) refractive index
n2=sqrt(-10.8 + 1j*1.47)    # medium 2 (gold) refractive index
n3=1.5            #  medium 3 (dielectrics) refractive index
n4=1.33            #   medium 4 (water) refractive index
ep1=n1**2         # medium 1 dielectric constant
ep2=n2**2         # medium 2 dielectric constant
ep3=n3**2         # medium 3 dielectric constant
ep4=n4**2         # medium 4 dielectric constant
d2=47             # thickness of medium 2 (nm)
d3=10             # thickness of medium 3 (nm)
WL=633            # vacuum wavelength (nm)
k0=2*pi/WL        # vacuum wavenumber

t1start=40         # start angle
t1end=70          #  end angle
t1points=300     # number of points
```

```

t1DegOut = linspace(t1start,t1end,t1points)    # array of incident angle
t1 = 0.25*pi+(1/n1)*arcsin((t1DegOut-45) /180*pi)      # angle change in radian
s1 = sin(t1)         # sin(t1)
c1 = cos(t1)         # cos(t1)
s2 = n1/n2*s1        # sin(t2)
c2 = sqrt(1-s2**2)   # cos(t2)
s3 = n1/n3*s1        # sin(t3)
c3 = sqrt(1-s3**2)   # cos(t3)
s4 = n1/n4*s1        # sin(t4)
c4 = sqrt(1-s4**2)   # cos(t4)

n1z=n1*c1          # n1z=k1z/k0
n2z=n2*c2          # n2z=k1z/k0
n3z=n3*c3          # n2z=k1z/k0
n4z=n4*c4          # n2z=k1z/k0

matT0=zeros((t1points,2,2),dtype=complex)          # initialization of T0 matrix
matT1=zeros((t1points,2,2),dtype=complex)          # initialization of T1 matrix
r0=zeros((t1points),dtype=complex) # initialization of reflection coefficient w/o dielectric layer
r1=zeros((t1points),dtype=complex) # initialization of reflection coefficient with dielectric layer

for i in range(t1points):

    matT0[i]=mMATp(n4z[i],n2z[i],n4,n2)@matFAI(n2z[i],d2,k0)@mMATp(n2z[i],n1z[i],n2,n1)
                                             # s-polarization transfer matrix T0
    matT1[i]=mMATp(n4z[i],n3z[i],n4,n3)@matFAI(n3z[i],d3,k0)@mMATp(n3z[i],n2z[i],n3,n2)@matFAI(n2z[i],d2,k0)@mMATp(n2z[i],n1z[i],n2,n1)
                                             # p-polarization transfer matrix T0

    r0[i]=-matT0[i,1,0]/matT0[i,1,1]     # reflection coefficient w/o dielectric layer
    r1[i]=-matT1[i,1,0]/matT1[i,1,1]     # reflection coefficient with dielectric layer

R0Abs=abs(r0)**2        # reflectivity w/o dielectric layer
R1Abs=abs(r1)**2        # reflectivity with dielectric layer

plt.figure(figsize=(8,6))
plt.figure(figsize=(8,6))
plot(t1DegOut,R1Abs, label="R1",linewidth = 3.0, color='gray')
plot(t1DegOut,R0Abs, label="R0",linewidth = 3.0, color='black')
xlabel(r"$\theta_1$ (deg.)",fontsize=20)
ylabel(r"Reflectivity",fontsize=20)
title("Surface Plasmon Resonance",fontsize=20)
grid(True)
legend(fontsize=20,loc='lower right')
```

```
plt.tick_params(labelsize=20)
tight_layout()
show()
```

8.2 Multilayer EMA calculation program

Program A.2 (multilayerEMA.py)

```
import scipy as sp
import matplotlib as mpl
import matplotlib.pyplot as plt

from scipy import pi,sin,cos,tan,arcsin,exp,linspace,sqrt,zeros,
    matrix,arrange
from matplotlib.pyplot import plot,show,xlabel,ylabel,title,
    legend,grid,axis,tight_layout


def func_nAg(WLs):
    ep=3.691-9.1522**2/((1240/WLs)**2+1j*0.021*(1240/WLs))
    index=sqrt(ep)
    return index

def func_nTiO2(WLs):
    ep=5.193 + 0.244/((WLs/1000)**2-0.0803)
    index=sqrt(ep)
    return index

def mMATs(n1z,n2z):
    return (1/(2*n1z))*matrix([[n1z+n2z,n1z-n2z],[n1z-n2z,n1z+n2z
        ]])
                                    # s-pol Mij matrix
def mMATp(n1z,n2z,n1,n2):
    return (1/(2*n1*n2*n1z))*matrix([[n1**2*n2z+n2**2*n1z,n1**2*
        n2z-n2**2*n1z],[n1**2*n2z-n2**2*n1z,n1**2*n2z+n2**2*n1z]])
                                    # p-pol Mij mat
def matFAI1(n1z,d1,k0):
    return matrix([[exp(1j*n1z*k0*d1), 0],[0,exp(-1j*n1z*k0*d1)
        ]])
                                     #  Phi matrix

def matPI1(n1,n1z):
    return matrix([[n1z/n1,n1z/n1,0,0],[n1,-n1
        ,0,0],[0,0,1,1],[0,0,n1z,-n1z]])

def matPI2(n2o,n2oz,n2ez,c2dash):
    return matrix([[c2dash,-c2dash,0,0],[n2o**2/n2ez*c2dash,n2o
        **2/n2ez*c2dash,0,0],[0,0,1,1],[0,0,n2oz,-n2oz]])

def matPI2inv(n2o,n2oz,n2ez,c2dash):
    return 0.5*matrix([[1/c2dash,n2ez/(n2o**2*c2dash),0,0],[-1/
```

```
        c2dash,n2ez/(n2o**2*c2dash),0,0],[0,0,1,1/n2oz],[0,0,1,-1/n2oz]])

def matPI3(n3,n3z):
    return matrix([[n3z/n3,n3z/n3,0,0],[n3,-n3,0,0],[0,0,1,1],[0,0,n3z,-n3z]])

def matPI3inv(n3,n3z):
    return 0.5*matrix([[n3/n3z,1/n3,0,0],[n3/n3z,-1/n3,0,0],[0,0,1,1/n3z],[0,0,1,-1/n3z]])

def matFAI2(k0,n2ez,n2oz,d2):
    return matrix([[exp(1j*n2ez*k0*d2), 0,0,0],[0,exp(-1j*n2ez*k0*d2),0,0],[0,0,exp(1j*n2oz*k0*d2),0],[0,0,0,exp(-1j*n2oz*k0*d2)]])

#############  Initialization  ##############
WLmin = 300             #  start wavelength
WLmax = 1000            #  end wavelength
WLperiod = 1            #  wavelength period
WLx = arrange(WLmin, WLmax+1, WLperiod) #  array of wavelength
NumWLx = int((WLmax-WLmin)/WLperiod)+1 #  number of wavelength
k0=2*pi/WLx             #  array of wavenumber

t1Deg = 45              #  angle of incidence
t1 = t1Deg /180*pi # Convert angle of incidence into radians

#############  Calculation of multilayer A model  ##############

n1=1
nA=1
n2=zeros(NumWLx, dtype=complex)
n3=zeros(NumWLx, dtype=complex)
n4=zeros(NumWLx, dtype=complex)
n5=zeros(NumWLx, dtype=complex)
n6=zeros(NumWLx, dtype=complex)
n7=zeros(NumWLx, dtype=complex)
n8=zeros(NumWLx, dtype=complex)
n9=zeros(NumWLx, dtype=complex)
d2=d4=d6=d8=10
d3=d5=d7=d9=10

for i in range(NumWLx):
     n2[i]=n4[i]=n6[i]=n8[i]=func_nAg(WLx[i])
     n3[i]=n5[i]=n7[i]=n9[i]=func_nTiO2(WLx[i])

s1 = sin(t1)
c1 = cos(t1)
s2, s3, s4, s5, s6, s7, s8, s9, sA = n1/n2*s1, n1/n3*s1, n1/n4*s1, n1/n5*s1, n1/n6*s1, n1/n7*s1, n1/n8*s1, n1/n9*s1, n1/nA*s1
c2, c3, c4, c5, c6, c7, c8, c9, cA = sqrt(1-s2**2), sqrt(1-s3**2), sqrt(1-s4**2), sqrt(1-s5**2), sqrt(1-s6**2), sqrt(1-s7**2), \
                                          sqrt(1-s8**2), sqrt(1-s9**2), sqrt(1-sA**2),
n1z, n2z, n3z, n4z, n5z, n6z, n7z, n8z, n9z, nAz = n1*c1, n2*c2, n3*c3, n4*c4, n5*c5, n6*c6, n7*c7, n8*c8, n9*c9, nA*cA
```

```

matTs=zeros((NumWLx,2,2),dtype=complex)  # initialization of s-pol transfer matrix
matTp=zeros((NumWLx,2,2),dtype=complex)  # initialization of p-pol transfer matrix
rsML1=zeros((NumWLx),dtype=complex)   # initialization of s-pol reflection coefficient
tsML1=zeros((NumWLx),dtype=complex)   # initialization of s-pol transmission coefficient
rpML1=zeros((NumWLx),dtype=complex)   # initialization of p-pol reflection coefficient
tpML1=zeros((NumWLx),dtype=complex)   # initialization of p-pol transmission coefficient

for i in range(NumWLx):

    matTs[i]= mMATs(nAz,n9z[i])@matFAI1(n9z[i],d9,k0[i])@mMATs(n9z[i],n8z[i])@matFAI1(n8z[i],d8,k0[i])@mMATs(n8z[i],n7z[i]) \
        @matFAI1(n7z[i],d7,k0[i])@mMATs(n7z[i],n6z[i])@matFAI1(n6z[i],d6,k0[i])@mMATs(n6z[i],n5z[i])@matFAI1(n5z[i],d5,k0[i]) \
        @mMATs(n5z[i],n4z[i])@matFAI1(n4z[i],d4,k0[i])@mMATs(n4z[i],n3z[i])@matFAI1(n3z[i],d3,k0[i])@mMATs(n3z[i],n2z[i]) \
        @matFAI1(n2z[i],d2,k0[i])@mMATs(n2z[i],n1z)
                                                                # s-pol transfer matrix
    matTp[i]= mMATp(nAz,n9z[i],nA,n9[i])@matFAI1(n9z[i],d9,k0[i])@mMATp(n9z[i],n8z[i],n9[i],n8[i])@matFAI1(n8z[i],d8,k0[i]) \
        @mMATp(n8z[i],n7z[i],n8[i],n7[i])@matFAI1(n7z[i],d7,k0[i])@mMATp(n7z[i],n6z[i],n7[i],n6[i])@matFAI1(n6z[i],d6,k0[i]) \
        @mMATp(n6z[i],n5z[i],n6[i],n5[i])@matFAI1(n5z[i],d5,k0[i])@mMATp(n5z[i],n4z[i],n5[i],n4[i])@matFAI1(n4z[i],d4,k0[i]) \
        @mMATp(n4z[i],n3z[i],n4[i],n3[i])@matFAI1(n3z[i],d3,k0[i])@mMATp(n3z[i],n2z[i],n3[i],n2[i])@matFAI1(n2z[i],d2,k0[i]) \
        @mMATp(n2z[i],n1z,n2[i],n1)
                                                               # p-pol transfer matrix
    rsML1[i]=-matTs[i,1,0]/matTs[i,1,1]     # reflection coefficient calculation for s-polarization
    tsML1[i]=matTs[i,0,0]-matTs[i,0,1]*matTs[i,1,0]/matTs[i,1,1]    # transmission coefficient calculation for s-polarization
    rpML1[i]=-matTp[i,1,0]/matTp[i,1,1]     # reflection coefficient calculation for p-polarization
    tpML1[i]=matTp[i,0,0]-matTp[i,0,1]*matTp[i,1,0]/matTp[i,1,1]    # transmission coefficient calculation for p-polarization

RsML1=abs(rsML1)**2   # s-pol reflectivity (Multilayer model A)
RpML1=abs(rpML1)**2   # p-pol reflectivity (Multilayer model A)
TsML1=abs(tsML1)**2   # s-pol transmittance (Multilayer model A)
TpML1=abs(tpML1)**2   # p-pol transmittance (Multilayer model A)

```

```
########## Calculation of multilayer B model  ############

n1=1
nA=1
n2=zeros(NumWLx, dtype=complex)
n3=zeros(NumWLx, dtype=complex)
n4=zeros(NumWLx, dtype=complex)
n5=zeros(NumWLx, dtype=complex)
n6=zeros(NumWLx, dtype=complex)
n7=zeros(NumWLx, dtype=complex)
n8=zeros(NumWLx, dtype=complex)
n9=zeros(NumWLx, dtype=complex)
d2=d4=d6=d8=10
d3=d5=d7=d9=10

for i in range(NumWLx):
     n2[i]=n4[i]=n6[i]=n8[i]=func_nTiO2(WLx[i])
     n3[i]=n5[i]=n7[i]=n9[i]=func_nAg(WLx[i])

s1 = sin(t1)
c1 = cos(t1)
s2, s3, s4, s5, s6, s7, s8, s9, sA = n1/n2*s1, n1/n3*s1, n1/n4*s1, n1/n5*s1, n1/n6*s1, n1/n7*s1, n1/n8*s1, n1/n9*s1, n1/nA*s1
c2, c3, c4, c5, c6, c7, c8, c9, cA = sqrt(1-s2**2), sqrt(1-s3**2), sqrt(1-s4**2), sqrt(1-s5**2), sqrt(1-s6**2), sqrt(1-s7**2), \
                                     sqrt(1-s8**2), sqrt(1-s9**2), sqrt(1-sA**2),
n1z, n2z, n3z, n4z, n5z, n6z, n7z, n8z, n9z, nAz = n1*c1, n2*c2, n3*c3, n4*c4, n5*c5, n6*c6, n7*c7, n8*c8, n9*c9, nA*cA

matTs=zeros((NumWLx,2,2),dtype=complex)   # initialization s-pol transfer matrix
matTp=zeros((NumWLx,2,2),dtype=complex)   # initialization p-pol transfer matrix
rsML2=zeros((NumWLx),dtype=complex)    #  initialization of rs
tsML2=zeros((NumWLx),dtype=complex)    #  initialization of ts
rpML2=zeros((NumWLx),dtype=complex)    #  initialization of rp
tpML2=zeros((NumWLx),dtype=complex)    # initialization of tp

for i in range(NumWLx):

    matTs[i]= mMATs(nAz,n9z[i])@matFAI1(n9z[i],d9,k0[i])@mMATs(n9z[i],n8z[i])@matFAI1(n8z[i],d8,k0[i])@mMATs(n8z[i],n7z[i]) \
        @matFAI1(n7z[i],d7,k0[i])@mMATs(n7z[i],n6z[i])@matFAI1(n6z[i],d6,k0[i])@mMATs(n6z[i],n5z[i])@matFAI1(n5z[i],d5,k0[i]) \
        @mMATs(n5z[i],n4z[i])@matFAI1(n4z[i],d4,k0[i])@mMATs(n4z[i],n3z[i])@matFAI1(n3z[i],d3,k0[i])@mMATs(n3z[i],n2z[i]) \
        @matFAI1(n2z[i],d2,k0[i])@mMATs(n2z[i],n1z)
                                                # s-pol transfer matrix
    matTp[i]= mMATp(nAz,n9z[i],nA,n9[i])@matFAI1(n9z[i],d9,k0[i]) @mMATp(n9z[i],n8z[i],n9[i],n8[i])@matFAI1(n8z[i],d8,k0[i]) \
        @mMATp(n8z[i],n7z[i],n8[i],n7[i])@matFAI1(n7z[i],d7,k0[i])
```

```
            @mMATp(n7z[i],n6z[i],n7[i],n6[i])@matFAI1(n6z[i],d6,k0[i]) \
        @mMATp(n6z[i],n5z[i],n6[i],n5[i])@matFAI1(n5z[i],d5,k0[i]) @mMATp(n5z[i],n4z[i],n5[i],n4[i])@matFAI1(n4z[i],d4,k0[i]) \
        @mMATp(n4z[i],n3z[i],n4[i],n3[i])@matFAI1(n3z[i],d3,k0[i]) @mMATp(n3z[i],n2z[i],n3[i],n2[i])@matFAI1(n2z[i],d2,k0[i]) \
        @mMATp(n2z[i],n1z,n2[i],n1)
                                                      # p-pol transfer matrix
    rsML2[i]=-matTs[i,1,0]/matTs[i,1,1]    # s-pol reflection coefficient
    tsML2[i]=matTs[i,0,0]-matTs[i,0,1]*matTs[i,1,0]/matTs[i,1,1]  # transmission coefficient calculation for s-polarization
    rpML2[i]=-matTp[i,1,0]/matTp[i,1,1]    # p-pol reflection coefficient
    tpML2[i]=matTp[i,0,0]-matTp[i,0,1]*matTp[i,1,0]/matTp[i,1,1]  # transmission coefficient calculation for p-polarization

RsML2=abs(rsML2)**2         # s-pol reflectivity (Multilayer model B)
RpML2=abs(rpML2)**2         # p-pol reflectivity (Multilayer model B)
TsML2=abs(tsML2)**2         #  s-pol transmittance (Multilayer model B)
TpML2=abs(tpML2)**2         #  p-pol transmittance (Multilayer model B)

##### EMA model calculations (3-layer problem for anisotropic thin films)######

n1=1
n3=1
d2=80

nTiO2=zeros((NumWLx),dtype=complex)
nAg=zeros((NumWLx),dtype=complex)

for i in range(NumWLx):
    nTiO2[i]=func_nTiO2(WLx[i])
    nAg[i]=func_nAg(WLx[i])

epx=0.5*(nTiO2**2 + nAg**2)
epz=2*(nTiO2**2)*(nAg**2)/((nTiO2**2)+(nAg**2))

no=sqrt(epx)
ne=sqrt(epz)

s1 = sin(t1)
c1 = cos(t1)
kappa=n1*s1
s3 = n1/n3*s1
c3 = sqrt(1-s3**2)

n1z=n1*c1
```

```
n3z=n3*c3

n2oz=sqrt(no**2-kappa**2)
n2ez=(no/ne)*sqrt(ne**2-kappa**2)
n2eEff=sqrt(kappa**2+n2ez**2)

matT=zeros((NumWLx,4,4),dtype=complex)  #  initialization of s-pol transfer matrix
rsEMA=zeros((NumWLx),dtype=complex)     # initialization of rs
tsEMA=zeros((NumWLx),dtype=complex)     # initialization of ts
rpEMA=zeros((NumWLx),dtype=complex)     # initialization of rp
tpEMA=zeros((NumWLx),dtype=complex)     # initialization of tp

for i in range(NumWLx):
    matT[i]=matPI3inv(n3,n3z)@matPI2(n2ez[i],n2eEff[i],n2oz[i])@matFAI2(k0[i],n2ez[i],n2oz[i],d2)@matPI2inv(n2ez[i],n2eEff[i],n2oz[i])@matPI1(n1,n1z)
    rsEMA[i]=-matT[i,3,2]/matT[i,3,3]
    tsEMA[i]=matT[i,2,2]-matT[i,2,3]*matT[i,3,2]/matT[i,3,3]
    rpEMA[i]=-matT[i,1,0]/matT[i,1,1]
    tpEMA[i]=matT[i,0,0]-matT[i,0,1]*matT[i,1,0]/matT[i,1,1]

RsEMA=abs(rsEMA)**2        # s-pol reflectivity(EMA model)
RpEMA=abs(rpEMA)**2        # p-pol reflectivity(EMA model)
TsEMA=abs(tsEMA)**2        # s-pol transmittance(EMA model)
TpEMA=abs(tpEMA)**2        # p-pol transmittance(EMAmodel)


############# PLOT ##############

plt.figure(figsize=(8,6))
plot(WLx,RsML1, label="RsML1",linewidth = 3.0, color='black')
plot(WLx,RsML2, label="RsML2",linewidth = 3.0, color='gray')
plot(WLx,RsEMA, label="RsEMA",linewidth = 3.0, color='black', linestyle='dashed')
xlabel(r"Wavelength(nm)",fontsize=22)
ylabel(r"reflectivity",fontsize=22)
title("",fontsize=22)
grid(True)
axis([300,1000,0,1.1])
legend(fontsize=20,loc='lower right')
plt.tick_params(labelsize=20)
tight_layout()
show()

plt.figure(figsize=(8,6))
plot(WLx,RpML1, label="RpML1",linewidth = 3.0, color='black')
plot(WLx,RpML2, label="RpML2",linewidth = 3.0, color='gray')
plot(WLx,RpEMA, label="RpEMA",linewidth = 3.0, color='black', linestyle='dashed')
xlabel(r"Wavelength(nm)",fontsize=22)
ylabel(r"reflectivity",fontsize=22)
title("",fontsize=22)
grid(True)
axis([300,1000,0,1.1])
legend(fontsize=20,loc='lower right')
plt.tick_params(labelsize=20)
```

```
tight_layout()
show()

plt.figure(figsize=(8,6))
plot(WLx,TsML1, label="TsML1",linewidth = 3.0, color=‘black’)
plot(WLx,TsML2, label="TsML2",linewidth = 3.0, color=‘gray’)
plot(WLx,TsEMA, label="TsEMA",linewidth = 3.0, color=‘black’,
    linestyle=‘dashed’)
xlabel(r"Wavelength(nm)",fontsize=22)
ylabel(r"transmittance",fontsize=22)
title("",fontsize=22)
grid(True)
axis([300,1000,0,1.1])
legend(fontsize=20,loc=‘lower right’)
plt.tick_params(labelsize=20)
tight_layout()
show()

plt.figure(figsize=(8,6))
plot(WLx,TpML1, label="TpML1",linewidth = 3.0, color=‘black’)
plot(WLx,TpML2, label="TpML2",linewidth = 3.0, color=‘gray’)
plot(WLx,TpEMA, label="TpEMA",linewidth = 3.0, color=‘black’,
    linestyle=‘dashed’)
xlabel(r"Wavelength(nm)",fontsize=22)
ylabel(r"transmittance",fontsize=22)
title("",fontsize=22)
grid(True)
axis([300,1000,0,1.1])
legend(fontsize=20,loc=‘lower right’)
plt.tick_params(labelsize=20)
tight_layout()
show()
```

8.3 Optical response of a bisphere

Program A.3 (Bisphere.py)

```
import scipy as sp
import scipy.special
import matplotlib as mpl
import matplotlib.pyplot as plt
import math
from scipy import pi,sin,cos,tan,arcsin,exp,linspace,arrange,sqrt
    ,zeros,array,matrix,asmatrix,real,imag,interpolate
from matplotlib.pyplot import plot,show,xlabel,ylabel,title,
    legend,grid, axis,tight_layout
from scipy.special import spherical_jn,spherical_yn, factorial
from RI import WLx, epAg, epAu, RIAu, RIAg

def kjo(k):
    return math.factorial(k)
```

```

def perpendi(k,j,r0,ep1,ep2,ep3):
    return ((ep1-ep3)*k*kjo(k+j))/(((k+1)*ep1+k*ep3)*kjo(k)*kjo(j)*(2*r0)**(k+j+1))

def paralleldi(k,j,r0,ep1,ep2,ep3):
    return -((ep1-ep3)*k*kjo(k+j))/(((k+1)*ep1+k*ep3)*kjo(k+1)*kjo(j-1)*(2*r0)**(k+j+1))


def uhen(ep1,ep2,ep3):
    return (ep1-ep3)/(2*ep1+ep3)

k0=2*pi/WLx  k0 = 2 * pi / WLx  # vacuum wavenumber
qq=15  # order of multipoles
r=50  # radius of sphere
gap=1/2 # half of gap
d=gap+r  #  d parameter
r0=d/r  #  r0 parameter

alpha_A=zeros(NumWLx, dtype=complex)  #  initialization of A
alpha_B=zeros(NumWLx, dtype=complex)  #  initialization of B

ep1=zeros(NumWLx, dtype=complex)
ep2=zeros(NumWLx, dtype=complex)
ep3=zeros(NumWLx, dtype=complex)
al=zeros([NumWLx,qq,qq], dtype=complex)
bl=zeros([NumWLx,qq,qq], dtype=complex)
fl=zeros([NumWLx,qq], dtype=complex)
Xal=zeros([NumWLx,qq], dtype=complex)
Xbl=zeros([NumWLx,qq], dtype=complex)
a11l=zeros([NumWLx,qq], dtype=complex)
b11l=zeros([NumWLx,qq], dtype=complex)


for i in range(NumWLx):
    ep1[i]=1                    # dielectric constant of ambient
    ep2[i]=RI.epAu[i]   # dielectric constant of sphere
    ep3[i]=RI.epAu[i]   # dielectric constant of substrate
    for k in range(qq):  # A coefficient
        for j in range(qq):
            if k==j:
                al[i,k,j]=1+perpendi(k+1,j+1,r0,ep1[i],ep2[i],ep3[i])
            else:
                al[i,k,j]=perpendi(k+1,j+1,r0,ep1[i],ep2[i],ep3[i])

    for k in range(qq):             #  B coefficient
        for j in range(qq):
            if k==j:
                bl[i,k,j]=1+paralleldi(k+1,j+1,r0,ep1[i],ep2[i],ep3[i])
            else:
                bl[i,k,j]=paralleldi(k+1,j+1,r0,ep1[i],ep2[i],ep3[i])

```

```
    for k in range(qq):
            if k==0:
                fl[i,k]=uhen(ep1[i],ep2[i],ep3[i])
            else:
                fl[i,k]=0

    Xal[i]=sp.linalg.solve(al[i],fl[i])   #  Solving simultaneous equations(A coefficient)
    Xbl[i]=sp.linalg.solve(bl[i],fl[i])   #  Solving simultaneous equations(B coefficient)

    alpha_A[i]=-4*pi*r**3*ep1[i]*Xal[i,0]  # polarizability(A coefficient)
    alpha_B[i]=-4*pi*r**3*ep1[i]*Xbl[i,0]  # polarizability(B coefficient)

Csca_A = k0**4/(6*pi)*abs(alpha_A)**2   # scattering cross-section(A coefficient)
Csca_B = k0**4/(6*pi)*abs(alpha_B)**2   # scattering cross-section(B coefficient)
Cabs_A = k0*imag(alpha_A)               # absorption cross-section(A coefficient)
Cabs_B = k0*imag(alpha_B)               # absorption cross-section(B coefficient)

Qsca_A = Csca_A / (2* (r**2) * pi) # scattering efficiency(A coefficient)
Qabs_A = Cabs_A / (2* (r**2) * pi) # absorption efficiency(A coefficient)
Qsca_B = Csca_B / ((r**2) * pi) # scattering efficiency(B coefficient)
Qabs_B = Cabs_B / ((r**2) * pi) # absorption efficiency(B coefficient)

plt.figure(figsize=(8,6))
plot(WLx,Qsca_A, label=r"$Q_{\rm sca}$",linewidth = 3.0, color='black')
plot(WLx,Qabs_A, label=r"$Q_{\rm abs}$",linewidth = 3.0, color='gray')
axis([400,700,0,12])
#xlabel("wavelength (nm)",fontsize=22)
#ylabel("efficiency",fontsize=22)
plt.tick_params(labelsize=20)
#legend(fontsize=20,loc='upper left')
tight_layout()
show()

plt.figure(figsize=(8,6))
plot(WLx,Qsca_B, label=r"$Q_{\rm sca}$",linewidth = 3.0, color='black')
plot(WLx,Qabs_B, label=r"$Q_{\rm abs}$",linewidth = 3.0, color='gray')
axis([400,700,0,4])
#xlabel("wavelength (nm)",fontsize=12)
#ylabel("efficiency",fontsize=12)
plt.tick_params(labelsize=20)
#legend(fontsize=20,loc='upper right')
```

```
tight_layout()
show()
```

8.4 Optical response of a truncated sphere

Program A.4 (truncated.py)

```
import scipy as sp
import scipy.special
import math
import matplotlib as mpl
import matplotlib.pyplot as plt

from scipy import pi,sin,cos,tan,arcsin,exp,linspace,arrange,sqrt,zeros,array,matrix,asmatrix,real,imag,interpolate,integrate
from matplotlib.pyplot import plot,show,xlabel,ylabel,title,legend,grid,axis,tight_layout
from scipy.special import spherical_jn,spherical_yn, factorial,lpmv,eval_legendre
from RI import WLx, NumWLx, epAg, epAu, RIAu, RIAg

def kjo(k):
    return math.factorial(k)

def perpen(k,j,r0,ep1,ep2,ep3):
    return ((ep2-ep1)*(ep1-ep3)*k*kjo(k+j))/((ep2+ep1)*((k+1)*ep1+k*ep3)*kjo(k)*kjo(j)*(2*r0)**(k+j+1))

def parallel(k,j,r0,ep1,ep2,ep3):
    return ((ep2-ep1)*(ep1 - ep3)*k*kjo(k+j))/((ep2+ep1)*((k+1)*ep1+k*ep3)*kjo(k+1)*kjo(j-1)*(2*r0)**(k+j+1))

def uhen(ep1,ep2,ep3):
    return (ep1-ep3)/(2*ep1+ep3)

def funcIMG(r, r0, t):
    return r*r-4*r*r0*t+4*r0*r0

def funcIMG2(r0, t):
    return 1+4*r0*r0-4*r0*t

def funcV(m, j, r, t, r0):
    return funcIMG(r,r0,t)**(-(j+1)/2)*lpmv(m,j,(r*t-2*r0)*funcIMG(r,r0,t)**(-1/2))

def funcV2(m, j, t, r0):
    return funcIMG2(r0,t)**(-0.5*(3+j))*(-(1+j)*funcIMG2(r0,t)*lpmv(m,j,(t-2*r0)/sqrt(funcIMG2(r0,t)))-2*(1+j-m)*r0*sqrt(funcIMG2(r0,t))*lpmv(m,j+1,(t-2*r0)/sqrt(funcIMG2(r0,t))))

def funcW(m, j, r, t, r0):
```

```
    return funcIMG(r,r0,t)**(j/2)*lpmv(m,j,(r*t-2*r0)*funcIMG(r,r0,t)**(-1/2))

def funcW2(m, j, t, r0):
    return funcIMG2(r0,t)**(j/2-1)*((j-4*j*r0*r0+2*r0*(t-2*r0))*lpmv(m,j,(t-2*r0)/sqrt(funcIMG2(r0,t)))-2*(1+j-m)*r0*sqrt(funcIMG2(r0,t))*lpmv(m,j+1,(t-2*r0)/sqrt(funcIMG2(r0,t))))

qq=11  # order of multipoles
theta_a = 90  #  theta_a = 180 - theta_sh   sphere: 0deg  hemisphere: 90 deg
theta_a = theta_a * pi/180
r0 = cos(theta_a)
rr=25  #  radius

k0=2*pi/WLx

ep1=zeros(NumWLx, dtype=complex)
ep2=zeros(NumWLx, dtype=complex)
ep3=zeros(NumWLx, dtype=complex)
ep4=zeros(NumWLx, dtype=complex)

matrixCinteg=zeros([qq,qq], dtype=float)
matrixDinteg=zeros([qq,qq], dtype=float)
matrixEinteg=zeros([qq], dtype=float)
matrixFinteg=zeros([qq,qq], dtype=float)
matrixGinteg=zeros([qq,qq], dtype=float)
matrixJinteg=zeros([qq,qq], dtype=float)
matrixKinteg=zeros([qq,qq], dtype=float)
matrixMinteg=zeros([qq,qq], dtype=float)
matrixNinteg=zeros([qq,qq], dtype=float)
matrixPinteg=zeros([qq], dtype=float)

matrixClist=zeros([NumWLx,qq,qq], dtype=complex)
matrixDlist=zeros([NumWLx,qq,qq], dtype=complex)
matrixElist=zeros([NumWLx,qq], dtype=complex)
matrixFlist=zeros([NumWLx,qq,qq], dtype=complex)
matrixGlist=zeros([NumWLx,qq,qq], dtype=complex)
matrixHlist=zeros([NumWLx,qq], dtype=complex)
matrixJlist=zeros([NumWLx,qq,qq], dtype=complex)
matrixKlist=zeros([NumWLx,qq,qq], dtype=complex)
matrixLlist=zeros([NumWLx,qq], dtype=complex)
matrixMlist=zeros([NumWLx,qq,qq], dtype=complex)
matrixNlist=zeros([NumWLx,qq,qq], dtype=complex)
matrixPlist=zeros([NumWLx,qq], dtype=complex)

matrixRA=zeros([NumWLx,2*qq,2*qq], dtype=complex)
matrixRB=zeros([NumWLx,2*qq,2*qq], dtype=complex)
matrixQA=zeros([NumWLx,2*qq], dtype=complex)
matrixQB=zeros([NumWLx,2*qq], dtype=complex)
vectorA=zeros([NumWLx,2*qq], dtype=complex)
vectorB=zeros([NumWLx,2*qq], dtype=complex)
alpha_A=zeros([NumWLx], dtype=complex)
alpha_B=zeros([NumWLx], dtype=complex)

ep1 = 1
ep2 = ep4 = 1.5**2
```

```
ep3 = epAu

for k in range(qq):
    matrixEinteg[k], dummy = integrate.quad(lambda t: lpmv(0,k+1,t)*(t-r0), -1, r0)
    matrixPinteg[k], dummy = integrate.quad(lambda t: lpmv(1,k+1,t)*lpmv(1,1,t), -1, r0)

    for j in range(qq):
        matrixCinteg[k,j], dummy = integrate.quad(lambda t: lpmv(0,k+1,t)*(lpmv(0,j+1,t)-(-1)**(j+1)*funcV(0,j+1,1,t,r0)), -1, r0)
        matrixDinteg[k,j], dummy = integrate.quad(lambda t: lpmv(0,k+1,t)*(lpmv(0,j+1,t)-(-1)**(j+1)*funcW(0,j+1,1,t,r0)), -1, r0)
        matrixFinteg[k,j], dummy = integrate.quad(lambda t: lpmv(0,k+1,t)*((j+2)*lpmv(0, j+1, t)-(-1)**(j+1)*funcV2(0, j+1, t, r0)), -1, r0)
        matrixGinteg[k,j], dummy = integrate.quad(lambda t: lpmv(0,k+1,t)*((j+1)*lpmv(0, j+1, t)+(-1)**(j+1)*funcW2(0, j+1, t, r0)), -1, r0)

        matrixJinteg[k,j], dummy = integrate.quad(lambda t: lpmv(1,k+1,t)*(lpmv(1,j+1,t)+(-1)**(j+1)*funcV(1,j+1,1,t,r0)), -1, r0)
        matrixKinteg[k,j], dummy = integrate.quad(lambda t: lpmv(1,k+1,t)*(lpmv(1,j+1,t)+(-1)**(j+1)*funcW(1,j+1,1,t,r0)), -1, r0)
        matrixMinteg[k,j], dummy = integrate.quad(lambda t: lpmv(1,k+1,t)*((j+2)*lpmv(1, j+1, t)+(-1)**(j+1)*funcV2(1, j+1, t, r0)), -1, r0)
        matrixNinteg[k,j], dummy = integrate.quad(lambda t: lpmv(1,k+1,t)*((j+1)*lpmv(1, j+1, t)-(-1)**(j+1)*funcW2(1, j+1, t, r0)), -1, r0)

for i in range(NumWLx):
    for k in range(qq):
        for j in range(qq):
            if k==j:
                matrixClist[i,k,j]=4*ep1/((ep1+ep2)*(2*(k+1)+1))-(ep1-ep2)/(ep1+ep2)*matrixCinteg[k,j]
            else:
                matrixClist[i,k,j]=-(ep1-ep2)/(ep1+ep2)*matrixCinteg[k,j]

    for k in range(qq):
        for j in range(qq):
            if k==j:
                matrixDlist[i,k,j]=-4*ep3[i]/((ep3[i]+ep4)*(2*(k+1)+1))+(ep3[i]-ep4)/(ep3[i]+ep4)*matrixDinteg[k,j]
            else:
                matrixDlist[i,k,j]=(ep3[i]-ep4)/(ep3[i]+ep4)*matrixDinteg[k,j]

    for k in range(qq):
        if k==0:
```

```
            matrixElist[i,k]=-2*ep1/(3*ep2)-(1-ep1/ep2)*matrixEinteg[k]
        else:
            matrixElist[i,k]=-(1-ep1/ep2)*matrixEinteg[k]

    for k in range(qq):
        for j in range(qq):
            if k==j:
                matrixFlist[i,k,j]=-4*ep1*ep2*(k+2)/((ep1+ep2)*(2*(k+1)+1))-(ep1*(ep1-ep2))/(ep1+ep2)*matrixFinteg[k,j]
            else:
                matrixFlist[i,k,j]=-(ep1*(ep1-ep2))/(ep1+ep2)*matrixFinteg[k,j]

    for k in range(qq):
        for j in range(qq):
            if k==j:
                matrixGlist[i,k,j]=-4*ep3[i]*ep4*(k+1)/((ep3[i]+ep4)*(2*(k+1)+1))-(ep3[i]*(ep3[i]-ep4))/(ep3[i]+ep4)*matrixGinteg[k,j]
            else:
                matrixGlist[i,k,j]=-(ep3[i]*(ep3[i]-ep4))/(ep3[i]+ep4)*matrixGinteg[k,j]

    for k in range(qq):
        if k==0:
            matrixHlist[i,k]=-2*ep1/3
        else:
            matrixHlist[i,k]=0


    for k in range(qq):
        for j in range(qq):
            if k==j:
                matrixJlist[i,k,j]=4*ep1*(k+1)*(k+2)/((ep1+ep2)*(2*(k+1)+1))-(ep1-ep2)/(ep1+ep2)*matrixJinteg[k,j]
            else:
                matrixJlist[i,k,j]=-(ep1-ep2)/(ep1+ep2)*matrixJinteg[k,j]

    for k in range(qq):
        for j in range(qq):
            if k==j:
                matrixKlist[i,k,j]=-4*ep3[i]*(k+1)*(k+2)/((ep3[i]+ep4)*(2*(k+1)+1))+(ep3[i]-ep4)/(ep3[i]+ep4)*matrixKinteg[k,j]
            else:
                matrixKlist[i,k,j]=(ep3[i]-ep4)/(ep3[i]+ep4)*matrixKinteg[k,j]

    for k in range(qq):
        if k==0:
            matrixLlist[i,k]=-4/3
        else:
            matrixLlist[i,k]=0
```

```

    for k in range(qq):
        for j in range(qq):
            if k==j:
                matrixMlist[i,k,j]=-4*ep1*ep2*(k+1)*(k+2)**2/((
                    ep1+ep2)*(2*(k+1)+1))-(ep1*(ep1-ep2))/(ep1+
                    ep2)*matrixMinteg[k,j]
            else:
                matrixMlist[i,k,j]=-(ep1*(ep1-ep2))/(ep1+ep2)*
                    matrixMinteg[k,j]

    for k in range(qq):
        for j in range(qq):
            if k==j:
                matrixNlist[i,k,j]=-4*ep3[i]*ep4*(k+1)**2*(k+2)
                    /((ep3[i]+ep4)*(2*(k+1)+1))-(ep3[i]*(ep3[i]-
                    ep4))/(ep3[i]+ep4)*matrixNinteg[k,j]
            else:
                matrixNlist[i,k,j]=-(ep3[i]*(ep3[i]-ep4))/(ep3[i
                    ]+ep4)*matrixNinteg[k,j]

    for k in range(qq):
        if k==0:
            matrixPlist[i,k]=-4*ep2/3-(ep1-ep2)*matrixPinteg[k]
        else:
            matrixPlist[i,k]=-(ep1-ep2)*matrixPinteg[k]

for i in range(NumWLx):
    for k in range(qq):
        for j in range(qq):
            matrixRA[i,k,j]=matrixClist[i,k,j]
            matrixRA[i,k,j+qq]=matrixDlist[i,k,j]
            matrixRA[i,k+qq,j]=matrixFlist[i,k,j]
            matrixRA[i,k+qq,j+qq]=matrixGlist[i,k,j]

for i in range(NumWLx):
    for k in range(qq):
          matrixQA[i,k]=matrixElist[i,k]
          matrixQA[i,k+qq]=matrixHlist[i,k]

for i in range(NumWLx):
    for k in range(qq):
        for j in range(qq):
            matrixRB[i,k,j]=matrixJlist[i,k,j]
            matrixRB[i,k,j+qq]=matrixKlist[i,k,j]
            matrixRB[i,k+qq,j]=matrixMlist[i,k,j]
            matrixRB[i,k+qq,j+qq]=matrixNlist[i,k,j]

for i in range(NumWLx):
    for k in range(qq):
          matrixQB[i,k]=matrixLlist[i,k]
          matrixQB[i,k+qq]=matrixPlist[i,k]

for i in range(NumWLx):
    vectorA[i]=sp.linalg.solve(matrixRA[i],matrixQA[i])
    vectorB[i]=sp.linalg.solve(matrixRB[i],matrixQB[i])

```

```
    alpha_A[i]=-4*pi*rr**3*ep1*vectorA[i,0]
    alpha_B[i]=-4*pi*rr**3*ep1*vectorB[i,0]

alpha_A_Re = real(alpha_A)
alpha_B_Re = real(alpha_B)
alpha_A_Im = imag(alpha_A)
alpha_B_Im = imag(alpha_B)
alpha_A_Abs = abs(alpha_A)
alpha_B_Abs = abs(alpha_B)

Csca_A = k0**4 / (6*pi) * abs(alpha_A)**2
Csca_B = k0**4 / (6*pi) * abs(alpha_B)**2
Cabs_A = k0 * imag(alpha_A)
Cabs_B = k0 * imag(alpha_B)

crossA = rr**2 * ((pi - theta_a) + 0.5 * sin(2 * theta_a))
crossB = rr**2 * pi

Qsca_A = Csca_A / crossA
Qabs_A = Cabs_A / crossA
Qsca_B = Csca_B / crossB
Qabs_B = Cabs_B / crossB

plt.figure(figsize=(8,6))
plot(WLx,Qabs_A, label=r"$Q_{\rm abs}^{z}$",linewidth = 3.0,
    color='black')
plot(WLx,Qabs_B, label=r"$Q_{\rm abs}^{||}$",linewidth = 3.0,
    color='gray')
axis([400,800,0,5])
xlabel("wavelength (nm)",fontsize=22)
ylabel("efficiency",fontsize=25)
plt.tick_params(labelsize=20)  # scale fontsize=18pt
legend(fontsize=20,loc='upper right')
tight_layout()
show()
```

8.5 Program of RCWA

Program A.5 (RCWA.py)

```
import scipy.interpolate, scipy.special, scipy.linalg
import math
import cmath
import numpy as np
import matplotlib.pyplot as plt

def Rcwa1d(pol, lambda0, kx0, period, layer, norder):
    """ RCWA for 1D binary grating
    pol: polarization, 'p' or 's'
    lambda0: wavelength of incident wave (μm)
    kx0: in-plane wave number of incident wave (1/μm)
```

```python
    period: period (μm)
    layer: layer structure
    norder: maximum diffraction order (2m+1 for ±m order) """

    nlayer = len(layer) # the number of layers including incident space and exiting space
    depth = np.zeros(nlayer) # thickness of each layer
    metal = np.array([False]*nlayer) # True, if at least one medium has dielectric constaant having imaginary part
    maxsect = max([len(v) for v in layer])//2 # the maximum number of elements composing 1 period
    nsect = np.zeros(nlayer, dtype=int) # the number of elements composing 1 period
    refra = np.zeros((nlayer, maxsect)) # (complex) refractive index of element
    filfac = np.zeros((nlayer, maxsect)) # width of elements normalized with period

    for j in range(nlayer): # retrieving of parameters from layer
        nsect[j] = len(layer[j])//2
        nsect[0] = 1
        nsect[nlayer-1] = 1
        depth[j] = layer[j][0]
        for i in range(nsect[j]):
            refra[j][i] = layer[j][i*2+1]
            if abs(refra[j][i].imag) > 1e-100:
                metal[j] = True
            filfac[j][i] = layer[j][i*2+2]
        filfac[0][0] = 1.
        filfac[nlayer-1][0] = 1.

    k0 = 2.0*math.pi/lambda0 # wave number in vacuum
    kc = k0*refra[nlayer-1][0] # wave number in incident space
    ks = k0*refra[0][0] # wave number in exiting space

    nmax = (norder-1)/2 # maximum diffraction order
    I = np.arrange(-nmax, nmax+1) # array of diffraction orders

    Zm = np.zeros([norder, norder]) # zero matrix
    p = norder//2 # location of zeroth order in matrix
    Eye = np.eye(norder) # identity matrix
    M = norder-1 # maximum order of Fourier series of dielectric constant

    K = 2.0*math.pi/period # grating vector
    kx = kx0+I*K # in-plane wave number of diffracted wave

    kzc = np.sqrt((kc**2-kx**2).astype(np.complex))
    # normal component of wave number of diffracted wave in incident space
    np.where((kzc.real+ kzc.imag)>0, kzc, -kzc) # correction of sign

    kzs = np.sqrt((ks**2-kx**2).astype(np.complex))
    # normal component of wave number of diffracted wave in exiting space
    if metal[0]:
```

```
        np.where((kzs.imag)>0, kzs, -kzs) # correction of sign
    else:
        np.where((kzs.real+ kzs.imag)>0, kzs, -kzs) # correction of sign

    Kx = np.diag(kx)/k0 # diagonal matrix of in-plane wave number of diffracted wave
    Kzc = np.diag(kzc)/k0
    # diagonal matrix of normal component of wave number of diffracted wave in incident space
    Kzs = np.diag(kzs)/k0
    # diagonal matrix of normal component of wave number of diffracted wave in exiting space

    EpsilonX = np.zeros([nlayer, norder, norder], dtype=np.complex)
    # Toeplitz matrix of Fourier coefficients of dielectric constant
    AlphaX = np.zeros([nlayer, norder, norder], dtype=np.complex)
    # Toeplitz matrix of Fourier coefficients of inverse of dielectric constant

    for kk in range(0, nlayer):
        if nsect[kk] > 1:
            vX = np.zeros(M*2+1) # array for Fourier coefficients of dielectric constant
            ivX = np.zeros(M*2+1) # array for Fourier coefficients of inverse of dielectric constant

            for jj in range(0, nsect[kk]): # calculation of Fourier coefficients
                disp = np.sum(filfac[kk][0:jj+1])-filfac[kk][jj]/2.0
                epsX = refra[kk][jj]**2 # permittivity
                asinc = filfac[kk][jj]*np.sinc(filfac[kk][jj]* np.arrange(1, M+ 1))
                vm = epsX*asinc[::-1]
                v0 = np.array([epsX*filfac[kk][jj]])
                vp = epsX*asinc
                vX = vX+ np.concatenate((vm, v0, vp)) \
                    * np.exp(-1j*2*math.pi*disp*np.arrange(-M, M+1))

                ivm = 1/epsX* asinc[::-1]
                iv0 = np.array([1/epsX*filfac[kk][jj]])
                ivp = 1/epsX* asinc
                ivX = ivX+np.concatenate((ivm, iv0, ivp)) \
                    * np.exp(-1j*2*math.pi*disp*np.arrange(-M, M+1))

            EpsilonX[kk, :, :] = scipy.linalg.toeplitz(vX[norder-1: 2*norder-1], \
                vX[norder-1::-1]) # generation of Toeplitz matrix of Fourier coefficients of dielectric constant
            AlphaX[kk, :, :] = scipy.linalg.toeplitz(ivX[norder-1: 2* norder-1], \
```

```
                ivX[norder-1::-1]) # generation of Toeplitz matrix of Fourier coefficients of inverse of dielectric constant
        else: # generation of Toeplitz matrix for homogeneous layer
            EpsilonX[kk, :, :] = Eye*(refra[kk][0]**2)
            AlphaX[kk, :, :] = Eye/(refra[kk][0]**2)

    if pol == "s": # in the case of s-polarization
        Rdu = Zm
        Rud = Zm
        Tuu = Eye
        Tdd = Eye
        for ii in range(0, nlayer):
            epsr = refra[ii][0]**2 # dielectric constant of exitting space
            if nsect[ii] > 1:
                A = Kx*Kx-EpsilonX[ii, :, :] # matrix in right-hand side of Eq. 5.14
                Eigen, W1 = np.linalg.eig(A) # eigenvalues and eigenvectors of above matrix
            else:
                W1 = Eye # eigenvectors for homogeneous layer
                Eigen = ((kx/k0)**2-epsr).astype(np.complex) # eigenvalues for homogeneous layer
            if ii == 0:
                W00 = W1
            Q = np.sqrt(-Eigen) # diagonal elements of matrix Q in Eq. 5.20
            if metal[ii]:
                Q = np.where(Q.imag>0.0, Q, -Q) # correction of sign
            else:
                Q = np.where((Q.real+ Q.imag)> 0.0, Q, -Q) # correction of sign
            V1 = np.dot(W1, np.diag(Q)) # Eq. 5.20
            if ii > 0:
                Q1 = np.dot(np.linalg.inv(W1), W0) # Eq. 5.118
                Q2 = np.dot(np.linalg.inv(V1), V0) # Eq. 5.118
                RudTilde = np.dot(Phip, np.dot(Rud, Phip)) # Eq. 5.110
                TddTilde = np.dot(Tdd, Phip) # Eq. 5.111
                F = np.dot(Q1, Eye+RudTilde) # Eq. 5.116
                G = np.dot(Q2, Eye-RudTilde) # Eq. 5.117
                Tau = np.linalg.inv(F+G) # Eq. 5.117
                Rud = Eye-2.0* np.dot(G, Tau) # Eq. 5.120
                Tdd = 2.0*np.dot(TddTilde, Tau) # Eq. 5.121
            if ii != nlayer-1:
                Phip = np.diag(np.exp(1j*k0*Q*depth[ii])) # Φ+ in Eq. 5.25
                W0 = W1
                V0 = V1
        Rud = np.dot(np.dot(W1, Rud), np.linalg.inv(W1))
            # right-hand side of Eq. 5.131 (except i)
        Tdd = np.dot(np.dot(W00, Tdd), np.linalg.inv(W1))
            # right-hand side of Eq. 5.132 (except i)
        Rs = Rud[:, p] # Eq. 5.131
```

```
        Ts = Tdd[:, p] # Eq. 5.132
        IR = (np.abs(Rs)**2)*np.real(kzc)/np.real(kzc[p]) # diffraction efficiencies of reflected waves
        IT = (np.abs(Ts)**2)*np.real(kzs)/np.real(kzc[p]) # diffraction efficiencies of transmitted waves

    else: # in the case of p-polarization
        Rdu = Zm
        Rud = Zm
        Tuu = Eye
        Tdd = Eye
        for ii in range(0, nlayer):
            epsr = refra[ii][0]**2 # dielectric constant of exitting space
            if nsect[ii] > 1:
                A = np.dot(Kx, np.dot(np.linalg.inv(EpsilonX[ii, :, :]), Kx)) \
                    - Eye # inside of parentheses in right-hand side of Eq. 5.39
                Eigen, W1 = np.linalg.eig(np.dot(np.linalg.inv(AlphaX[ii, :, :]), A))
                    # eigen values and eigenvectors of matrix in right-hande side of Eq. 5.39
            else:
                W1 = Eye # eigenvectors for homogeneous layer
                Eigen = ((kx/k0)**2-epsr).astype(np.complex) # eigenvalues for homogeneous layer
            if ii == 0:
                W00 = W1
            Q - np.sqrt(-Eigen) # diagonal elements of matrix Q in Eq. 5.39
            if metal[ii]:
                Q = np.where(Q.imag>0.0, Q, -Q) # correction of sign
            else:
                Q = np.where((Q.real+Q.imag)>0.0, Q, -Q) # correction of sign
            if nsect[ii] > 1:
                V1 = np.dot(np.dot(AlphaX[ii, :, :], W1), np.diag(Q)) # Eq. 5.47
            else:
                V1 = np.diag(Q)/epsr # Eq.~5.47
            if ii > 0:
                Q1 = np.dot(np.linalg.inv(W1), W0) # Eq. 5.118
                Q2 = np.dot(np.linalg.inv(V1), V0) # Eq. 5.118
                RudTilde = np.dot(np.dot(Phip, Rud), Phip) # Eq. 5.110
                TddTilde = np.dot(Tdd, Phip) # Eq. 5.111
                F = np.dot(Q1, (Eye+RudTilde)) # Eq. 5.116
                G = np.dot(Q2, (Eye-RudTilde)) # Eq. 5.117
                Tau = 2.0*np.linalg.inv(F+G) # Eq. 5.117
                Rud = Eye-np.dot(G, Tau) # Eq. 5.120
                Tdd = np.dot(TddTilde, Tau) # Eq. 5.121
            if ii != nlayer-1:
                Phip = np.diag(np.exp(1j*k0*Q*depth[ii])) # Φ+ in Eq. 5.25
                W0 = W1
```

```
                V0 = V1
        Rud = np.dot(np.dot(W1, Rud), np.linalg.inv(W1)) # right-hand side of Eq. 5.131 (except i)
        Tdd = np.dot(np.dot(W00, Tdd), np.linalg.inv(W1)) # right-hand side of Eq.~5.132 (except i)
        Rp = Rud[:, p] # Eq. 5.131
        Tp = Tdd[:, p] # Eq. 5.132
        IR = (np.abs(Rp)**2)*np.real(kzc)/np.real(kzc[p]) # diffraction efficiencies of reflected waves
        IT = (np.abs(Tp)**2)*np.real(kzs/refra[0][0]**2) \
            /np.real(kzc[p]/refra[nlayer-1][0]**2) # diffraction efficiencies of transmitted waves
    return IR, IT

if __name__ == "__main__":
    layer = ((0, 1.5, 1.0), (0.25, 1.5, 1/2, 1.0, 1/3, 1.5, 1/6), \
        (0.25, 1.5, 1/3, 1.0, 2/3), (0, 1.0, 1.0)) # layer structure
    pitch = 1. # period (μm)
    norder = 21 # diffraction order taken into account (2m+1)
    disporder = range(-2,3) # diffraction order to be displayed (2m+1)
    angle = 30*math.pi/180 # angle of incidence (rad)
    wl_start = 0.5+1e-10 # starting wavelength (μm)
    wl_end = 1.5 # finishing wavelength (μm)
    wl = np.linspace(wl_start, wl_end, 200) # array if wavelengths
    imax = len(wl)
    ir = np.zeros([imax, norder]) # array for storing diffraction efficiencies of reflected waves
    it = np.zeros([imax, norder]) # array for storing diffraction efficiencies of transmitted waves
    for i in range(0, imax):
        ir[i,:], it[i,:] = Rcwa1d(‘p’, wl[i], 2*math.pi*math.sin(angle)/wl[i], \
            pitch, layer, norder) # calling function RCWA

    plt.figure(1) # display of diffraction efficiencies of transmitted waves
    lines= (‘solid’, ‘dashed’, ‘dashdot’, ‘dotted’, ‘solid’)
    for m in disporder:
         plt.plot(wl, it[:, m+norder//2], label="m = {0}".format(m), \
             linewidth=3, linestyle=lines[m-disporder[0]])
    plt.xlim(wl_start, wl_end)
    plt.xlabel(‘Wavelength (μm)’, fontsize=16)
    plt.ylabel(‘Transmittance’, fontsize=16)
    plt.legend(loc=‘center’, frameon=False, fontsize=16)

    plt.figure(2) # display of diffraction efficiencies of reflected waves
    for m in disporder:
         plt.plot(wl, ir[:, m+norder//2], label="m = {0}".format(m), \
             linewidth=3, linestyle=lines[m-disporder[0]])
    plt.xlim(wl_start, wl_end)
```

```
    plt.xlabel('Wavelength (μm)', fontsize=16)
    plt.ylabel('Reflectance', fontsize=16)
    plt.legend(loc='center', frameon=False, fontsize=16)

    plt.show()
```

8.6 Program of FDTD

Program A.6.1 (runfdtd.py)

```
import time
from collections import namedtuple
from fdtd import *

if __name__ == "__main__":

    regionx = 200.0e-9  # object region
    regiony = 200.0e-9  # object region
    regionz = 200.0e-9  # object region
    dxtarget = 2.5e-9  # dx [m]
    dytarget = 2.5e-9  # dy [m]
    dztarget = 2.5e-9  # dz [m]

    source = 'dipole'  # 'dipole' or 'plane' wave source
    pulse = 'pulse'  # 'pulse' or 'cw' source

    lambda0 = 0.561e-6  # center wavelength in vacuum [m]
    courantfac = 0.98  # Courant factor
    mt = 2**15  # number of iterations, must be integer power of 2
    mfft = 2**9  # number of sampling for FFT, must be integer power of 2
    extrapol = 4  # zero-filling factor before FFT

    msf = 3  # width for scattering field region (>=3)
    mpml = 8  # number of perfectly matched layers
    kappamax = 100.0  # parameter for CFS-CPML
    amax = 10.0  # parameter for CFS-CPML
    mpow = 3  # parameter for CFS-CPML

    r1 = 25.0e-9  # radius of inner sphere
    Obj = namedtuple('Obj', ('shape', 'material', 'position', 'size'))
    objs = (
        Obj('background', 'vacuum', 0, 0),
        Obj('substrate', 'SiO2', (0, 0, r1), 0),
        Obj('sphere', 'Au', (0, 0, 0), r1)
        )

    Dipole = namedtuple('Dipole', ('pol', 'phase', 'x', 'y', 'z'))
```

```
    # phase: ‘in’ in-phase, ‘anti’ antiphase
    dipoles = (
        Dipole(‘z’, ‘in’, 0, 0, -30e-9),
        )

    # field monitors
    savenum = 32  # total number of data saving
    saveint = mt//savenum  # interval for data saving
    Fmon= namedtuple(‘Fmon’, (‘ehfield’, ‘axis’, ‘position’))
    fieldmons = (savenum, saveint,
        Fmon(‘Ex’, ‘y’, 0),
        Fmon(‘Ex’, ‘z’, 0),
        Fmon(‘Ez’, ‘y’, 0),
        Fmon(‘Hy’, ‘x’, 0)
        )

    # epsilon monitors
    Epsmon = namedtuple(‘Epsmon’, (‘pol’, ‘axis’, ‘position’))
    epsmons = (
        Epsmon(‘x’, ‘z’, 0), \
        Epsmon(‘x’, ‘y’, 0), \
        Epsmon(‘z’, ‘z’, 0))

    r1 = 25.0e-9  # radius of sphere
    Dtct = namedtuple(‘Dtct’, (‘pol’, ‘x’, ‘y’, ‘z’))
    detectors = (
        Dtct(‘x’, 0, 0, 0),
        Dtct(‘x’, r1 + 5.0e-9, 0, 0),
        Dtct(‘z’, r1 + 5.0e-9, 0, 0),
        Dtct(‘x’, r1, 0, r1),
        Dtct(‘z’, r1, 0, r1),
        )

    em = Fdtd(\
        source, pulse, lambda0, courantfac, mt, mfft, extrapol, \
        regionx, regiony, regionz, dxtarget, dytarget, dztarget, \
        mpml, msf, kappamax, amax, mpow, \
        objs, fieldmons, epsmons, detectors, dipoles)
    start = time.time()
    em.sweep()
    print(‘Elapsed time = %f s’ % (time.time() - start))
```

ProgramA.6.2 (fdtd.py)

```
import sys
import math
import os
import numpy as np
from preprocess import *

class Fdtd(Preprocess):

    def sweep(self):
```

```
        """ Time development with CFS-PML and ADE """

        self.save_idv()
        numt = 0
        for jt in range(self.mt):
            # update E-field
            self.sweep_isolate_e()
            self.sweep_boundary_e()
            # E-field source injection
            if self.source == ‘plane’:
                self.normalinc_p_e(jt)
            else:
                self.dipole_source(jt)
            # auxiary E-field update
            self.develop_pcurrent()

            # update H-field
            self.sweep_isolate_h()
            self.sweep_boundary_h()
            # H-field source injection
            if self.source == ‘plane’:
                self.normalinc_p_h()

            # store H and E fields
            if (jt+1)%self.saveint == 0 and numt < self.savenum:
                self.save_ehfield(numt)
                numt = numt+1
            self.detect_efield(jt)

            # update arrays
            self.update_field()

        # calculate spectra
        self.detect_spectra()

    def dipole_source(self, jt):
        """ Dipole source """

        env_factor = 1.0/4.0

        tau = math.pi/self.omega0
        if self.pulse == ‘pulse’:
            t0 = 5.0*tau
        else:
            t0 = 0.0
            omega_env = self.omega0*env_factor
        tempe = (jt-1)*self.dt - t0

        if self.pulse == ‘pulse’:
            campe = math.sin(self.omega0*tempe)
            j00 = math.exp(-tempe*tempe/tau/tau) * campe
        else:
            tempe2 = tempe - math.pi/omega_env
            campe = math.cos(self.omega0*tempe2)
            if tempe2 <  -math.pi/omega_env:
                j00 = 0
            elif tempe2 < 0:
```

```
                j00 = 0.5 * (1+math.cos(omega_env*tempe2)) * campe
            else:
                j00 = campe

        for dipole in self.idipoles:
            if dipole.pol == ‘x’:
                self.Ex2[dipole.iz, dipole.iy, dipole.ix] = \
                    self.Ex2[dipole.iz, dipole.iy, dipole.ix] - dipole.phase* j00
            elif dipole.pol == ‘y’:
                self.Ey2[dipole.iz, dipole.iy, dipole.ix] = \
                    self.Ey2[dipole.iz, dipole.iy, dipole.ix] - dipole.phase* j00
            elif dipole.pol == ‘z’:
                self.Ez2[dipole.iz, dipole.iy, dipole.ix] = \
                    self.Ez2[dipole.iz, dipole.iy, dipole.ix] - dipole.phase* j00
            else:
                print(‘Error at dipole_source!’)

        self.esource[jt] = j00

    def normalinc_p_e(self, jt):
        """
            Source: x-polarized and z-propagating plane wave
            TF/SF compensation for E
        """

        # generation of the temporal shape of the source wave
        iz00 = self.mz1   # origin for incident wave
        env_factor = 1.0/4.0
        tau = math.pi/self.omega0
        if self.pulse == ‘pulse’:
            t0 = 5.0*tau
        else:
            t0 = 0.0
            omega_env = self.omega0*env_factor

        tempe = (jt+ 0.5)*self.dt - t0 \
            - (self.izst- iz00)*self.dz*math.sqrt(self.epsr[self.bgmater])/self.cc
        temph = jt*self.dt - t0- (self.izst-iz00-0.5) \
            * self.dz*math.sqrt(self.epsr[self.bgmater])/self.cc
        campe = math.sin(self.omega0*tempe)
        camph = math.sin(self.omega0*temph)
        if self.pulse == ‘pulse’:
            SEx00 = math.exp(-tempe*tempe/tau/tau)*campe
            SHy00 = math.exp(-temph*temph/tau/tau)*camph \
                / (self.zz0/math.sqrt(self.epsr[self.bgmater]))
        else:
            if tempe < 0.0:
                SEx00 = 0.0
            elif tempe < math.pi/omega_env:
                SEx00 = 0.5 * (1.0-math.cos(omega_env*tempe)) * campe
            else:
```

```
                SEx00 = campe

            if temph < 0.0:
                SHy00 = 0.0
            elif temph < math.pi/omega_env:
                SHy00 = 0.5*(1.0-math.cos(omega_env* temph))*camph \
                    / (self.zz0/math.sqrt(self.epsr[self.bgmater]))
            else:
                SHy00= camph / (self.zz0/math.sqrt(self.epsr[self.bgmater]))

        # store source E field
        self.esource[jt] = SEx00

        # store source E and H fields for FFT
        if jt%self.sampint == 0:
            jfft = jt//self.sampint

        #   Ex development
        for iz in range(1, self.mzz):
            imater = self.isdx[iz]
            self.SEx2[iz] = self.SEx1[iz]*self.ce1[imater] \
                - self.spx2[iz]*self.ce3[imater] \
                - (self.SHy1[iz]-self.SHy1[iz-1])*self.ckez[iz]*self.ce2[imater]

        #   -z pml
        for iz in range(1, self.mz1):
            self.SpsiExz2m[iz] = self.SpsiExz1m[iz]*self.cbze[iz] \
                + (self.SHy1[iz]-self.SHy1[iz-1])*self.ccze[iz]
            self.SEx2[iz] = self.SEx2[iz] - self.SpsiExz2m[iz]*self.ce2[self.isdx[iz]]

        #   +z pml
        for iz in range(self.mz2+ 1, self.mzz):
            izz = iz - self.mz2
            izzr = self.mzz - iz
            self.SpsiExz2p[izz] = self.SpsiExz1p[izz]*self.cbze[izzr] \
                + (self.SHy1[iz]-self.SHy1[iz-1])*self.ccze[izzr]
            self.SEx2[iz] = self.SEx2[iz] - self.SpsiExz2p[izz]*self.ce2[self.isdx[iz]]

        # source compensation for E
        self.SEx2[self.izst] = self.SEx2[self.izst] + self.cez2[self.isdx[self.izst]]*SHy00

        self.SEx2[0] = 0.0
        self.SEx2[self.mzz] = 0.0

        #   Hy development
        for iz in range(self.mzz):
            self.SHy2[iz] = self.SHy1[iz] - (self.SEx2[iz+1]-self.SEx2[iz])*self.ckhz1[iz]
```

```

        #   -z pml
        for iz in range(self.mz1):
            self.SpsiHyz2m[iz] = self.SpsiHyz1m[iz]*self.cbzh[iz] \
            + (self.SEx2[iz+1]-self.SEx2[iz])*self.cczh[iz]
            self.SHy2[iz] = self.SHy2[iz]- self.SpsiHyz2m[iz]*self.coefh

        #   +z pml
        for iz in range(self.mz2, self.mzz):
            izz = iz - self.mz2
            izzr = self.mzz - iz- 1
            self.SpsiHyz2p[izz] = self.SpsiHyz1p[izz]*self.cbzh[izzr] \
                + (self.SEx2[iz+1]-self.SEx2[iz])*self.cczh[izzr]
            self.SHy2[iz] = self.SHy2[iz] - self.SpsiHyz2p[izz]*self.coefh

        # source compensation for H
        self.SHy2[self.izst-1] = self.SHy2[self.izst-1] + self.ckhz1[self.izst]*SEx00

        iy1 = self.mox1
        iy2 = self.moy2
        iz1 = self.moz1
        iz2 = self.moz2 - 1

        # -x boundary
        ix = self.mxx1
        for iz in range(iz1, iz2):
            self.Ez2[iz,iy1:iy2,ix] = self.Ez2[iz,iy1:iy2,ix] \
                - self.cex2[self.isdz[iz]]*self.SHy2[iz]

        # +x boundary
        ix = self.mox2
        for iz in range(iz1, iz2):
            self.Ez2[iz,iy1:iy2,ix] = self.Ez2[iz,iy1:iy2,ix] \
                + self.cex2[self.isdz[iz]]*self.SHy2[iz]

        ix1 = self.mox1
        ix2 = self.mox2 - 1

        # -z boundary
        iz = self.moz1
        self.Ex2[iz,iy1:iy2,ix1:ix2] = self.Ex2[iz,iy1:iy2,ix1:ix2] \
            + self.cez2[self.isdx[iz]]*self.SHy2[iz-1]

        # +z boundary
        iz = self.moz2 - 1
        self.Ex2[iz,iy1:iy2,ix1:ix2] = self.Ex2[iz,iy1:iy2,ix1:ix2] \
            - self.cez2[self.isdx[iz]]*self.SHy2[iz]

        # develop spx2 for ADE
        iz1 = 1
```

```
        iz2 = self.mzz
        self.spx2[iz1:iz2] = self.cj1[self.isdx[iz1:iz2]]*self.spx2[iz1:iz2] \
            + self.cj3[self.isdx[iz1:iz2]]*(self.SEx2[iz1:iz2]+self.SEx1[iz1:iz2])

        #  update
        self.SEx1[:] = self.SEx2[:]
        self.SHy1[:] = self.SHy2[:]
        self.SpsiExz1m[:] = self.SpsiExz2m[:]
        self.SpsiExz1p[:] = self.SpsiExz2p[:]
        self.SpsiHyz1m[:] = self.SpsiHyz2m[:]
        self.SpsiHyz1p[:] = self.SpsiHyz2p[:]

    def normalinc_p_h(self):
        """
          Source: x-polarized and z-propagating plane wave
          TF/SF compensation for H
        """

        ix1 = self.mox1
        ix2 = self.mox2 - 1
        iz1 = self.moz1
        iz2 = self.moz2

        # -y boundary
        iy = self.my1+ self.msf
        for iz in range(iz1, iz2):
            for ix in range(ix1, ix2):
                self.Hz2[iz,iy-1,ix] = self.Hz2[iz,iy-1,ix] \
                    - self.ckhy1[iy-1]*self.SEx1[iz]

        # +y boundary
        iy = self.my2- self.msf
        for iz in range(iz1, iz2):
            for ix in range(ix1, ix2):
                self.Hz2[iz,iy,ix] = self.Hz2[iz,iy,ix] + self.ckhy1[iy]*self.SEx1[iz]

        iy1 = self.moy1
        iy2 = self.moy2

        # -z boundary
        iz = self.moz1
        for iy in range(iy1, iy2):
            for ix in range(ix1, ix2):
                self.Hy2[iz-1,iy,ix] = self.Hy2[iz-1,iy,ix] \
                    + self.ckhz1[iz-1]*self.SEx1[iz]

        # +z boundary
        iz = self.moz2 - 1
        for iy in range(iy1, iy2):
            for ix in range(ix1, ix2):
                self.Hy2[iz,iy,ix] = self.Hy2[iz,iy,ix]- self.ckhz1[iz]*self.SEx1[iz]

    def sweep_isolate_h(self):
```

```

        ix1 = 0
        iy1 = 0
        iz1 = 0

        #     Hx development

        ix2 = self.mxx+ 1
        iy2 = self.myy
        iz2 = self.mzz

        for iy in range(iy1, iy2):
            self.Hx2[iz1:iz2,iy,ix1:ix2] = self.Hx1[iz1:iz2,iy,ix1:ix2] \
                - (self.Ez2[iz1:iz2,iy+1,ix1:ix2]-self.Ez2[iz1:iz2,iy,ix1:ix2]) \
                * self.ckhy1[iy]
        for iz in range(iz1, iz2):
            self.Hx2[iz,iy1:iy2,ix1:ix2] = self.Hx2[iz,iy1:iy2,ix1:ix2] \
                + (self.Ey2[iz+1,iy1:iy2,ix1:ix2]-self.Ey2[iz,iy1:iy2,ix1:ix2]) \
                * self.ckhz1[iz]

        #     Hy development

        ix2 = self.mxx
        iy2 = self.myy+ 1
        iz2 = self.mzz

        for iz in range(iz1, iz2):
            self.Hy2[iz,iy1:iy2,ix1:ix2] = self.Hy1[iz,iy1:iy2,ix1:ix2] \
                - (self.Ex2[iz+1,iy1:iy2,ix1:ix2]-self.Ex2[iz,iy1:iy2,ix1:ix2]) \
                * self.ckhz1[iz]
        for ix in range(ix1, ix2):
            self.Hy2[iz1:iz2,iy1:iy2,ix] = self.Hy2[iz1:iz2,iy1:iy2,ix] \
                + (self.Ez2[iz1:iz2,iy1:iy2,ix+1]-self.Ez2[iz1:iz2,iy1:iy2,ix]) \
                * self.ckhx1[ix]

        #     Hz development

        ix2 = self.mxx
        iy2 = self.myy
        iz2 = self.mzz+ 1

        for ix in range(ix1, ix2):
            self.Hz2[iz1:iz2,iy1:iy2,ix] = self.Hz1[iz1:iz2,iy1:iy2,ix] \
                - (self.Ey2[iz1:iz2,iy1:iy2,ix+1]-self.Ey2[iz1:iz2,iy1:iy2,ix]) \
                * self.ckhx1[ix]
        for iy in range(iy1, iy2):
```

```
            self.Hz2[iz1:iz2,iy,ix1:ix2] = self.Hz2[iz1:iz2,iy,ix1:ix2] \
                + (self.Ex2[iz1:iz2,iy+1,ix1:ix2]-self.Ex2[iz1:iz2,iy,ix1:ix2]) \
                * self.ckhy1[iy]

    def sweep_boundary_h(self):

        # -x boundary

        # Hy PML
        ix1 = 0
        iy1 = 0
        iz1 = 0
        ix2 = self.mx1
        iy2 = self.myy+ 1
        iz2 = self.mzz

        for ix in range(ix1, ix2):
            self.psiHyx2m[iz1:iz2,iy1:iy2,ix] \
                = self.psiHyx1m[iz1:iz2,iy1:iy2,ix]*self.cbxh[ix] \
                + (self.Ez2[iz1:iz2,iy1:iy2,ix+1]-self.Ez2[iz1:iz2,iy1:iy2,ix]) \
                * self.ccxh[ix]
            self.Hy2[iz1:iz2,iy1:iy2,ix]= self.Hy2[iz1:iz2,iy1:iy2,ix] \
                + self.psiHyx1m[iz1:iz2,iy1:iy2,ix]* self.coefh

        # Hz PML
        ix2 = self.mx1
        iy2 = self.myy
        iz2 = self.mzz+ 1

        for ix in range(ix1, ix2):
            self.psiHzx2m[iz1:iz2,iy1:iy2,ix] \
                = self.psiHzx1m[iz1:iz2,iy1:iy2,ix]*self.cbxh[ix] \
                + (self.Ey2[iz1:iz2,iy1:iy2,ix+1]-self.Ey2[iz1:iz2,iy1:iy2,ix]) \
                * self.ccxh[ix]
            self.Hz2[iz1:iz2,iy1:iy2,ix]= self.Hz2[iz1:iz2,iy1:iy2,ix] \
                - self.psiHzx1m[iz1:iz2,iy1:iy2,ix]*self.coefh

        # +x boundary

        # Hy PML
        ix1 = self.mx2
        iy1 = 0
        iz1 = 0
        ix2 = self.mxx
        iy2 = self.myy+ 1
        iz2 = self.mzz

        for ix in range(ix1, ix2):
            ixx = ix- self.mx2
```

```
            ixxr = self.mxx- ix- 1
            self.psiHyx2p[iz1:iz2,iy1:iy2,ixx] \
                = self.psiHyx1p[iz1:iz2,iy1:iy2,ixx]* self.cbxh[ixxr] \
                + (self.Ez2[iz1:iz2,iy1:iy2,ix+1]-self.Ez2[iz1:iz2,iy1:iy2,ix]) \
                * self.ccxh[ixxr]
            self.Hy2[iz1:iz2,iy1:iy2,ix]= self.Hy2[iz1:iz2,iy1:iy2,ix] \
                + self.psiHyx1p[iz1:iz2,iy1:iy2,ixx]*self.coefh

        # Hz PML
        ix2 = self.mxx
        iy2 = self.myy
        iz2 = self.mzz+ 1

        for ix in range(ix1, ix2):
            ixx = ix- self.mx2
            ixxr = self.mxx- ix- 1
            self.psiHzx2p[iz1:iz2,iy1:iy2,ixx] \
                = self.psiHzx1p[iz1:iz2,iy1:iy2,ixx]*self.cbxh[ixxr] \
                + (self.Ey2[iz1:iz2,iy1:iy2,ix+1]-self.Ey2[iz1:iz2,iy1:iy2,ix]) \
                * self.ccxh[ixxr]
            self.Hz2[iz1:iz2,iy1:iy2,ix]= self.Hz2[iz1:iz2,iy1:iy2,ix] \
                - self.psiHzx1p[iz1:iz2,iy1:iy2,ixx]*self.coefh

        # -y boundary

        # Hx PML
        ix1 = 0
        iy1 = 0
        iz1 = 0
        ix2 = self.mxx+ 1
        iy2 = self.my1
        iz2 = self.mzz

        for iy in range(iy1, iy2):
            self.psiHxy2m[iz1:iz2,iy,ix1:ix2] \
                = self.psiHxy1m[iz1:iz2,iy,ix1:ix2]*self.cbyh[iy] \
                + (self.Ez2[iz1:iz2,iy+1,ix1:ix2]-self.Ez2[iz1:iz2,iy,ix1:ix2]) \
                * self.ccyh[iy]
            self.Hx2[iz1:iz2,iy,ix1:ix2]= self.Hx2[iz1:iz2,iy,ix1:ix2] \
                - self.psiHxy1m[iz1:iz2,iy,ix1:ix2]*self.coefh

        # Hz PML
        ix2 = self.mxx
        iy2 = self.my1
        iz2 = self.mzz+ 1

        for iy in range(iy1, iy2):
            self.psiHzy2m[iz1:iz2,iy,ix1:ix2] \
```

```
                = self.psiHzy1m[iz1:iz2,iy,ix1:ix2]*self.cbyh[iy] \
                + (self.Ex2[iz1:iz2,iy+1,ix1:ix2]-self.Ex2[iz1:iz2,iy,ix1:ix2]) \
                * self.ccyh[iy]
            self.Hz2[iz1:iz2,iy,ix1:ix2]= self.Hz2[iz1:iz2,iy,ix1:ix2] \
                + self.psiHzy1m[iz1:iz2,iy,ix1:ix2]*self.coefh

        # +y boundary

        # Hx PML
        ix1 = 0
        iy1 = self.my2
        iz1 = 0
        ix2 = self.mxx+ 1
        iy2 = self.myy
        iz2 = self.mzz

        for iy in range(iy1, iy2):
            iyy = iy- self.my2
            iyyr = self.myy- iy- 1
            self.psiHxy2p[iz1:iz2,iyy,ix1:ix2] \
                = self.psiHxy1p[iz1:iz2,iyy,ix1:ix2]*self.cbyh[iyyr] \
                + (self.Ez2[iz1:iz2,iy+1,ix1:ix2]-self.Ez2[iz1:iz2,iy,ix1:ix2]) \
                * self.ccyh[iyyr]
            self.Hx2[iz1:iz2,iy,ix1:ix2]= self.Hx2[iz1:iz2,iy,ix1:ix2] \
                - self.psiHxy1p[iz1:iz2,iyy,ix1:ix2]*self.coefh

        # Hz PML
        ix2 = self.mxx
        iy2 = self.myy
        iz2 = self.mzz+ 1

        for iy in range(iy1, iy2):
            iyy = iy- self.my2
            iyyr = self.myy- iy- 1
            self.psiHzy2p[iz1:iz2,iyy,ix1:ix2] \
                = self.psiHzy1p[iz1:iz2,iyy,ix1:ix2]*self.cbyh[iyyr] \
                + (self.Ex2[iz1:iz2,iy+1,ix1:ix2]-self.Ex2[iz1:iz2,iy,ix1:ix2]) \
                * self.ccyh[iyyr]
            self.Hz2[iz1:iz2,iy,ix1:ix2]= self.Hz2[iz1:iz2,iy,ix1:ix2] \
                + self.psiHzy1p[iz1:iz2,iyy,ix1:ix2]*self.coefh

        # -z boundary

        # Hx PML
        ix1 = 0
        iy1 = 0
        iz1 = 0
        ix2 = self.mxx+ 1
```

```
        iy2 = self.myy
        iz2 = self.mz1

        for iz in range(iz1, iz2):
            self.psiHxz2m[iz,iy1:iy2,ix1:ix2] \
                = self.psiHxz1m[iz,iy1:iy2,ix1:ix2]*self.cbzh[iz] \
                + (self.Ey2[iz+1,iy1:iy2,ix1:ix2]-self.Ey2[iz,iy1:iy2,ix1:ix2]) \
                * self.cczh[iz]
            self.Hx2[iz,iy1:iy2,ix1:ix2]= self.Hx2[iz,iy1:iy2,ix1:ix2] \
                + self.psiHxz1m[iz,iy1:iy2,ix1:ix2]*self.coefh

        # Hy PML
        ix2 = self.mxx
        iy2 = self.myy+ 1
        iz2 = self.mz1

        for iz in range(iz1, iz2):
            self.psiHyz2m[iz,iy1:iy2,ix1:ix2] \
                = self.psiHyz1m[iz,iy1:iy2,ix1:ix2]*self.cbzh[iz] \
                + (self.Ex2[iz+1,iy1:iy2,ix1:ix2]-self.Ex2[iz,iy1:iy2,ix1:ix2]) \
                * self.cczh[iz]
            self.Hy2[iz,iy1:iy2,ix1:ix2]= self.Hy2[iz,iy1:iy2,ix1:ix2] \
                - self.psiHyz1m[iz,iy1:iy2,ix1:ix2]*self.coefh

        # +z boundary

        # Hx PML
        ix1 = 0
        iy1 = 0
        iz1 = self.mz2
        ix2 = self.mxx+ 1
        iy2 = self.myy
        iz2 = self.mzz

        for iz in range(iz1, iz2):
            izz = iz- self.mz2
            izzr = self.mzz- iz- 1
            self.psiHxz2p[izz,iy1:iy2,ix1:ix2] \
                = self.psiHxz1p[izz,iy1:iy2,ix1:ix2]*self.cbzh[izzr] \
                + (self.Ey2[iz+1,iy1:iy2,ix1:ix2]-self.Ey2[iz,iy1:iy2,ix1:ix2]) \
                * self.cczh[izzr]
            self.Hx2[iz,iy1:iy2,ix1:ix2]= self.Hx2[iz,iy1:iy2,ix1:ix2] \
                + self.psiHxz1p[izz,iy1:iy2,ix1:ix2]*self.coefh

        # Hy PML
        ix2 = self.mxx
        iy2 = self.myy+ 1
        iz2 = self.mzz
```

```
        for iz in range(iz1, iz2):
            izz = iz- self.mz2
            izzr = self.mzz- iz- 1
            self.psiHyz2p[izz,iy1:iy2,ix1:ix2] \
                = self.psiHyz1p[izz,iy1:iy2,ix1:ix2]*self.cbzh[izzr] \
                + (self.Ex2[iz+1,iy1:iy2,ix1:ix2]-self.Ex2[iz,iy1:iy2,ix1:ix2]) \
                * self.cczh[izzr]
            self.Hy2[iz,iy1:iy2,ix1:ix2]= self.Hy2[iz,iy1:iy2,ix1:ix2] \
                - self.psiHyz1p[izz,iy1:iy2,ix1:ix2]*self.coefh

    def sweep_isolate_e(self):

        ix2 = self.mxx
        iy2 = self.myy
        iz2 = self.mzz
        """---------------------
          Ex development
        ---------------------"""
        ix1 = 0
        iy1 = 1
        iz1 = 1
        for iy in range(iy1, iy2):
            self.Ex2[iz1:iz2,iy,ix1:ix2] \
                = self.Ex1[iz1:iz2,iy,ix1:ix2]*self.ce1[self.idx[iz1:iz2,iy,ix1:ix2]] \
                - self.px2[iz1:iz2,iy,ix1:ix2]*self.ce3[self.idx[iz1:iz2,iy,ix1:ix2]] \
                + (self.Hz1[iz1:iz2,iy,ix1:ix2]-self.Hz1[iz1:iz2,iy-1,ix1:ix2]) \
                * self.ckey[iy]*self.ce2[self.idx[iz1:iz2,iy,ix1:ix2]]

        for iz in range(iz1, iz2):
            self.Ex2[iz,iy1:iy2,ix1:ix2]= self.Ex2[iz,iy1:iy2,ix1:ix2] \
                - (self.Hy1[iz,iy1:iy2,ix1:ix2]-self.Hy1[iz-1,iy1:iy2,ix1:ix2]) \
                * self.ckez[iz]*self.ce2[self.idx[iz,iy1:iy2,ix1:ix2]]

        """---------------------
          Ey development
        ---------------------"""
        ix1 = 1
        iy1 = 0
        iz1 = 1
        for iz in range(iz1, iz2):
            self.Ey2[iz,iy1:iy2,ix1:ix2] \
                = self.Ey1[iz,iy1:iy2,ix1:ix2]*self.ce1[self.idy[iz,iy1:iy2,ix1:ix2]] \
                - self.py2[iz,iy1:iy2,ix1:ix2]*self.ce3[self.idy[iz,iy1:iy2,ix1:ix2]] \
                + (self.Hx1[iz,iy1:iy2,ix1:ix2]-self.Hx1[iz-1,iy1:iy2,ix1:ix2]) \
```

```
                * self.ckez[iz]*self.ce2[self.idy[iz,iy1:iy2,ix1:ix2]]
        for ix in range(ix1, ix2):
            self.Ey2[iz1:iz2,iy1:iy2,ix] = self.Ey2[iz1:iz2,iy1:iy2,ix] \
                - (self.Hz1[iz1:iz2,iy1:iy2,ix]-self.Hz1[iz1:iz2,iy1:iy2,ix-1]) \
                * self.ckex[ix]*self.ce2[self.idy[iz1:iz2,iy1:iy2,ix]]

        """---------------------
          Ez development
        ---------------------"""
        ix1 = 1
        iy1 = 1
        iz1 = 0
        for ix in range(ix1, ix2):
            self.Ez2[iz1:iz2,iy1:iy2,ix] \
                = self.Ez1[iz1:iz2,iy1:iy2,ix]*self.ce1[self.idz[iz1:iz2,iy1:iy2,ix]] \
                - self.pz2[iz1:iz2,iy1:iy2,ix]*self.ce3[self.idz[iz1:iz2,iy1:iy2,ix]] \
                + (self.Hy1[iz1:iz2,iy1:iy2,ix]-self.Hy1[iz1:iz2,iy1:iy2,ix-1]) \
                * self.ckex[ix]*self.ce2[self.idz[iz1:iz2,iy1:iy2,ix]]
        for iy in range(iy1, iy2):
            self.Ez2[iz1:iz2,iy,ix1:ix2]= self.Ez2[iz1:iz2,iy,ix1:ix2] \
                - (self.Hx1[iz1:iz2,iy,ix1:ix2]-self.Hx1[iz1:iz2,iy-1,ix1:ix2]) \
                * self.ckey[iy]*self.ce2[self.idz[iz1:iz2,iy,ix1:ix2]]

    def sweep_boundary_e(self):

        ix2 = self.mxx
        iy2 = self.myy
        iz2 = self.mzz

        # -x-side boundary
        # Ey PML
        iy1 = 0
        iz1 = 1
        for ix in range(1, self.mx1):
            self.psiEyx2m[iz1:iz2,iy1:iy2,ix] \
                = self.psiEyx1m[iz1:iz2,iy1:iy2,ix]*self.cbxe[ix] \
                + (self.Hz1[iz1:iz2,iy1:iy2,ix]-self.Hz1[iz1:iz2,iy1:iy2,ix-1]) \
                * self.ccxe[ix]
            self.Ey2[iz1:iz2,iy1:iy2,ix] = self.Ey2[iz1:iz2,iy1:iy2,ix] \
                - self.psiEyx2m[iz1:iz2,iy1:iy2,ix] \
                * self.ce2[self.idy[iz1:iz2,iy1:iy2,ix]]
        self.Ey2[:,:,0] = 0.0

```

```
        # Ez PML
        iy1 = 1
        iz1 = 0
        for ix in range(1, self.mx1):
            self.psiEzx2m[iz1:iz2,iy1:iy2,ix] \
                = self.psiEzx1m[iz1:iz2,iy1:iy2,ix]*self.cbxe[ix] \
                + (self.Hy1[iz1:iz2,iy1:iy2,ix]-self.Hy1[iz1:iz2,iy1:iy2,ix-1]) \
                * self.ccxe[ix]
            self.Ez2[iz1:iz2,iy1:iy2,ix]= self.Ez2[iz1:iz2,iy1:iy2,ix] \
                + self.psiEzx2m[iz1:iz2,iy1:iy2,ix] \
                * self.ce2[self.idz[iz1:iz2,iy1:iy2,ix]]
        self.Ez2[:,:,0] = 0.0

        # +x-side boundary
        # Ey PML
        iy1 = 0
        iz1 = 1
        for ix in range(self.mx2+ 1, self.mxx):
            ixx = ix- self.mx2
            ixxr = self.mxx- ix
            self.psiEyx2p[iz1:iz2,iy1:iy2,ixx] \
                = self.psiEyx1p[iz1:iz2,iy1:iy2,ixx]*self.cbxe[ixxr] \
                + (self.Hz1[iz1:iz2,iy1:iy2,ix]-self.Hz1[iz1:iz2,iy1:iy2,ix-1]) \
                * self.ccxe[ixxr]
            self.Ey2[iz1:iz2,iy1:iy2,ix] = self.Ey2[iz1:iz2,iy1:iy2,ix] \
                - self.psiEyx2p[iz1:iz2,iy1:iy2,ixx] \
                * self.ce2[self.idy[iz1:iz2,iy1:iy2,ix]]
        self.Ey2[:,:,self.mxx] = 0.0

        # Ez PML
        iy1 = 1
        iz1 = 0
        for ix in range(self.mx2+ 1, self.mxx):
            ixx = ix- self.mx2
            ixxr = self.mxx- ix
            self.psiEzx2p[iz1:iz2,iy1:iy2,ixx] \
                = self.psiEzx1p[iz1:iz2,iy1:iy2,ixx]*self.cbxe[ixxr] \
                + (self.Hy1[iz1:iz2,iy1:iy2,ix]-self.Hy1[iz1:iz2,iy1:iy2,ix-1]) \
                * self.ccxe[ixxr]
            self.Ez2[iz1:iz2,iy1:iy2,ix] = self.Ez2[iz1:iz2,iy1:iy2,ix] \
                + self.psiEzx2p[iz1:iz2,iy1:iy2,ixx] \
                * self.ce2[self.idz[iz1:iz2,iy1:iy2,ix]]
        self.Ez2[:,:,self.mxx] = 0.0

        # -y-side boundary
        # Ex PML
        ix1 = 0
        iz1 = 1
```

```
        for iy in range(1, self.my1):
            self.psiExy2m[iz1:iz2,iy,ix1:ix2] \
                = self.psiExy1m[iz1:iz2,iy,ix1:ix2]*self.cbye[iy]
                    \
                + (self.Hz1[iz1:iz2,iy,ix1:ix2]-self.Hz1[iz1:iz2,
                    iy-1,ix1:ix2]) \
                * self.ccye[iy]
            self.Ex2[iz1:iz2,iy,ix1:ix2] = self.Ex2[iz1:iz2,iy,
                ix1:ix2] \
                + self.psiExy2m[iz1:iz2,iy,ix1:ix2] \
                * self.ce2[self.idx[iz1:iz2,iy,ix1:ix2]]
        self.Ex2[:,0,:] = 0.0

        # Ez PML
        ix1 = 1
        iz1 = 0
        for iy in range(1, self.my1):
            self.psiEzy2m[iz1:iz2,iy,ix1:ix2] \
                = self.psiEzy1m[iz1:iz2,iy,ix1:ix2]*self.cbye[iy]
                    \
                + (self.Hx1[iz1:iz2,iy,ix1:ix2]-self.Hx1[iz1:iz2,
                    iy-1,ix1:ix2]) \
                * self.ccye[iy]
            self.Ez2[iz1:iz2,iy,ix1:ix2] = self.Ez2[iz1:iz2,iy,
                ix1:ix2] \
                - self.psiEzy2m[iz1:iz2,iy,ix1:ix2] \
                * self.ce2[self.idz[iz1:iz2,iy,ix1:ix2]]
        self.Ez2[:,0,:] = 0.0

        # +y-side boundary
        # Ex PML
        ix1 = 0
        iz1 = 1
        for iy in range(self.my2+ 1, self.myy):
            iyy = iy - self.my2
            iyyr = self.myy - iy
            self.psiExy2p[iz1:iz2,iyy,ix1:ix2] = \
                self.psiExy1p[iz1:iz2,iyy,ix1:ix2]*self.cbye[iyyr
                    ] \
                + (self.Hz1[iz1:iz2,iy,ix1:ix2]-self.Hz1[iz1:iz2,
                    iy-1,ix1:ix2]) \
                * self.ccye[iyyr]
            self.Ex2[iz1:iz2,iy,ix1:ix2] = self.Ex2[iz1:iz2,iy,
                ix1:ix2] \
                + self.psiExy2p[iz1:iz2,iyy,ix1:ix2] \
                * self.ce2[self.idx[iz1:iz2,iy,ix1:ix2]]
        self.Ex2[:,self.myy,:] = 0.0

        # Ez PML
        ix1 = 1
        iz1 = 0
        for iy in range(self.my2+ 1, self.myy):
            iyy = iy - self.my2
            iyyr = self.myy-  iy
            self.psiEzy2p[iz1:iz2,iyy,ix1:ix2] \
                = self.psiEzy1p[iz1:iz2,iyy,ix1:ix2]*self.cbye[
                    iyyr] \
```

```
                + (self.Hx1[iz1:iz2,iy,ix1:ix2]-self.Hx1[iz1:iz2,iy-1,ix1:ix2]) \
                * self.ccye[iyyr]
            self.Ez2[iz1:iz2,iy,ix1:ix2] = self.Ez2[iz1:iz2,iy,ix1:ix2] \
                - self.psiEzy2p[iz1:iz2,iyy,ix1:ix2] \
                * self.ce2[self.idz[iz1:iz2,iy,ix1:ix2]]
        self.Ez2[:,self.myy,:] = 0.0

        # -z-side boundary
        # Ex PML
        ix1 = 0
        iy1 = 1
        for iz in range(1, self.mz1):
            self.psiExz2m[iz,iy1:iy2,ix1:ix2] \
                = self.psiExz1m[iz,iy1:iy2,ix1:ix2]*self.cbze[iz] \
                + (self.Hy1[iz,iy1:iy2,ix1:ix2]-self.Hy1[iz-1,iy1:iy2,ix1:ix2]) \
                * self.ccze[iz]
            self.Ex2[iz,iy1:iy2,ix1:ix2] = self.Ex2[iz,iy1:iy2,ix1:ix2] \
                - self.psiExz2m[iz,iy1:iy2,ix1:ix2] \
                * self.ce2[self.idx[iz,iy1:iy2,ix1:ix2]]
        self.Ex2[0,:,:] = 0.0

        # Ey PML
        ix1 = 1
        iy1 = 0
        for iz in range(1, self.mz1):
            self.psiEyz2m[iz,iy1:iy2,ix1:ix2] \
                = self.psiEyz1m[iz,iy1:iy2,ix1:ix2]*self.cbze[iz] \
                + (self.Hx1[iz,iy1:iy2,ix1:ix2]-self.Hx1[iz-1,iy1:iy2,ix1:ix2]) \
                * self.ccze[iz]
            self.Ey2[iz,iy1:iy2,ix1:ix2]= self.Ey2[iz,iy1:iy2,ix1:ix2] \
                + self.psiEyz2m[iz,iy1:iy2,ix1:ix2] \
                * self.ce2[self.idy[iz,iy1:iy2,ix1:ix2]]
        self.Ey2[0,:,:] = 0.0

        # +z-side boundary
        # Ex PML
        ix1 = 0
        iy1 = 1
        for iz in range(self.mz2+ 1, self.mzz):
            izz = iz- self.mz2
            izzr = self.mzz- iz
            self.psiExz2p[izz,iy1:iy2,ix1:ix2] \
                = self.psiExz1p[izz,iy1:iy2,ix1:ix2]*self.cbze[izzr] \
                + (self.Hy1[iz,iy1:iy2,ix1:ix2]-self.Hy1[iz-1,iy1:iy2,ix1:ix2]) \
                * self.ccze[izzr]
            self.Ex2[iz,iy1:iy2,ix1:ix2] = self.Ex2[iz,iy1:iy2,ix1:ix2] \
```

```
                - self.psiExz2p[izz,iy1:iy2,ix1:ix2] \
                * self.ce2[ self.idx[iz,iy1:iy2,ix1:ix2]]
        self.Ex2[self.mzz,:,:] = 0.0

        # Ey PML
        ix1 = 1
        iy1 = 0
        for iz in range(self.mz2+ 1, self.mzz):
            izz = iz- self.mz2
            izzr = self.mzz- iz
            self.psiEyz2p[izz,iy1:iy2,ix1:ix2] \
                = self.psiEyz1p[izz,iy1:iy2,ix1:ix2]*self.cbze[izzr] \
                + (self.Hx1[iz,iy1:iy2,ix1:ix2]-self.Hx1[iz-1,iy1:iy2,ix1:ix2]) \
                * self.ccze[izzr]
            self.Ey2[iz,iy1:iy2,ix1:ix2]= self.Ey2[iz,iy1:iy2,ix1:ix2] \
                + self.psiEyz2p[izz,iy1:iy2,ix1:ix2] \
                * self.ce2[ self.idy[iz,iy1:iy2,ix1:ix2]]
        self.Ey2[self.mzz,:,:] = 0.0

    def develop_pcurrent(self):

        ix2 = self.mxx
        iy2 = self.myy
        iz2 = self.mzz

        #  px2 development
        ix1 = 0
        iy1 = 1
        iz1 = 1
        self.px2[iz1:iz2,iy1:iy2,ix1:ix2] \
            = self.cj1[self.idx[iz1:iz2,iy1:iy2,ix1:ix2]] \
            * self.px2[iz1:iz2,iy1:iy2,ix1:ix2] \
            + self.cj3[self.idx[iz1:iz2,iy1:iy2,ix1:ix2]] \
            * (self.Ex2[iz1:iz2,iy1:iy2,ix1:ix2]+ self.Ex1[iz1:iz2,iy1:iy2,ix1:ix2])

        #  py2 development
        ix1 = 1
        iy1 = 0
        iz1 = 1
        self.py2[iz1:iz2,iy1:iy2,ix1:ix2] \
            = self.cj1[self.idy[iz1:iz2,iy1:iy2,ix1:ix2]] \
            * self.py2[iz1:iz2,iy1:iy2,ix1:ix2] \
            + self.cj3[self.idy[iz1:iz2,iy1:iy2,ix1:ix2]] \
            * (self.Ey2[iz1:iz2,iy1:iy2,ix1:ix2]+ self.Ey1[iz1:iz2,iy1:iy2,ix1:ix2])

        #  pz2 development
        ix1 = 1
        iy1 = 1
        iz1 = 0
        self.pz2[iz1:iz2,iy1:iy2,ix1:ix2] \
            = self.cj1[self.idz[iz1:iz2,iy1:iy2,ix1:ix2]] \
            * self.pz2[iz1:iz2,iy1:iy2,ix1:ix2] \
```

```
                + self.cj3[self.idz[iz1:iz2,iy1:iy2,ix1:ix2]] \
                * (self.Ez2[iz1:iz2,iy1:iy2,ix1:ix2]+ self.Ez1[iz1:
                    iz2,iy1:iy2,ix1:ix2])

    def update_field(self):

        self.Ex1[:,:,:] = self.Ex2[:,:,:]
        self.Ey1[:,:,:] = self.Ey2[:,:,:]
        self.Ez1[:,:,:] = self.Ez2[:,:,:]
        self.Hz1[:,:,:] = self.Hz2[:,:,:]
        self.Hx1[:,:,:] = self.Hx2[:,:,:]
        self.Hy1[:,:,:] = self.Hy2[:,:,:]

        self.psiEzx1m[:,:,:] = self.psiEzx2m[:,:,:]
        self.psiEyx1m[:,:,:] = self.psiEyx2m[:,:,:]
        self.psiHzx1m[:,:,:] = self.psiHzx2m[:,:,:]
        self.psiHyx1m[:,:,:] = self.psiHyx2m[:,:,:]
        self.psiHzx1p[:,:,:] = self.psiHzx2p[:,:,:]
        self.psiHyx1p[:,:,:] = self.psiHyx2p[:,:,:]
        self.psiEzx1p[:,:,:] = self.psiEzx2p[:,:,:]
        self.psiEyx1p[:,:,:] = self.psiEyx2p[:,:,:]

        self.psiEzy1m[:,:,:] = self.psiEzy2m[:,:,:]
        self.psiExy1m[:,:,:] = self.psiExy2m[:,:,:]
        self.psiHzy1m[:,:,:] = self.psiHzy2m[:,:,:]
        self.psiHxy1m[:,:,:] = self.psiHxy2m[:,:,:]
        self.psiEzy1p[:,:,:] = self.psiEzy2p[:,:,:]
        self.psiExy1p[:,:,:] = self.psiExy2p[:,:,:]
        self.psiHzy1p[:,:,:] = self.psiHzy2p[:,:,:]
        self.psiHxy1p[:,:,:] = self.psiHxy2p[:,:,:]

        self.psiEyz1m[:,:,:] = self.psiEyz2m[:,:,:]
        self.psiExz1m[:,:,:] = self.psiExz2m[:,:,:]
        self.psiHyz1m[:,:,:] = self.psiHyz2m[:,:,:]
        self.psiHxz1m[:,:,:] = self.psiHxz2m[:,:,:]
        self.psiEyz1p[:,:,:] = self.psiEyz2p[:,:,:]
        self.psiExz1p[:,:,:] = self.psiExz2p[:,:,:]
        self.psiHyz1p[:,:,:] = self.psiHyz2p[:,:,:]
        self.psiHxz1p[:,:,:] = self.psiHxz2p[:,:,:]

    def save_idv(self):
        """ save material index distribution """

        for epsmon in self.iepsmons:
            if epsmon.pol == ‘x’:
                if epsmon.axis == ‘x’: # normal to x-axis
                    ieps2d = self.idx[:self.mzz+1, \
                        :self.myy+1, epsmon.position]
                elif epsmon.axis == ‘y’:
                    ieps2d = self.idx[:self.mzz+1, \
                        epsmon.position, :self.mxx]
                else:
                    ieps2d = self.idx[epsmon.position, \
                        :self.myy+1, :self.mxx]
            elif epsmon.pol == ‘y’:
                if epsmon.axis == ‘x’:
                    ieps2d = self.idy[:self.mzz+1, \
```

```
                        :self.myy, sepsmon.position]
                elif epsmon.axis == ‘y’:
                    ieps2d = self.idy[:self.mzz+1, \
                        epsmon.position, :self.mxx+1]
                else:
                    ieps2d = self.idy[epsmon.position, \
                        :self.myy, :self.mxx+1]
            else:
                if epsmon.axis == ‘x’:
                    ieps2d = self.idz[:self.mzz, \
                        :self.myy+1, epsmon.position]
                elif epsmon.axis == ‘y’:
                    ieps2d = self.idz[:self.mzz, \
                        epsmon.position, :self.mxx+1]
                else:
                    ieps2d = self.idz[epsmon.position, \
                        :self.myy+1, :self.mxx+1]

            if not os.path.exists(‘./field’):
                os.mkdir(‘./field’)
            np.savetxt(epsmon.fname, ieps2d, fmt= ‘%d’, delimiter = ‘ ’)

    def save_ehfield(self, numt):
        """ save electric field and magnetic field """

        for ifieldmon in self.ifieldmons:
            location = ifieldmon.position
            ehfield = ifieldmon.ehfield

            # normal to x-axis
            if ifieldmon.axis == ‘x’:
                if ehfield == ‘Ex’:
                    field2d = self.Ex2[0:self.mzz+1,0:self.myy+1, location]
                elif ehfield == ‘Ey’:
                    field2d = self.Ey2[0:self.mzz+1,0:self.myy, location]
                elif ehfield == ‘Ez’:
                    field2d = self.Ez2[0:self.mzz,0:self.myy+1, location]
                elif ehfield == ‘Hx’:
                    field2d = self.Hx2[0:self.mzz,0:self.myy, location]
                elif ehfield == ‘Hy’:
                    field2d = self.Hy2[0:self.mzz,0:self.myy+1, location]
                elif ehfield == ‘Hz’:
                    field2d = self.Hz2[0:self.mzz+1,0:self.myy, location]

            # normal to y-axis
            elif ifieldmon.axis == ‘y’:
                if ehfield == ‘Ex’:
                    field2d = self.Ex2[0:self.mzz+1,location,0:self.mxx]
                elif ehfield == ‘Ey’:
```

```
                    field2d = self.Ey2[0:self.mzz+1,location,0:self.mxx+1]
                elif ehfield == 'Ez':
                    field2d= self.Ez2[0:self.mzz,location,0:self.mxx+1]
                elif ehfield == 'Hx':
                    field2d= self.Hx2[0:self.mzz,location,0:self.mxx+1]
                elif ehfield == 'Hy':
                    field2d= self.Hy2[0:self.mzz,location,0:self.mxx]
                elif ehfield == 'Hz':
                    field2d= self.Hz1[0:self.mzz+1,location,0:self.mxx]

            # normal to z-axis
            elif ifieldmon.axis == 'z':
                if ehfield == 'Ex':
                    field2d = self.Ex2[location,0:self.myy+1,0:self.mxx]
                elif ehfield == 'Ey':
                    field2d = self.Ey2[location,0:self.myy,0:self.mxx+1]
                elif ehfield == 'Ez':
                    field2d = self.Ez2[location,0:self.myy+1,0:self.mxx+1]
                elif ehfield == 'Hx':
                    field2d = self.Hx2[location,0:self.myy,0:self.mxx+1]
                elif ehfield == 'Hy':
                    field2d = self.Hy2[location,0:self.myy+1,0:self.mxx]
                elif ehfield == 'Hz':
                    field2d = self.Hz2[location,0:self.myy,0:self.mxx]

            if not os.path.exists('./field'):
                os.mkdir('./field')
            fname = ifieldmon.prefix + '{0:0>3}'.format(numt) + '.txt'
            np.savetxt(fname, field2d, fmt= '%e', delimiter=' ')

    def detect_efield(self, jt):
        """ detection of E field """

        for i, detector in enumerate(self.idetectors):
            ix = detector.x
            iy = detector.y
            iz = detector.z
            if detector.pol == 'x':
                self.edetect[i][jt] = self.Ex1[iz,iy,ix]
            elif detector.pol == 'y':
                self.edetect[i][jt] = self.Ey1[iz,iy,ix]
            else:
                self.edetect[i][jt] = self.Ez1[iz,iy,ix]

    def detect_spectra(self):
```

```
        """ Fourier Transformation to obtain E-field spectra """

        if not os.path.exists('field'):
            os.mkdir('field')
        fname = 'field/Response.txt'
        col = 'Time(ps) Source'
        for i in range(len(self.idetectors)):
            col= col+ ' Detector['+ str(i)+ ']'
        atime = np.arrange(0, self.mt)*self.dt*1.0e12
        atime = np.append([atime], [self.esource], axis=0)
        atime = np.append(atime, self.edetect, axis=0)
        np.savetxt(fname, atime.T, fmt='%.4e', delimiter=' ', \
            header=col, comments='')

        esource2 = self.esource[::self.sampint]
        esourceft = np.absolute(np.fft.rfft(esource2, n=self.
            mfft2))** 2
        edetect2 = self.edetect[:,::self.sampint]
        edetectft = np.absolute(np.fft.rfft(edetect2, n=self.
            mfft2))** 2
        col = 'Frequency(THz) Wavelength(um) Source'
        for i in range(len(self.idetectors)):
            col = col + ' Detector[' + str(i) + ']'
        thz = np.arrange(self.mfft2//2+1, dtype=np.float64 ) \
            * 1.0e-12/(self.dt*self.sampint*self.mfft2)
        wavelength= np.ones(self.mfft2//2+1) * self.cc * 1.0e-6
        wavelength[1:] = wavelength[1:] / thz[1:]
        wavelength[0] = wavelength[1]
        thz = np.append([thz], [wavelength], axis=0)
        thz = np.append(thz, [esourceft], axis=0)
        thz = np.append(thz, edetectft, axis=0)
        if not os.path.exists('./field'):
            os.mkdir('./field')
        fname = './field/Spectra.txt'
        np.savetxt(fname, thz.T, fmt='%.4e', delimiter=' ', \
            header=col, comments='')
```

8.7 Visualize shapes (for DDSCAT)

Program A.7 (ShapePlot.py)

```
import matplotlib.pyplot as plt
import numpy as np
import matplotlib.pyplot as plt
from mpl_toolkits.mplot3d import Axes3D
from scipy import zeros, array
from matplotlib.pyplot import plot,show,xlabel,ylabel,title,
    legend,grid, axis,subplot

num=1

```

```
xmin = -100      # Calculation range setting
xmax = 100
ymin = -100
ymax = 100
zmin = -100
zmax = 100

numx = xmax-xmin+1  #  Number of points in x direction
numy = ymax-ymin+1  #  Number of points in y direction
numz = zmax-zmin+1  #   Number of points in z direction
num = numx*numy*numz  #  Number of total points in x direction

p = np.zeros([numx,numy,numz],dtype=int)  #  initialization of flag p(x,y,z)

iii=0
xorigin=0  #  initialization of gravity center in x-direction
yorigin=0  #  initialization of gravity center in y-direction
zorigin=0  #  initialization of gravity center in z-direction

for z in range(zmin, zmax):
    for y in range(ymin, ymax):
        for x in range(xmin, xmax):
            if (x/10)**2 + (y/25)**2 + (z/10)**2 <= 1:   # determine whether the coordinates constitute a shape
                p[x-xmin,y-ymin,z-zmin] = 1   # p=1 for the coordinates that make up the shape
                      #  Since the array of p is an integer greater than or equal to 0, it is shifted by xmin
                xorigin=xorigin+x   #   Sum the x-coordinates to find the gravity center
                yorigin=yorigin+y   #   Sum the y-coordinates to find the gravity center
                zorigin=zorigin+z   #   Sum the z-coordinates to find the gravity center
                iii+=1
            else:
                p[x-xmin,y-ymin,z-zmin] = 0     #       p=0 if the coordinates do not constitute a shape

Xorigin=xorigin/iii   # the gravity center x component
Yorigin=yorigin/iii   # the gravity center y component
Zorigin=zorigin/iii   # the gravity center z component

xx=zeros(iii, dtype=int)
yy=zeros(iii, dtype=int)
zz=zeros(iii, dtype=int)

i=0
for z in range(zmin, zmax):
    for y in range(ymin, ymax):
        for x in range(xmin, xmax):
            if p[x-xmin,y-ymin,z-zmin] == 1:
                xx[i]=x
                yy[i]=y
```

```
                    zz[i]=z
                    i+=1

fig = plt.figure()
ax = Axes3D(fig)
ax.scatter3D(np.ravel(xx),np.ravel(yy),np.ravel(zz))     #3D plot

plt.show()
```

Bibliography

[1] M. Born and E. Wolf. *Principles of Optics.* Pergamon Press, Oxford, 1959.

[2] G. R. Fowles. *Introduction to Modern Optics.* Dover, New York, 1968.

[3] B. E. A. Saleh and M. C. Teich. *Fundamentals of Photonics.* Wiley, New York, 2007.

[4] D. S. Bethune. Optical harmonic generation and mixing in multilayer media: Analysis using optical transfer matrix techniques. *J. Opt. Soc. Am. B*, 6:910–916, 1989.

[5] D. S. Bethune. Optical harmonic generation and mixing in multilayer media: Extension of optical transfer matrix approach to include anisotropic materials. *J. Opt. Soc. Am. B*, 8:367–373, 1991.

[6] A. Poddubny, I. Iorsh, P. Belov, and Kivshar Y. Optical harmonic generation and mixing in multilayer media: Extension of optical transfer matrix approach to include anisotropic materials. *Nat. Photon.*, 7:958–967, 2013.

[7] K.-H. Kim, Y.-S. No, S. Chang, J.-H. Choi, and H.-G. Park. Invisible hyperbolic metamaterial nanotube at visible frequency. *Sci. Rep.*, 5:16027, 2015.

[8] C. F. Bohren and D. R. Huffman. *Absorption and Scattering of Light by Small Particles.* Wiley, New York, 1983.

[9] van de Hulst. *Light Scattering by Small Particles.* Dover, New York, 1957.

[10] J. A. Stratton. *Electromagnetic Theory.* McGraw-Hill, New York, 1941.

[11] A. E. Neeves and M. H. Birnboim. Composite structures for the enhancement of nonlinear-optical susceptibility. *J. Opt. Soc. Am. B*, 6:787–796, 1989.

[12] R. B. Johnson and R. W. Christy. Optical constants of the noble metals. *Phys. Rev. B*, 6:4370–4379, 1972.

[13] M. Kerker and E. Matijević. Scattering of electromagnetic waves from concentric infinite cylinders. *J. Opt. Soc. Am.*, 51:506–508, 1961.

[14] M. M. Wind, J. Vliger, and D. Bedeaux. The polarizability of a truncated sphare on a substrate I. *Physica A*, 141A:33–57, 1987.

[15] T. Okamoto and Yamaguchi I. Optical absorption study of the surface plasmon resonance in gold nanoparticles immobilized onto a gold substrate by self-assembly techniqe. *J. Phys. Chem. B*, 107:10321–10324, 2003.

[16] R. Ruppin. Optical absorption of two spheres. *J. Phys. Soc. Jpn.*, 58:1446–1451, 1989.

[17] G. Gupta, D. Tanaka, Y. Ito, D. Shibata, M. Shimojo, K. Furuya, K. Mitsui, and K. Kajikawa. Absorption spectroscopy of gold nanoisland films: Optical and structural characterization. *Nanotechnology*, 20:025703, 2009.

[18] G. Gupta, Y. Nakayama, K. Furuya, K. Mitsuishi, M. Shimojo, and K. Kajikawa. Cross-sectional transmission electron microscopy and optical characterization of gold nanoislands. *Jpn. J. Appl. Phys.*, 48:080207, 2009.

[19] M. G. Moharam and T. K. Gaylord. Rigorous coupled-wave analysis of planar-grating diffraction. *J. Opt. Soc. Am.*, 71:811–818, 1981.

[20] M. G. Moharam and T. K. Gaylord. Diffraction analysis of dielectric surface-relief gratings. *J. Opt. Soc. Am.*, 72:1385–1392, 1982.

[21] M. G. Moharam and T. K. Gaylord. Rigorous coupled-wave analysis of grating diffraction– E-mode polarization and losses. *J. Opt. Soc. Am.*, 73:451–455, 1983.

[22] M. G. Moharam and T. K. Gaylord. Rigorous coupled-wave analysis of metallic surface-relief gratings. *J. Opt. Soc. Am. A*, 3:1780–1787, 1986.

[23] L. Li and C. W. Haggans. Convergence of the coupled-wave method for metallic lamellar diffraction gratings. *J. Opt. Soc. Am. A*, 10:1184–1189, 1993.

[24] M. G. Moharam, E. B. Grann, and D. A. Pommet. Formulation for stable and efficient implementation of the rigorous coupled-wave analysis of binary grating. *J. Opt. Soc. Am. A*, 12:1068–1076, 1995.

[25] G. Granet and B. Guizal. Efficient implementation of the coupled-wave method for metallic lamellar gratings in tm polarization. *J. Opt. Soc. Am. A*, 13:1019–1023, 1996.

[26] P. Lalanne and G. M. Morris. Highly improved convergence of the coupled-wave method for tm polarization. *J. Opt. Soc. Am. A*, 13:779–784, 1996.

[27] L. Li. Formulation and comparison of two recursive matrix algorithms for modeling layered diffraction gratings. *J. Opt. Soc. Am. A*, 13:1024–1035, 1996.

[28] P. Lalanne. Improved formulation of the coupled-wave method for two-dimensional gratings. *J. Opt. Soc. Am. A*, 14:1592–1598, 1997.

[29] E. Popov, M. Nevière, B. Gralak, and G. Tayeb. Staircase approximation validity for arbitrary-shaped gratings. *J. Opt. Soc. Am. A*, 19:33–42, 2002.

[30] M. Nevière and E. Popov. *Light Propagation in Periodic Media.* Marcel Dekker, New York, 2003.

[31] L. Li. Note on the S-matrix propagation algorithm. *J. Opt. Soc. Am. A*, 20:655–661, 2003.

[32] M. G. Moharam, D. A. Pommet, E. B. Grann, and T. K. Gaylord. Stable implementation of the rigorous coupled-wave analysis for surface-relief gratings: Enhanced transmittance matrix approach. *J. Opt. Soc. Am. A*, 12:1077–1086, 1995.

[33] L. Li. Bremmer series, R-matrix propagation algorithm, and numerical modeling of diffraction gratings. *J. Opt. Soc. Am. A*, 11:2829–2836, 1994.

[34] T. Vallius, J. Tervo, P. Vahimaa, and J. Turunen. Electromagnetic field computation in semiconductor laser resonators. *J. Opt. Soc. Am. A*, 23:906–911, 2006.

[35] L. Li. New formulation of the fourier modal method for crossed surface-relief gratings. *J. Opt. Soc. Am. A*, 14:2758–2767, 1997.

[36] K. S. Yee. Numerical solution of initial boundary value problems involving maxwell's equations in isotropic media. *IEEE Trans. Antennas Propagat.*, 14:302–307, 1966.

[37] A. Taflove and S. C. Hagness. *Computational Electrodynamics: The Finite-Difference Time-Domain Method*, 3rd ed. Artech House, Boston, 2005.

[38] M. Okoniewski, M. Mrozowski, and M. A. Stuchly. Simple treatment of multi-term dispersion in FDTD. *IEEE Microwave Guided Wave Lett.*, 7:121–123, 1997.

[39] V. G. Padalka and I. N. Shklyarevskii. Determination of the microcharacteristics of silver and gold from the infrared optical constants and the conductivity at 82 and 295°k. *Opt. Spectrosc.*, 11:285–288, 1961.

[40] J.-P. Berenger. A perfectly matched layer for the absorption of electromagnetic waves. *J. Comput. Phys.*, 114:185–200, 1994.

[41] S. D. Gedney. An anisotropic PML absorbing media for the FDTD simulation for fields in lossy and dispersive media. *Electromagnetics*, 16:399–415, 1996.

[42] W. C. Chew and W. H. Weedon. A 3D perfectly matched medium from modified Maxwell's equations with stretched coordinates. *Microwave Opt. Tech. Lett.*, 7:599–604, 1994.

[43] S. D. Gedney. An anisotropic perfectly matched layer-absorbing medium for truncation of FDTD lattices. *IEEE Trans. Antennas and Propagation*, 44:1630–1639, 1996.

[44] J. A. Roden and S. D. Gedney. Convolution PML (CPML): An efficient FDTD implementation of the CFS-PML for arbitrary media. *Microwave Opt. Tech. Lett.*, 27:334–339, 2000.

[45] M. Kuzuoglu and R. Mittra. Frequency dependence of the constitutive parameters of causal perfectly matched anisotropic absorbers. *IEEE Microwave Guided Wave Lett.*, 6:447–449, 1996.

[46] S. D. Gedney. *CIntroduction to the Finite-Difference Time-Domain (FDTD) Method for Electromagnetics.* Mprgan & Claypool, 2011.

[47] L. Zhang and T. Seideman. Rigorous formulation of oblique incidence scattering from dispersive media. *Phys. Rev. B*, 82:155117, 2010.

[48] Y.-N. Jiang, D.-B. Ge, and S.-J. Ding. Analysis of TF-SF boundary for 2D-FDTD with plane P-wave propagation in layered dispersive and lossy media. *Prog. Electromagn. Res.*, 83:157–172, 2008.

[49] I. R. Çapoğlu and G. S. Smith. A total-field/scattered-field plane-wave source for the FDTD analysis of layered media. *IEEE Tran. Anttenas Propag.*, 56:158–169, 2008.

[50] F. Yang, J. Chen, R. Qiang, and A. Elsherbeni. A simple and efficient FDTD/PBC algorithm for scattering analysis of periodic structures. *Radio Sci.*, 42:RS4004, 2007.

[51] P. Harms, R. Mittra, and W. Ko. Implementation of the periodic boundary condition in the finite-difference time domain algorithm for FSS structures. *IEEE Trans. Antennas Propag.*, 42:1317–1324, 1994.

[52] J. A. Roden, S. D. Gedney, M. P. Kesler, J. G. Maloney, and P. H. Harms. Time-domain analysis of periodic structures at oblique incidence: Orthogonal and nonorthogonal FDTD implementations. *IEEE Trans. Microwave Theory Tech.*, 46:420–427, 1998.

[53] A. Aminian and Y. Rahmat-Samii. Spectral FDTD: A novel computational technique for the analysis of periodic structures. *Proc. IEEE Antennas Propag. Soc. Symp.*, 3:3139–3142, 2004.

[54] D. Schurig. Off-normal incidence simulations of metamaterials using FDTD. *Int. J. Numer. Model.*, 19:215–228, 2006.

[55] K. Umashankar and A. Taflove. A novel method to analyze electromagnetic scattering of complex objects. *IEEE Trans. Electromagn. Compa.*, EMC-24:397–405, 1982.

[56] P. W. Zhai, C. H. Li, G. W. Kattawar, and P. Yang. FDTD far-field scattering amplitudes: Comparison of surface and volume integration methods. *J. Quant. Spectrosc. Radiat. Transfer*, 106:590–594, 2007.

[57] R. J. Luebbers, K. S. Kunz, M. Shneider, and F. Hunsberger. A finite-difference time-domain near zone to far zone transformation. *IEEE Trans. Antennas Propagat.*, 39:429–433, 1991.

[58] D. J. Robinson and J. B. Schneider. On the use of the geometric mean in FDTD near-to-far-field transformations. *IEEE Trans. Antennas Propagat.*, 55:3204–3211, 2007.

[59] S. A. Maier. *Plasmonics: Fundamentals and Applications.* Springer, New York, 2007.

[60] A. H. Nuttall. Some windows with very good sidelobe behavior. *IEEE Trans. Acoust. Speech Signal Process.*, ASSP-29:84–91, 1981.

[61] B. T. Draine and P. J. Flatau. Discrete-dipole approximation for scattering calculations. *J. Opt. Soc. Am. A*, 11:1491–1499, 1994.

[62] M. A. Yurkin, V. P. Maltsevb, and A. G. Hoekstra. The discrete dipole approximation for simulation of light scattering by particles much larger than the wavelength. *J. Quantitive Specroscopy & Radiative Transfer*, 106:546–557, 2007.

[63] B. T. Draine and J. Goodman. Beyond clausius-mossotti: Wave propagation on a polarizable point lattice and the discrete dipole approximation. *Astrophys. J.*, 405:685–697, 1993.

Index

FSC
www.fsc.org
MIX
Papier | Fördert gute Waldnutzung
FSC® C083411

Kolibri 360 GmbH
Ferdinand-Jühlke-Straße 7
99095 Erfurt, Deutschland
produktsicherheit@kolibri360.de